SEX

SEX

A Natural History

———

J OANN E LLISON R ODGERS

A W.H. FREEMAN BOOK

TIMES BOOKS · HENRY HOLT AND COMPANY · NEW YORK

Times Books
Henry Holt and Company, LLC
Publishers since 1866
115 West 18th Street
New York, New York 10011

Portions of chapter 11 are taken from Joann Ellison Rodgers, *Drugs and Sexual Behavior*
(New York: Chelsea House, 1988).

Library of Congress Cataloging-in-Publication Data
Rodgers, Joann Ellison.
    Sex : a natural history / Joann Ellison Rodgers.—1st ed.
    p. cm.
Includes bibliographical references and index.
ISBN: 0-7167-3744-2 (hc.)
1. Sex (Biology) I. Title.

QP251 .R577 2002
576.8 '55—dc21   2001054409

First Edition 2002

Designed by Fritz Metsch

Printed in the United States of America

1   3   5   7   9   10   8   6   4   2

For my granddaughter,
Margaret Ellison "Ellie" Rodgers,
and her generations to come

CONTENTS

SECARE AND *EROS*: SEX, LOVE, SCIENCE, AND CIRCUMSTANCES

" 'It is an old maxim of mine,' said Holmes, 'that when you have excluded the impossible, whatever remains, however improbable, must be the truth.' Answered Watson, 'Perhaps, you may have convinced me as to the motive, but you are yet to explain how it is done.' "

—ARTHUR CONAN DOYLE

People love sex. We have it every chance we get, in every position and season. We will take incredible risks, exhaust ourselves, even self-destruct to get it, do it, keep it. Once we experience its power and its pleasure—even when we only can imagine it—we seek it with the intensity of an addict after a fix. If nature were into efficient engineering of reproductive systems, interest in sex would end with a woman's last ovulation. Any self-respecting MIT graduate cum Harvard MBA would insist on a just-in-time inventory system: When you're out of eggs, you're out of sex. Instead, we, like every other living thing, are throbbing collections of protoplasm whose energies are ever in screaming search of sex. We want sex not just for reproducing, not just on purpose, but for pleasure and even just for the pleasure of its pursuit.

Consequently, if sex were a novel, it would be an Everyman story. The heroes and heroines would be every one of our genes, cells, tissues, hormones, organs, and most of all the chemistry of the brain and mind. The plot would track the quest to bring them together in harmony against all odds to merge a heavily guarded set of gametes while simultaneously bringing pleasure to participants. At the end, we would respect the characters if not altogether like them. A friend and colleague, two-time Pulitzer Prize–winning science journalist Jon Franklin, once said that truly heroic stories are full of "aha!" moments that show how ordinary characters resolve to imagine, then successfully execute creative, resourceful, lasting, and most of all practical solutions to fundamental, serious, life and death complications. He could have been summarizing the story of sex. There is no magical deus ex machina, no metaphysical or divine act that makes sex *work*. Just diligent, tenacious, randomly acquired, hard-earned tangles of internal constructions and functions, hardwired yet flexible enough in each one of us to meet unique circumstances.

Paradoxically, for most of us, *how* these elemental facts of life happen is as poorly understood as they are compelling. This is partly, at least, because Rube Goldberg couldn't reengineer a more clumsy, less "efficient" process. Sex in humans and other animals is an almost ludicrously complicated, infinitely resourceful, and varied network of anatomical, chemical, social, biological, and emotional signals and schemes for regenerating and perpetuating chromosomes, genes, and DNA.

But our lack of understanding is also because we seem squeamishly reluctant to think too hard about sex. Sex, according to Johns Hopkins University psychologist and sexologist John Money, is a thoroughly unloved human behavior. "Isn't it supremely ironic," he once told me, "that the very thing responsible for every parent and every child is a philosophical orphan? People thirst for a sexonomy, a factual set of fundamental principles that lets us talk and think about sex as a reflexive, built-in thing we do, no different than coughing, thinking, sneezing, or singing. Sex is here because we are members of the human species, not because of any act of yours or mine. And most people can't bring themselves to talk about it."

We prefer "doin' what comes naturally" to considering what "naturally" means. While excavating the secrets of life is notoriously difficult, those whose livelihood it is to try have become adept at peeking under the planet's blankets to find out just what "naturally" does mean. Sometimes hilariously so. En route to finding ways to get rid of a citrus crop pest called *Diaprepes abbreviatus*, an inch-long black beetle, University of Florida zoologist H. Jane Brockmann and Ally Harari of Ben-Gurion University in Israel spent long hours sorting out the mating rites of this bug and concluded simply that "everybody mounted everybody." Males mount females, males mount males and even mating couples. Females mount females and lure alpha males in the process. It turns out that the beetles have a hard time sorting out who is who. But nature doesn't play dice with anything so important as survival of a species, leading Brockmann, with classic understatement, to note that "when you see something like that, it demands an explanation."

Well, yes, but to those on the lookout for the natural signals of sex, these beetles' mating habits may also explain something about why Hugh Hefner made a splendid living exploiting the flirtatious dance of animal attraction between wealthy men and big-breasted young blondes. Or between men and women in general. The signals are different only in kind, not in category. As the beetle experts learned, the females who tried to mate with other females tended to wind up with the largest males, the insect equivalent of the men with wealth, which helped the girls in terms of their survival, because bigger, in the insect world, is indeed often better. The larger the male, the more likely he is to bring nuptial gifts, such as more food and a set of genes that helped him get them. It also turns out that while it's hard even for the beetles to sort it out, females *are* a bit bigger than males. So the males first follow their noses to the females, tracking a particular scent the girls give off; but because the boys still can't always tell who wears pink from who wears blue, they frequently mount another guy or, mistaking size for sex, head for a beetle couple already mating. Without any better navigational cue the males depend on sight, moving toward "big" beetles, which are more likely female but may well be a mating pair. And, if such a male is big enough, he can always push a smaller male off the back of the female of the pair he mistook for a big girl and have his way. In a male sexist world, that would be the end of it. But to the keen observers of these beetles, such visual signals suggest a far more important theme in the nature of sex: that it takes two to tango, and to choose to dance in the first place. It seems the females are also doing the choosing. How? By mounting each other. The bigger female, who mounts another female, attracts the biggest guys who, myopic as they are, think the female-female pair is one *giant* lady bug. Result: The biggest male gets the shrewdest female. To test all this further, the scientists spent more hours and days observing the reaction of male beetles to sets of large and small female-mounted pairs. The larger beetles went straight for the larger females. By attracting the biggest males, the most "fit" females got good genes (i.e., those from bigger, presumably more potent and wealthy males) and nuptial gifts such as food that male bugs tend to pass on during copulation. If all this doesn't remind us of the behavior of some two-legged animals, we haven't been paying attention. And if this isn't the nature of sex, and natural, nothing is.

Just why we so willingly endorse the mystique of sex but avoid confronting its mystery may rest in part in the difficulty of doing what the beetle watchers do—that is, spend a lot of time as objective voyeurs—but also in our universal vulnerability to its force. We adore telling stories about sex, but we don't want sex to tell us too much about ourselves.

Sex *scares* us. And why shouldn't it? "Making love," says psychologist David Buss, author of *The Evolution of Desire,* "is one of the most important, complex and perilous cooperative exchanges that any of us engage in during our lives. Loaded with promise and fraught with dangerous pitfalls, love affairs tax our ability at deal-making to the fullest, requiring the complete repertoire of psychological specializations that evolved for cooperation." University of Texas zoologist David Crews, who has studied sex behavior for decades, is more blunt. "Sexuality," he declares, "is all in your head and that's the most dangerous place it can be. It has so many components, but we only have one thought at a time. So usually we act on it and set down a course. We get confused, forgetting all the other aspects. We get into a *lot* of predicaments."

Why bother to parse any of this? Beyond idle curiosity, or concern about sexually transmitted disease, why probe? After all, no one needs an instruction manual; the mechanics aren't *that* hard to figure out, or even master. If the goal is to get, beget, or be gotten, even the sexually challenged get help from a vast inventory of how-to and advice books, medical texts, illustrated guides, devices, and therapists.

One reason to look further is that information at the clinical level is to a genuine knowledge of sex what a list of grapes is to a master wine maker—as remote from creation of a great or even a good vintage as a 1945 Chateau Mouton Rothschild is from fermented grape juice.

Another is that because sex is essential to collective survival, societies are obsessed with it, creating rules that irrationally inhibit and too often cruelly torment people. Debates over political "solutions" to sexual "problems" are well-served by information about the biology and psychology of sex. As archaeologist Timothy Taylor noted in his book *The Prehistory of Sex: Four Million Years of Human Sexual Culture,* sixteenth-century Spaniards were so outraged at the homosexuality and transvestism they encountered among the indigenous people they conquered that they destroyed almost all of the sculpture, pottery, monuments, and jewelry that depicted such practices, distorting greatly our view of sexual behavior and morality. "We cannot assume," Taylor adds, "that our modern way of thinking about sex—either biologically or socioculturally—is necessarily any more objective than any other way of thinking about sex. Even within the rather narrow Western tradition . . . from Plato to Shere Hite, it is clear that no one has ever had a monopoly on the truth about our bodies."

The most powerful and useful reason, however, is that whereas the evolutionary "goal" of sex is straightforward—a union between two suitable and compatible sets of chromosomes that assures survival of the species—its *nature* may be the clearest, brightest window through which to look, and mirror in which to reflect, the meandering and sometimes misleading connections between bedrock biology and behavior. Sex is, on the whole, the organizing theme, the ulterior motive of our inner and outer lives. So much is involved in the way we attach to others and reproduce ourselves that no other human process is likely to teach us more about the links between what we are and what to do.

Taken from the Latin *secare,* meaning to divide or cut (dividing in two implies reproduction), the word "sex" evokes not only copulation but the two sexes and their genitals. Yet sex is also about *eros* (from the Greek word for love), and thus about desire, motives, and intimacy. All of the kissing, touching, tickling, biting, teasing, positioning, penetrating, prolonging, and climaxing is the wrapper; the contents are at least as compelling.

Sex as we know and practice it today, in all of its expressions, from prostitution to bestiality, is part and parcel of our long, long, prehistoric legacy. In his *Prehistory of Sex,* Taylor exposes the hidden sexual treasures of museums and archaeological collections worldwide to support his contention that modern humans, from their earliest versions two hundred thousand to three hundred thousand years ago, had substantially invented everything there was to invent about sex, much about which we're *still* afraid to ask. Early man and woman, for instance, not only understood where babies come from but easily separated sex from reproduction, using contraception and abortion to control fertility. They developed animal husbandry, "oppressed" women to maximize population (harems *are* productive, after all), and probably practiced artificial insemination. They cross-dressed, recognized drag queens, established brothels, and created elaborately pornographic statues, fetishes, and paintings. "Starting around 5000 years ago, it is possible to document great variation in human sexuality in Eurasia: bestiality, homosexuality, prostitution . . . transvestism (male and female), transsexuality, hormone treatments, sadomasochism, autoerotic asphyxia . . . sex as an acrobatic and competitive pastime and sex as a transcendental spiritual discipline," he says. The artifacts that survived to fuel his research, all just a few thousands years old, "represent just a tiny part of the four-million-year saga of its prehistory," but include such compelling items as golden penis sheaths, the full erection of a Dordogne cave painter's depiction of a shaman in a trance, and other paintings of body piercing, anal sex, sadomasochism, and masturbation.

Our ancestors' sexual *culture*, then, is part of our sexual repertoire, too. But if at times surreal, sex is never *unreal*. Penises, personalities, vaginas, nipples, skin, neurons, phobias, temperament, facts, fallacies, and families all get in the act, and sometimes in the way. Our ovaries, testes, and gametes play out their complicated roles as long as we do our part to get them coordinated.

Predictably, it's the coordination of things *right now* that consumes our interest, our energy, and our time. Each thought of love and longing, each act of courtship and copulation, each flush of desire and arousal represents the sum and synergy of biological ingredients, always, always, in tune with particular circumstances and with those paramount drivers of sex, the mind and brain. Each sexual engagement is a moment in time that reveals what it is to be quintessentially and quite literally alive.

"Ideas," wrote novelist Henry Miller, whose *Tropic of Cancer* and other stories were once literally banned in Boston, "have to be wedded to action; if there is no sex, no vitality in them, there is no action." And it is in the *process* of sex that all the action takes place—the process, millions of years in the making, that snakes and lurches its way through an intricate dance beginning with the elegant, unheard intercourse of DNA molecules and concluding with what even poets and artists depict as a relatively graceless and disheveled display. "The act of procreation," wrote Leonardo Da Vinci in *Dell'Anatomia*, "and the members employed therein are so repulsive, that if it were not for the beauty of the faces and the adornments of the actors, and the pent-up impulse, nature would lose the human species." (Not likely, Leonardo.)

As choreographed by humans over the past 4 million years, sex vigorously endures. What's new is what's been learned about the nature of sex, the inner sex lives of our cells and ourselves. Understanding sex's nature lends meaning to the knowledge that both men and women have orgasms while dreaming, and that men universally are far more occupied with a woman's physical attractiveness than women are with a man's. To know sex at such levels is to better know ourselves.

Happily, the growing number of serious scientists engaged in the study of sex have a lot to tell us. For the best of them the story of sex isn't about mechanics featured in "sex education" film strips or some arcane account of genes and molecules, but indeed a way of organizing and thinking about all of the things we know or think we know about what is most central to human nature and behavior. If there is an overarching principle informing the story of sex, it's some version of what Martin Daly and Margo Wilson write in their book *Sex, Evolution, and Behavior:* "All social organization is in principle interpretable as the outcome of the sexual strategies by which animals try to reproduce themselves."

Over the past decade, those seeking sex's nature have overcome conceptual, social, political, and scientific obstacles and moved the field to a golden age of collaboration, sharing imaginative and even revolutionary ways of looking at the information they gather. To an astonishing degree, they have learned how to "ask" minds, brains, bodies, cells, and molecules—in creatures great and small—how sex works. To an equally astonishing degree, they are getting answers by learning how events that happen at one level of biology—in brain cells that make and react to chemicals, for example—influence, control, start, stop, and account for action at very distant levels. They are teasing out long hidden facts that tie sexuality to its biology *and* its circumstances.

Building on dramatic discoveries of the twentieth century, scientists are linking genes and the chemical blueprints they carry to both *secare* and *eros.* Their turf now ranges widely, from neuroscience and psychology to genetics, cell biology, endocrinology, and evolution—because nothing, no part of our conscious or unconscious lives, stays untouched by sex. When we're sexually attracted to someone, the allure plays itself out in our hormones, genes, and behavior, in our "hearts" and in our minds. When we encounter sex, it is simultaneously present in our bodies and brains, our memories, even in our culture. No wonder then that "Sex research covers everything from A to Z, anthropology to zoology," says Howard Ruppell, executive director of the Society for the Scientific Study of Sex. It embraces the enormous variation in human and animal sex behavior, not only between species but also between individuals in any one species. It brings psychiatrists and evolutionary biologists to the same meetings as zoologists and archaeologists to track each others' research

and build on what each has learned. More than a half century ago, embry-
ologist Frank Rattray Lillie of the University of Chicago first suggested
that male hormones secreted in the womb were responsible for the fact
that when cows have twins of opposite sexes, the female is sterile and
behaves more like her brother the bull than other cows. Since then, scientists have systematically looked beneath the blankets and the skin to
where the biochemical action of sex takes place. There is biological music
that accompanies sex and the goal of science is to offer at least eight bars
everybody can hum.

What universally characterizes the best of those contributing to our
understanding of sex is their amazing ability to develop ways to even ask
such questions as whether there might be molecules of monogamy or
hormones for love. I am humbled by how hard it is to scientifically demonstrate what we lust after without a second—or even a first—thought.
Most of us glibly throw out such lines as "men and women are different
below the belt and in the brain" or "the brain is responsible for behavior,"
but scientific support for these sweeping generalizations required intellectual tours de force. Teasing out the secrets of the body-mind-brain-
emotion-hormone-gonad circuits is a task that brings weak scholars to
their knees. Just exactly how *does* a biologist ask a male bird or bird-
watcher to objectively reveal what propels him to love or make love to a
particular female?

He does it by exploiting the theoretical knowledge and technological
means available now for the first time in history to map the brain, measure
and regulate hormones, manipulate behavior, track DNA, and link specific complex traits to individual genes. With a discriminating eye on the
work of Alfred Kinsey, William Masters, Virginia Johnson, and even Dr.
Ruth, scientists have brought to sex research the tools of molecular biology, anthropology, social biology, brain imaging, and genetic engineering.
Now they are moving quickly to apply these tools to mating in all its
forms and functions, integrating discoveries from songbird brains and peacocks, one-celled creatures and fruit flies, worms, parasites, mice, and
men. Riding the wave of technological and conceptual advances in their
investigations of "inner sex," they've devised a way to collect and track
individual daily urine samples to test for sex hormones and sexual cycles
among gorillas in the wild. ("They won't pee in the bottle," one quipped.)
They've knocked out individual genes to identify biological causes of sexual aggression and parenting, designed glass bottles so that everyone can

watch fruit flies with a certain mutated gene seek same-sex liaisons, constructed cages that let virgin female mice sniff but not mate with males
to see if the mere presence of the male within smell range triggers brain
changes in the female (it does). They've fashioned toy dummy wasps to
fool males into mating, strapped artificial weights to beetles to test the
notion that males give females a substantial amount of body weight during
sperm transfer and that females prefer heavier males. They've placed tiny
microphones near insects to measure their "bedroom" sounds and made
digital recordings of novel copulatory screams by macaques to analyze
how the brain organizes to perceive and react to sex signals.

By tracing connections among brain activities, hormones, emotions,
and anatomical structure, investigators are sorting out with some certainty
how oxytocin, the natural hormone that triggers the uterine contractions
of labor, keeps female prairie voles faithful to one mate; how it is that a
woman in love can have orgasms simply by *imagining* a tongue on her
clitoris or breasts; how Harlequin paperback romances and fantasies are
so powerfully arousing; and why, for a nursing mother, the oxytocin,
secreted by cells in a primitive part of her brain, makes her nipples tingle
with pleasure as her milk "lets down" or at the thought of a lover's touch.

At the benches and in the field, sex scientists have learned that if you
stare intensely into the eyes of an attractive girl you've just met in a bar,
she'll likely avert her gaze because it's *physically* uncomfortable to do
otherwise; and that when love is lost, it *does* feel like a punch in the gut.
They also have learned that if you become her lover, she'll stare back;
that what makes our hearts pound as we flee a mugger or build to a
sexual climax is the same as what gives a long distance runner her "high";
and that the same muscles that constrict the anus of a novice bungee
jumper are called into play to intensify the pleasure of orgasm. (Indeed,
it may be the "thrill" of that constriction that impels the jumper's two
hundredth plunge.) They've even amassed evidence that brain chemistry
governs pair-bonding in neurochemical cycles during reproductive years,
generating emotional attachment peaks and valleys that roughly correspond to prehistoric time requirements for child rearing as well as Americans' patterns of divorce.

Even if some of these findings fail to hold up to further scrutiny, they
already have forged a perspective that has enriched sex research. Scientists
are thinking "out of the box," searching for what David Crews calls "com-
plementarities" instead of singularities. If it's true that like other animals,

men and women have subliminal odors that attract the opposite sex, the pheromones we've been hearing so much about these days, that's interesting. But if it's also true that for any particular man, only *some* women attract him, then clearly something *more* enterprising is going on between the two sexes.

The story of sex is crowded with such accounts, some as irresistible as the sex drive: of what makes semen smell musty and taste sweet, and the sensual behavior these things *create* as well as reflect; of the erotic lure of the deep, thumping bass rhythms of a rock 'n' roll band; of the literally wilting effects of depression on the messages between brain and penis.

In telling the story of sex, others have done so from a particular point of view (feminism, for instance) or singled out a particular phenomenon (e.g., sperm competition) or expert theory as the dominant theme or basis of sex and sex behavior. There are also those that explore solely the evolutionary origins of sex. In this category is Matt Ridley's *The Red Queen*. Sex exists, Ridley writes, because it is the only reproductive design that can constantly mix genes and provide a necessary supply treadmill of potentially advantageous, disease-resistance genes that keep us from succumbing to the endless onslaught of bugs determined to kill us off. The process, like the subjects of the Red Queen in Lewis Carroll's *Through the Looking Glass*, "takes all the running you can do to keep in the same place." But at the end of the day, scientists don't really know *precisely why* sex exists.

Still others focus single-mindedly on hormonal chemistry or brain anatomy to fully explain some aspect of human behavior. The present story of sex, however, attempts a broader sweep at the rich stash of scientific discoveries about how sex gets done each step along the way to genetic survival. Astute scientific readers will note, and may even be disturbed by, the book's deliberate shifts between the evolutionary biologist's ultimate time tests of human behavior, those measured by persistence over eons; and the physiologist's moment-to-moment reconstruction of the biochemistry of *now*, the ebb and flow of action and reaction, brain receptors emptied and filled in response to some proximate event. ("Physiology" is derived from the Greek words for the science of natural events and causes.) Certainly it would be a mistake for the reader to conclude from these shifts that, with only a few exceptions, physiological research in today's *Homo sapiens* has been, or ever can be, done to show that the latter absolutely supports the former, or vice versa. Nevertheless, I've moved without a stamped and valid passport between the two sciences, a liberty taken in the interest of keeping the story going. Trying to describe sex from so many different perspectives opens the text to attack from all sides. Indeed, thoughtful scientific readers, including Sue Carter Porges and James Weinrich, cautioned me properly against it. I fully recognize the danger of such speculative time travel, but risked it to offer fluid possibilities, a view of both camps and how they might be integrated.

Journalists, including myself, like to construct books at least partly out of a particular point of pique, and mine was founded on damage done to human sexual inquiry by the Freudian century. This book about sex contains only a single mention of Sigmund Freud, this one, a mention made only to say that it is *possible*, at last, to write a whole book about sex and never mention him at all. It's not so much a whine about his implausible impact as a testimonial to what rigorous scientists know about sex. A reporter's romp through the scientific literature of sex ought to tell a *knowledgeable* story about monogamy *and* its molecules; on courtship *and* its hormones; on gender *and* its genes; on sex behavior *and* its brain connections; on scholarly sex *and* good old fashioned "dirty" sex.

This account of sex also gives importance to the observation that the legendary battle of the sexes both begs the question of how we get to resolution of sexual conflict and denies the cooperative aspects of sex. And it appreciates that perhaps the largest differences in sexuality may not be between species so much as between individual males and females of all species. Especially the human species. There is, says Crews, who has exhaustively studied the role of hormones in the sexual behavior of animals, "enormous plasticity" in how each of us plays our sex scenes, based on novel circumstances and our physiological ability to adapt to such endless variety. "The variation among individuals within a sex is usually greater than the difference between the averages for each of the sexes," Crews says, and "[a]lthough the sexes differ in many traits, this often is a statistical phenomenon. Even as adults, each individual is capable of displaying the behaviors of the opposite sex and it is rare to find a truly sex-specific or sex-exclusive *behavior* at all." Indeed, it is altogether possible, even likely, that the complex string of traits we call sexual behavior came *before* the evolution of egg and sperm, penises and vaginas, the division into male and female. And more, that males emerged only after so-called hermaphrodite organisms—females that could reproduce themselves—first materialized.

Researchers may never fully pin down specific molecular tracks left by sexual exclusivity, sexual accessibility, sexual attractiveness, and emotional devotion, the strung-together quartet of human sexual strategies painstakingly identified by evolutionary psychologist David Schmitt. But, there is logic in the concept that the brain houses the mind, and the mind functions as interpreter of the slim space between our inner selves and the outside world we deal with in every moment of our lives. For biology's parishioners, the scientists in pursuit of those molecules, the mantra is "I wouldn't have seen it if I hadn't believed it!" rather than "I won't believe it until I see it." What makes the science of sex so alluring and so challenging is what makes biology, not physics, the twenty-first century's hottest and most complex science. Because unlike in physics, where a knowledge of mechanical and particle characteristics and behaviors leads to working models and provable causes (e.g., we stay rooted to the Earth because of gravitational force), in biology causes are never the sum of the parts, but rather the sum of parts interacting in particular organisms at a particular time in a particular environment. No one has seriously challenged Copernicus for centuries; evolutionary biology is still a "theory" to a majority of American citizens. Indeed, scientific information can arguably never tell us what is always or absolutely true; it can only, as biologist Ernst Mayr concluded after eight decades of inquiry, approximate what is possibly or probably *believable*.

That scientists get, for now, believable, even cogent answers that occasionally stand the test of challenge and analysis, time and reason is amazing, and I have tried to faithfully tell their stories.

They in turn have relied on the mostly involuntary subjects of the story of sex: people and animals on whom experiments are made or who themselves are experiments of nature—the diseased and disordered and "different" across a zoo of species. Besides the humans, homage must be paid to prairie voles, whiptail lizards, red-sided garter snakes, Japanese quail, songbirds, possums, chimps, lobsters, blue-headed wrasse, one-celled algae, wild and inbred mice, rats, roaches, and dozens of other species who have "donated" their brains and bodies to science. After all, evolution has no direction toward perfection or superiority, so there's no way to say what is the best or more advanced model from which to draw lessons about the biology of human behavior or human nature. But I am persuaded by those whose work dominates sexual biology that there has perhaps been too *little* appreciation of how parallel are animal strategies— biological and emotional—for working out the "problem" of sex and too much emphasis on our self-consciousness to explain our sexual selves.

Just as we did not evolve opiate receptors in the brain to enable addiction to heroin, we have not evolved as sexual beings to enjoy sex. Nature's driving force does not particularly "care" if we have love with our sex, or multiple orgasms, only that we *successfully* replicate. "If Arnold Schwarzenegger never has a child, in fact if he does not have a lot of children, then in terms of evolutionary fitness, he is a 90-pound weakling," notes molecular biologist Shirley Tilghman, now the president of Princeton University.

A further statement about sex and sensibilities seems in order, especially if, like me, you grew up on Woody Allen's neurotic reflections on sex.

Almost universally, sex, even in fish, lizards, parasites, and peacocks, is a research orphan, or worse. Conducting human sex research requires a fighter pilot's skill and courage, and there's no parachute to safety. A potent piece of evidence for that statement comes from Diane deMauro, in her 1995 report "Sexuality Research in the United States: An Assessment of the Social and Behavioral Sciences." She notes that the work of the late Alfred Kinsey is nearly a half century out of date. Very few besides William Masters and Virginia Johnson, whose groundbreaking work is decades old as well, are lining up to make sexuality a "primary focus" of their research. (I found it remarkable that a bibliography of thirty-six key publications listed on the Web site of the Kinsey Institute was dominated by vintage books and articles. Only eight were published in the 1990s and six in the 1980s.)

Scholars know, often from bitter experience, that without the instincts of a cold warrior, studying sex can be dangerous and getting published punitive. Nearly every scientist I interviewed (more than fifty) volunteered that she or he had felt the threat of career-killing disapproval by public and private funding agencies, politicians, institutions, and the public. They prefaced every conversation with a plea that I take special care not to tie their observations too closely to human sex, even though they believe that, generally speaking, what happens in a mouse *is* relevant to what happens in a man.

Sometimes, the information I sought in published materials was deliberately buried by the authors in the arcane code language of their specialties, curbing visibility by all but their savvy peers. Without their help with the code, I might never have looked for, much less found, important information on female orgasm in articles about pupillary dilation as a measure of vagus nerve response to electrical stimulation.

However we get them, the facts summon us to become explorers of sex, to look "under the hood" for those events at which the brain meets behavior, the mind's capacity for love meets the body's compulsion to act, chemistry meets charisma and romance reproduction. When such information is melded with observations among our near and distant ancestors, and within the strict terms of evolution—if things *work* they tend to be copied—scientists can simultaneously train the microscope and the telescope on our most primal activity and create a picture that is both detailed and in perspective. They're giving us the forest *and* the trees.

No part of the human experience falls outside the purview of sex. Aggression and violence, culture and commerce all have an impact on, or are influenced by, sex. There's a sexual role for the senses and symmetry, dance and disease, fantasy, feelings, feminism and film, folklore and gender, genetics and humor, IQ and inhibition, language and love, medicine and memory, morality, pain, painting and perversion, chemistry and politics, racism and rape, religion and science. We *evolved* to think Renoir's paintings beautiful. The romanticism and eroticism they evoke for people everywhere are no random consequence of the processes that help to assure sex and reproduction.

Arguably, we invented cities, clothes, art, history, the work ethic, the family, and every human investment and venture because of sex, as a result of sex, to recruit sex, to be good at it, to master it, capture it, and ultimately to try and contain and control it. An economist looking at a societal organization chart might well conclude that while sex is freely distributed, it is never without cost, and that it may well be the stable currency required for every meaningful transaction. For sex is forever and completely tied to our identity, to the best and worst of who and what we are and to our stake in the future.

At a different level of discourse, we might consider sex research the way we think about the study of cave paintings at Lascaux, Chauvet, and Cosquer, which altered forever people's perspective on the development of modern human beings. Like the French cave art, vivid experiments of nature and modern science are extending our vision of sex and its role in all of life. To apply wishful or wistful political, social, or religious agendas to sex, to put facts into a frame that limits how we think people *ought* to behave, or *once* behaved, is to commit the same mistakes as those who argued that cave art was a prank, that early man was too "primitive" to have made art so compelling, creative, symbolic, and abstract. Believing—

or not—that twenty-first-century human beings have or should have evolved beyond what "prehistoric savages" did adds nothing to our knowledge or understanding of what sex is, or how it works, any more than judging a drawing tells us what art is or how it came to exist. "There is great risk," evolutionary biologist David Haig warned me, "intellectually, scientifically, and socially in trying to draw moral lessons from sexual phenomena. All attempts to do so are a form of power grab or manipulation."

What else we can say with certainty is that sexual reproduction is so successful, so much the winning strategy in the survival sweepstakes that everything we see and understand about animal and human sex reflects solutions to quite literally life-threatening problems. If human sex were a Michelin guide for survival, every route and signpost; every superhighway and footpath; every mountain, valley, overlook, stream, and city; every detour, fork in the road, and attraction would represent field-tested, three-starred journeys chosen and endured by generations of tourists who never planned anything, including a trip home. There are no package tours, no guarantees, only origins and altered destinies. Getting there is not half the adventure, it's all of it. When we know *how* it all works, we increase our appreciation of the journey.

In any case, after nearly five years of sifting the work of scholars interested in the matter of sex, I am mostly still full of the wonder of it all. (Also, I am numb to double entendres gleefully exposed by friends and loved ones—and even my ceaselessly patient editor—after every innocent utterance, including "I'm coming!")

Meanwhile, the only certainty in our sexual futures is change wrought by failure and success, death, and adaptation of cosmic dimensions. Like every biological event, sex and the mechanisms that support it are subject to ongoing experiments of nature, some surely destined to be "dead ends" and some adaptive. Perhaps the "two-sex" system as the dominant animal mating strategy, and all "acceptable" sexual behaviors, are merely way stations en route to extinction, making way for different strategies, perhaps some that can better reconcile or embrace our spiritual, thoughtful, intelligent, emotional traits. Conceivably, heterosexual sex, and two genders, are a transitional set, much like the dinosaur archaeopteryx Melvin Konner wrote of in *The Tangled Wing,* "a piss-poor reptile and . . . not very much of a bird." Are we a "piss-poor mammal" and not very much of a "spiritual creature" working our way to something else?

Contrary to our egocentric wishes and vague memories of grade school Darwin, we cannot, with any ease, consciously or willfully change how our bodies and minds *do* sex to meet the whims of Western civilization, urban culture, religious beliefs, and environmental destruction. Our uniquely human influence on the process can be profound, of course. We perform in vitro fertilization and genetic engineering of single traits; we contracept and prohibit, regulate and rule-make. Nevertheless, *secare* and *eros* are our once and (foreseeable) future destinies. Nature's game rules still trump our will, and our personal version of sex cannot override biological or psychological legacy. Through sex and love comes our only immortality, through our children. (Or, more precisely, through our grandchildren, since from nature's perspective, parents' only real assurance of a genetic future is the fertility of our *offspring*, not ourselves.) The only sensible challenge—and real reward—is to catch nature's own act.

SEX

SEX THROUGH THE EONS

"Nothing in biology makes sense except in the light of evolution."
—THEODORE DOBZHINSKY

"Is sex dirty? Only if it's done right."
—WOODY ALLEN

*"Amoebas at the start were not complex. They tore themselves apart
and started Sex."*
—ARTHUR GUITERMAN IN *THE LIGHT GUITAR*

Unconcerned and unhampered by those watching them, Malko and Lana focus only on each other and what will happen now. Tired of the chase, the teasing, the indecision, and of shopping around, of arguments and jealousies, the two find retreat in the small world that lovers find, even in a crowd. They lean toward each other, Lana peering deeply into Malko's face. They touch, shyly at first, then almost ferociously, gazing at each other. Malko, almost tripping on Lana's heels, pushes her ahead, their dinner left uneaten. Indifferent, for the moment, to the envy of those around them, their glancing pinches, rubs, taps become firm and rhythmic strokes across each other's breasts, genitals, and buttocks. Their arms and legs are everywhere, nudges accompanied by flesh-tingling nips and muzzled, urgent sounds. They make louder sounds, their mouths open and gaping. The closest observers watch their arousal with excitement, sensing the sexual charge in their shudders and twitches, the concentration in their posture. Malko pulls at Lana and she pulls at him, and they tumble around like playful children, their exposed and naked bodies grabbing, reaching, and wrestling. Now they hug, fondling the tenderest spots. Upping the tempo, they grope, massage, rub, lick, suck, and tug at each other until their genitals, torsos, faces ache from the contact.

Finally, after maneuvering themselves face-to-face, Lana rocks back on her shoulders as Malko mounts, inserting his stiffened penis into her, both of them concentrating on the penetration. They seem to prefer the eye-to-eye encounter spurred by the "missionary position" they clearly have mastered. Quickly, the compelling and predictable climax proceeds with rapidly tightening and relaxing movement, sometimes synchronized and sometimes uneven, but always faster and exhausting. With Malko astride, Lana's limbs askew, their bodies grind, tense, and then abruptly relax. Almost immediately, a female onlooker impatiently moves alongside them.

Lana turns her attention to the newcomer at once and soon she and Lana—tired or not—push their pelvises together in a fierce genital-to-genital rubbing, giving louder and louder voice to Lana's renewed delight. For his part, Malko, a willing accomplice in this ménage à trois, quickly makes it a foursome, as he invites closer a male friend just as eager for sex play. As Malko and his companion grab themselves and each other in the motions of masturbation, Malko spends a brief moment to glance at his two younger brothers nearby sucking on each other's penises, while Lana and her girlfriend move toward a circle of others.

This is sex. *Normal and routine*, but at the same time breathtaking sex. The kind of sex some know firsthand and many more of us want. Empty of inhibition, full of dirty dancing, irresistible urges, variety, and endurance. Genital-swelling, chest-heaving, fantasy-pushing, pulse-quickening, nerve-tingling, muscle-contracting sex. Dirty sex done right.

What makes it so normal, so *natural*, is that all of it occurred not among adventurous or oversexed Homo sapiens, but among our closest biological relatives, the bonobo apes, so-called pygmy chimps, whose face-to-face couplings, along with their irrepressible, carefree, homophilic, and even pedophilic bouts, are in the animal kingdom so far as we know shared with very few species besides humans.

Bonobos, with presumably no choice in the matter, have left it to the forces of natural history—of natural selection and time—the task of shaping, determining, and defining their sexual bodies and brains. With this sexual physiology comes the behavior that only their human cousins call free love: "perversions," and an open prostitution of sorts, in which there is a frequent exchange of sex for food and other favors within their small troops. And to some enduring if indeterminate extent have we, their descendants, also made use of such forces—unless, uniquely in all the animal kingdom, evolution boycotted our human forebears.

In our minds and in our myths, human sex, through all of its preludes to intercourse, orgasm, and the physical mechanics of begetting and safeguarding the next generation, may appear to involve more sophistication, grace, nuance, and complexity than a roll around the forest floor in exchange for a ripe fruit snack. All those dates, chaperoned courtships, prenuptial agreements. All that *romance* and *love* and *commitment*. As it turns out, the scientific study of sex in all of its variety has uncovered the reliable presence of a sexual arc, a predictable biological and behavioral continuum in every species, including our own and our nearest evolutionary relatives.

In each species, the arc may be comprised of segments that differ in their duration, their expression, their extravagance, their subtleties, or their lack thereof. But in whatever animal scientists have examined, unmistakable biological and behavioral counterparts exist for such segments of the human sexual arc as attraction, courtship, arousal, and allegiance. If nature abhors a vacuum, it certainly adores a pattern. As in the preliminary sketches for medieval tapestries, the patterns of sex emerge regardless of colorful variations and shimmery silken threads applied by innovative weavers to the designer's basic, woolen plan for warp and weft.

We can find it fun to pay attention to the variety. Bonobos nuzzle, bite, eat, and fondle, apparently enjoying the tease and tickle when breeding and when not. Among mandrills and orangutans, by contrast, males completely ignore both females and the young except between June and November, their breeding seasons, and otherwise don't even socialize with other males. Little red flying foxes copulate while both he and she hang by their feet, heads down, supporting their own weight and in complete silence. They begin with twenty seconds of oral sex, he then thrusts for precisely twenty seconds and ejaculates for nine. After exactly 175 seconds of passion and 35 seconds of rest, they start all over, always for these same durations. Among lions, pairs have intercourse every fifteen minutes for three days during the mating season. Antelopes and elephant shrews take about three seconds to complete the act, while harlequin toads of Latin America, whose male members grab females from behind, hang on sometimes for *months*. For penguins in Patagonia, monogamous old flames return each year to have sex outside the family home, while male Australian marsupial mice go at it for twelve hours and literally die of exhaustion. Bat rays do it with the females on top—and only in moonlight. A pig's orgasm—at least the physiological elements that conclude coitus—

is said to last half an hour. Meanwhile female baboons give in to the dominant male's demand for sex, but often sneak off with a bachelor "stranger" who mates, then leaves with her blessing.

As for romance, is there an animal equivalent of dinner and drinks, flowers and foreplay? Male bower birds build elaborate constructions, decorated with particular color schemes, impressing females with their expensive and energy-consuming habits as prelude to luring them into sex (but never in the bower, by the way). Male peacocks, splendid tails in full-fanned graceful elegance, prance while the plain peahen makes up her mind about his bejeweled seduction. Once she accedes, the sex is almost violent, and later he has essentially nothing more to do with her at all.

Vive la difference. But the important thing is not to be distracted by *la difference.* The patterns, the sexual arc, remain beneath the glitz and goofiness. We humans may think we are the sexiest, most inventive and urbane creatures—or the most debased and debauched. But we are nei-ther. Consider a yuppie couple in hot pursuit the way a bonobo ethnol-ogist or biologist would. In moves that social scientists have observed so many times they seem scripted, such a couple, having dinner and drinks in a swank restaurant, surrounded by a crowd, tune out the world and start their presexual "dance." It begins with the couple sitting forward in the restaurant chairs, heads nearly touching across the small table. In moves the bonobo observers would instantly recognize as presexual "grooming," she sweeps imaginary lint from his suit jacket (the tentative touch). He pushes a small wave of hair from her face (the eager response).

Soon enough, fingers mingle and linger together, hands cup each other's. They may feed each other from their plates, but will often leave most of their dinner untouched as Malko and Lana did. They stare into each other's faces as if looking for engraved micro-messages on lips. They whisper. They chirp. They laugh with *open* mouths. They smile Mona Lisa smiles. They walk from the table to the door hip to hip, almost in lockstep. Soon they, too, fondle, massage, rub, suck, lick, tug. They too tumble around, nip each other, squeeze and pull at each other's genitals, buttocks, and faces. And they, too, have oral sex before intercourse.

What bonobos, people, mandrills, foxes, lions, peacocks, and all the others are telling us is that sex as we know it, sex with each of its com-ponents and all of its complexities, contradictions, rituals, and all-consuming appeal, has a natural and vital process and purpose deeply rooted in the origin and persistence of life itself. In the origin and devel-opment of species, no surviving component of sex, can be considered *unnatural* or unnecessary. *All* aspects of sex observable in animals today, no less than sexual reproduction itself, are what biologists and psychologists call "highly conserved." All aspects of sex are the evolutionary winners across eons of natural selection, of trial and error. They persist in us and every other creature precisely because of their importance in survival.

What else this conservation tells us is that the critical components of human sexual natures and acts are so important to survival that they must be determined, in some part, by mechanisms that assure their transition from one generation to the next. Scientists hunting for the foundations of sex have done well to search in genes, genders, brains, bodies, minds, and chemistries.

Sex as people know it did not spring forth in some biological "Eureka!" Nor is sexual reproduction—the fusion of male and female gametes held apart by separate sexes—a human birthright. Both evolved, taking shape in changing circumstances, by fits and starts. Whatever sex looks like, it represents and reflects the triumph of *sexual* strategies for reproduction. Sex and all its accouterments are with us for one reason: the process of sexual selection, first described in detail by Charles Darwin. As with natural selection, in which traits that stick around are those that confer survival benefits, sexual selection favors any physical or mental characteristic that helps sexes compete within members of their own sex and the other to get the best mates or best access to choice of mates. In short, to win the reproductive lottery. Every trait, from strength, size, and beauty to cleverness in recruiting allies or neutralizing enemies, is our sexually selected legacy. Says evolutionary psychologist Geoffrey F. Miller, "natural selection is about living long enough to reproduce. Sexual selection is about convincing others to mate with you. It's about animal minds, our ancestors' minds, influencing other minds. . . ."

Does this mean that human sex and sexuality today are all in our biology, fixed in genomic cement? Not at all. There are clearly distinctive aspects brought to us by the wonders of the cerebral cortex and millennia of events that only we, with our big brains and appetites for social organization, could have wrought. Since the Stone Age at least, sexual behaviors have been influenced by social and cultural heritage as well as genes and biochemicals. But in the distant past, in places the yuppie couple are unlikely to find even on a map, our "sophisticated" repertoire of sexual behavior—getting to "yes, let's"—was pretty much worked out to the mutual satisfaction of generations of he's and she's of countless species whose physiological operations responded to opportunity. Other males and females may have somewhat different versions of what it means to flirt, seduce, conquer, submit, and (presumably) "come." But all of them got to the present the same way, through a natural history.

How do we know? What is the evidence that sex, in all its permutations, even ours, is (1) essential to the life of species and (2) originated and developed naturally over time? The answer turns out to be less straightforward than one might think. For while we might view sex as inevitable, and immutable, the scientific record shows us nothing of the sort.

To begin with, sex is both costly and inefficient, which is no doubt why various creatures periodically have abandoned it. Some aphids and water fleas, for example, breed without sex on occasion and sexually at other times. And there are enough non-fertilizing self-sustaining forms of life (grass never needs fertilizing unless you stop cutting it—although lawn mowers are not a part of the evolutionary process, grazers certainly are) and enough parthenogenetic species (in which all the members are females) to suggest that one "sex" is probably enough, as long as it's female. Even among vertebrates, at least twenty-seven species of reptiles, mostly lizards, are all female and produce clones of female babies. Moreover, those who would conclude that sexual reproducers rely only on two sexes would be wrong, too. One slime mold studied by biologists has, according to one estimate, fifty-eight thousand genders, if by "gender" we mean a distinct variant of sex strategy and behavior.

Theoretically, at least, virgin birth is possible, even among humans. All that's needed is a woman in whom an egg arises with mutant genes and begins making extra copies of its DNA and then divides and develops. Domesticated turkeys do this more than infrequently. All offspring produced this way would be female, too, of course (X chromosomes only), and therefore "parthenogenetic." If such a woman's daughters were able to carry on this family trait, her genes would really thrive compared to her neighbor who needed men's sperm, because parthenogenetic offspring carry twice as many of their parents' genes (100 percent) as do sexually produced offspring. Within a few generations, this female clan could easily overwhelm the sexually reproducing neighborhood.

Yet 99.9 percent of the species of life on planet Earth don't operate this way. Asexual reproduction just isn't the norm and virgin births just aren't popping up these days. Even in those turkeys, it turns out that when they are healthy, virginal births don't happen. In fact, in turkeys, wasps, and other creatures, sex can be switched off by parasitic or viral diseases.

So why is there almost universal two-sex, male-and-female, egg-and-sperm sex? How did it get established? (Even in plants, where hermaph-roditism, or single-sex sex, is the norm, gender dimorphism is a common evolutionary trend in so-called angiosperm species.) Is sex some leftover event, something that arose randomly in the deep past, then stuck around because it worked and faced no reason to go away? And even if sexual reproduction proved to be a useful way to join the genetic material of two individuals and produce offspring with some combination of parental DNA, what could possibly account for all of the courtship and mating rituals, stratagems and contortions that people and other animals go through to get the job done? Wouldn't random, mindless copulation be a lot easier on everyone and everything, with every animal of a species interested automatically in every other animal of that species? Why have to search for Mr. or Ms. "Right"? If sex is the answer to life's prayers, then why the disruptive and confounding "battle of the sexes"? Why two sexes with vastly different ideas about what is sexy and what is sexually proper?

There is not *an* answer to these questions. No simple "Because that way . . ." response that can make it all come clear in a moment. This is what we have, and to understand why, it's useful to think about how a rational, self-perpetuating, safe, and effective system for ensuring survival of members of a species might be designed by natural engineers. The specifications would be pretty clear. You'd need a system that would assure flexible ability to adapt to new situations, to threats, predators, and disease. You'd need one that ultimately would involve more than conflict, since war is hell on the survival of any species. You'd need a system that was, in and of itself, rewarding and "good" for individuals in a group, to keep individuals at it. And one that would assure that each of us could "win," which means each would succeed in sending his or her own par-ticular DNA into the next generation and beyond. Finally, you'd need a system that had built into it assurances that not all offspring conceived or born would live forever (genetically speaking), as that would over-whelm resources, conserve bad genes along with good, and kill off the entire population. When scientists go looking for the origins, purpose, and processes of sex, they have these suppositions as guidelines to organ-izing their treks through psyche and soma, and to what they find in the human sexual enterprise. Much of what they've found pretty much fits the specifications. Notes James Weinrich of the University of California, San Diego, "Natural selection . . . acts in a way so that over time, individuals which come into being are nearly all very good at trying to win, so most of them do well to some extent."

Consider, for example, that men across all cultures and ages are relatively more promiscuous than women in choosing sex partners. Yet women, although in our recorded history more restrained, are no less interested in sex, even if they are less interested in sex that makes babies. A male's mating "job" is essentially over after fertilization, and at first blush, it seems one-night stands are all he needs to send his genes into the next generation at little or no further expense to himself. A female, on the other hand, must always consider the need for resources to rear any offspring, and it's hard to be a single mom. It's true that once he knows his efforts will produce a baby, Dad might have more interest in making sure his genes survive to impregnate another generation, so he may be more willing to stay home in the cave and hunt the bison. But often, that's not altogether so. Mom is relatively always more interested in rearing the child (it carries her genes, too, after all). She is aware that her lifetime production of babies is far more limited than his: all the eggs she'll ever have are with her at birth and from then on she begins losing them faster than she can use them; he, on the other hand, can produce sperm into his nineties and beyond. Thus, she needs to conserve her strength in order to send lots of copies of her genes into the future. And though Dad may lust for the bimbo, once he gets invested in offspring at all, he is just as likely as she has always been to want someone who can think and solve problems and stick around to care for junior for ten or twenty years.

Nevertheless, the male of the species *Homo sapiens* is far more upset if the female fools around than she is at his one-night stands. According to a study performed by University of Michigan evolutionary psychologist David Buss, when men are asked to just *imagine* their wives in bed with another man, their heart rates speed up to the equivalent of having had three cups of caffeine-rich coffee.

How does all this fit those specifications? What's partly thought to be behind this heart-pounding response may well be the fact that because a woman's egg is fertilized inside her, hidden from him, a man can't ever be sure that child is his, so his need to watch and guard his mate is urgent, and jealousy is *her* proof that he's truly invested in this infant. Likewise, women need to spend decades of time and energy in pregnancy, birth, and child rearing, so emotional infidelity is far more threatening for them than a man's one-night stands.

Similarly, if men are more likely to choose a sex partner for her big breasts and symmetrical face, and women are more likely to choose a partner who has wealth, power, and status to father her children, both choices only make sense from a biological and psychological standpoint as the outcome of sexual selection. If men are more enthusiastic about sex with *anyone*, and women are less swayed by a handsome face and a great body; if women want a sensitive male and men could care less if a woman is smart at a first meeting, these sexual traits and behaviors, no less than intercourse and orgasm, have been selected and entrenched in our natures.

Of course, it doesn't hurt if her choice is handsome, but ugly kings can be as potent as gorgeous gigolos and have more coins in the coffer. If only pretty people mated, humanity would soon be extinguished because, like highly inbred racehorses or show dogs, selection for "beauty" traits alone can lead to exaggerations that benight the species. Some evolutionary psychologists studying college students (i.e., young, mostly fertile, unattached, and free-to-roam men and women) have found evidence that the evolution of human brain and mind was heavily influenced by the need to assure the transmission of a person's genes into the next generation, and that means men want nineteen-year-old Barbies and women want somewhat older, well-heeled, and stable sex partners. But to suggest that any man who fails to look for a Playboy model lacks a truly male brain, and that any female who fails to fall for some combination of James Bond and Bill Gates is equally a biological loser, misses the big point.

In a revealing essay in *The New York Times*, Meredith F. Small, a Cornell anthropologist and an authority on how biology and culture shape our behavior, pointed out, "I have . . . chosen to spend the last decade with an artist, a man with no money and few goods beyond a bunch of paint brushes. Silly me. But then he's a loser as well. . . . He has chosen me, a woman much older than himself, one with low fertility who hasn't seen a decent hip-to-waist ratio in years. Are we really losers?" The answer is a resounding "no," Small says, not because these patterns don't exist, but because they don't tell the whole story.

In most surveys of college kids' perceptions of an ideal mate, the list of traits does indeed include looks and largesse. But what also shows up at the top of the list for both college men and women in these surveys, says Small, are such things as "kind and understanding," "intelligent," and "exciting personality." Yes, says Small, in similar surveys done by David Buss, men do rate looks somewhat higher than do women and women rate wealth a bit higher than do men. But they "ignore the obvious and higher rated results and men and women surprisingly want the same thing." Adds Small, "we have a long evolutionary path during which our brains and bodies adapted and are still adapting."

In historic and traditional cultures, women do "look for status in men because they can't get it alone. And given the choice between an old wealthy guy of high status and a young one, any sane woman would take the young one. What person wouldn't want a partner judged ideal by the cultural standards. At the moment, skinny girls are offered up as the ideal and so men want that type." But in the real world, she notes, the vast majority of men don't get Gwyneth Paltrow or Kate Moss, just as a generation ago they didn't get Liz or Sophia, either. Concluded Small, "Genes are passed along in the real world where attraction, love, mating and marriage are not always linked. In reality, people choose mates (or have them chosen for them) in myriad ways. Our brains are not designed specifically to desire this or that, but to *weigh options* [italics mine], and then get down to the business of having babies." In this scenario, Small and her artist aren't losers but typical, embracing both the traits they share and those that may conflict.

At the level of DNA and molecules, or even sperm and egg, similar conflicts—and cooperatives—of interest serve the system well, with the conflicts in need of eventual resolution and tolerance. In this scenario as well, evolution has produced various means of molecular management that are the match of what goes on outside the skin, and sex is forever bound to a remarkable panorama of behaviors, ideas, strategies, and traits. From Barbie doll figures to peacock tails and rooster combs, from monogamy to infidelity, from the war between the sexes to the most romantically drawn intimacies ever imagined, we *evolved* all of the processes that drive sex. Women have eggs and men have sperm and only when the two *negotiate and agree* to get together is there the best chance of survival for each. "Secondary" sex traits and behaviors—the bigger breasts, the hairy chests, the flirting and the flash that favor connections, courtship, and commitment are as much a portrait of sex's origins and evolution as sperm and egg.

For the past half century or so, evolution's mechanisms have been uncovered by platoons of biologists, psychologists, and anthropologists searching systematically for the sustained, sustainable, perpetuated, and preserved underpinnings of courtship, flirting, promiscuity, commitment, and the whole range of human workings and behaviors involved in sex and reproduction. Even rape, fetishes, bondage, and other so-called aberrant sexual behaviors are almost certainly biologically predisposed, if not adaptive, and may therefore be what biologists call "conserved" traits, attributes or properties useful or essential to life across all cultures and genomes. In mammals, from the tree shrew to the lion to the bonobo to us, these patterns of behavior, these traits, or their counterparts, exist.

In a Midwestern town of forty thousand, a high school sophomore cries herself to sleep longing for the attentions of the handsome football star whose backseat tumbles with "bad" girls makes him forbidden by the girl's family and all the more appealing to her, as she foments a plan to wear something "special" to school and lure him into a few stolen kisses, if not a tumble of her own. Her physiological need, her reproductive status, and her strategies are not altogether removed from that of the Florida black beetle, Lara the bonobo, or the castle-bound Guinevere longing for Lancelot. Athleticism and body building, one-night stands, romantic love, and jealousy, along with infidelity, monogamy, and homosexuality, are so universally demonstrable across species and cultures that they have long been presumed in large measure to have been drawn through the filter of sexual evolution and biology. What's new is that in recent decades, researchers have the means to reveal them, predict them, manipulate them experimentally and expose them to strict analysis.

While these "means" have generally been developed to extract the biological foundations of behaviors and traits linked to such socially critical issues as learning, drug abuse, and the development of drugs tailored to individual biological makeup, they have begun to be exploited by those interested in revealing the genetic and biochemical foundations of profoundly complex behaviors, including those we collectively call "sex." Genetic engineering methods such as transgenic and null mutant mice can now be used to identify and study single genes that play a role in complex behaviors. By inserting or knocking out genes in mice, scientists create models in which they can assess the impact of a given mutation on behaviors involving multiple genes. Further, by studying the impact of insertions or deletions in rodent strains that carry measurably and naturally different behavioral backgrounds, and crossbreeding such strains, they have been able to begin to make educated guesses about the amount of variation in behavior that can be attributed to genetics, and the limits to such genetic contributions.

Jeanne M. Wehner of the University of Colorado's Institute for Behavioral Genetics is one who has modeled such complex behavioral traits in mice as fear, learning, and anxiety, traits that play a role in such aspects of sex as attraction, flirtation, competition for mates, and courtship. "What we already have learned is that there is no such thing as a 'learning gene' or an 'anxiety gene,' but there are genes that regulate behaviors and activities involved in these complex behaviors. There are without doubt genetic components to behavior and we know from animal experiments that what causes variation of a behavioral trait is variation in genes and the overall genetic background on which individual genes express themselves." In her laboratory, Wehner has used drugs and genetic means to manipulate glutamate receptors in a portion of the brain of two different strains of mice with very different learning and anxiety traits, and has shown she can dramatically alter each strain's innate response to fear and anxiety. By crossbreeding and analyzing their genetic makeup, she has found five genetic regions in the brains of these mice linked to the differences in these animals' response to a complex behavior called "fear conditioning." And by creating new strains made by moving hunks of genes in these regions from one strain to the other, she has identified a cluster of genes that do indeed enhance learning by increasing sensitivity to fear. What's relevant to the story of other complex behaviors, such as those found in mating, is that the fear region in these mice seems to track to differences in the animals' emotionality, to what is in a sense their "mind."

What also is relevant, Wehner notes, is that even when scientists can completely control the genetic variability, as in the mouse studies, only 40 percent of the variation in individual behavior can be attributed to genetics. Studies of identical twins reared apart have shown some traits to be 70 percent heritable, "but that means," she emphasizes, "that 30 percent is not. Genes are clearly not the whole pictures, and single genes never are. It's the variability that is important in human behavior and interactions among genes that determine this variation." The lesson, perhaps, is that at the level of whole species and populations, patterns of behavior and certain overall traits are conserved, while in individual members of these populations, variation is what keeps the evolutionary process healthy.

What we think is attractive or ugly; why we want more or fewer sex partners at different times in our lives; why testicles and penises are the sizes they are; the differences in male and female arousals; women's buttocks; hairy armpits; rates of orgasm; and fads and trends in human beauty and fashion: these all exist as they do because they have at some point in our past been favored by those of our ancestors who succeeded in winning the reproductive fitness and survival lottery. We are their descendants. Those who favored less winning traits never survived to wind up in *us*.

Of course, the origin of sex is of no more interest to our sophomore or to a couple in bed than it is to bonobos, penguins, peacocks, or bat rays. But every couple's interest in and preoccupation with sex, the energy they expend to get it and get it with the "right" person, are traceable to sex's origins. Evolutionary biology is the framework for these origins, and sexual selection is that framework's blueprint.

Charles Darwin introduced the concept of sexual selection in 1871 when he needed to explain biological and behavioral traits that could not very easily be explained by natural selection. Natural selection, of course, was the mechanism Darwin proposed to account for the process of evolution: animals that grew longer, sharper fangs tended to win fights or kill more prey and thus passed on these traits to their descendants. Similarly, females able to secure and keep more resources passed those traits to their descendants. And those whose coloring tended to match more closely the surrounding foliage were not as vulnerable to predators.

But what about a peacock's tail or a frog's mating call? Clearly, the one's flamboyant feathery baggage and the other's loud attempts to draw attention to himself did nothing to help either hide from predators. But these things did assure survival in another way: Darwin proposed that by helping males seduce the most biologically desirable females or win out over other males, peacock plumage, deer antlers, frog calls, squirrel lines, and all of the bright coloration so often seen in the males of various species were all ways that males could compete successfully with other males for access to the most fit females. The bigger the male, the more extravagant his look, the more likely he was to reproduce his genes and assure his line, his immortality.

Darwin had it pretty much right. In an experiment that Darwin himself first suggested 120 years ago, a team of scientists clipped the "eyes" from some peacocks' magnificent tail feathers and learned that males missing as few as 20 out of an average 170 eyes per tail experienced a dramatic drop in their mating games. Clearly, those tail eyes played a role in attracting a mate. In the last fifty or sixty years, scientists have come through with overwhelming evidence that the biological, psychological, genetic, and chemical events going on in any act of sex fit the same overall tenets of natural and sexual selection, albeit in ways more subtle and complicated than even someone as observant as Darwin could imagine.

Consider, for example, an experiment conducted by Peter Staats, a pain researcher at Johns Hopkins University School of Medicine. He asked forty college students to plunge one of their hands in a tank of ice water and keep it there until they could no longer tolerate the pain. During a second dip, they were randomly assigned to different groups. Some were asked to think of a preferred sexual fantasy such as kissing, flirting, or other enjoyable activities with a favorite partner. The others were asked to envision either a non-preferred sexual fantasy or a neutral fantasy (such as walking to class), or were not given any special instructions. Those who could fantasize a sexual scene they liked were able to keep their hands in the ice water more than twice as long as their counterparts in the other groups (three minutes versus a bit more than one minute).

It doesn't take a neuroscientist to imagine the value of natural links and even close alignment of our built-in responses to sex, pleasure, and pain. Those links make great sense in the context of a major theme in biology and evolution: traits that are highly conserved tend to be multi-purpose. Their expression varies with circumstance and serve many goals. Just as human brains developed opiate receptors, not to advance the cause of the Colombian drug cartel but to block pain and encourage the "natural highs" of sexual arousal and orgasm, so, too, it would seem, our other pain relief–pleasure enhancement systems are plugged into our sexual natures.

It also should be noted that in this case, *both* sexes possess the systems. And herein lies a major theme in the evolution of our sexual natures: that in order for a trait or set of traits to make it into the genomic hard drive and thus the culture forever, it has to benefit both sexes, even if the expressions are dimorphic or *exclusive* to one sex or the other. Such traits coevolved in the sexes, each sex encouraging and affirming a particular trait acted on by natural selection. The peacock example suggests not only that the male evolved a certain number of these colorful eyes in an ever-enlarging tail to attract the peahen, but also that the peahen, no less than the human female, was doing her own selecting, buying into or even driving the male's increasing expenditure of precious and scarce energy into his feathered display. It seems evident that she has over time developed an eye for the peacock equivalent of tight buns and broad pecs, a sense of aesthetics to accommodate *her* evolving sexual and reproductive needs. She has some say about *her* sexual preferences and is far from a passive player in the mating game. She appears to be able to *choose* the most beautifully feathered males to fulfill her biological destinies, if not fantasies. This is true not only for the peahen: other scientists report evidence of similar active female choice in 177 of 186 species investigated.

Combat, conflict, and female preferences, intriguing as they may be, are only a part of the sexual selection story. They alone can never fill out the natural engineering specifications for species survival mentioned earlier. In many species, it seems, traits not typically associated with sexual selection—for example, mobility, navigational skills, and a sense of relationship to things and spaces—are equally or even more important to mating success. If, for example, your are a male thirteen-lined squirrel, and the females you long for are in estrus for only four hours over a two-week period in any one year and widely spaced over acres of territory, finery and ferocity won't help you much. What will help is your spatial learning ability and the size of the part of your brain that helps form memories, identify shapes and smells, and separate important sensory information from noise—the hippocampus. Experiments with these squirrels confirm that the studs traveled over 7 hectares and seduced three to five females, while the duds journeyed over only 3.5 hectares and found only one to three willing females. Is this species an aberration in the animal kingdom? Some accident of nature? Not at all. Other rodents, notably meadow and pine voles, have a similar strategy. They, too, must roam wide spaces to find true love, and like the thirteen-lined squirrel, the more polygamous the vole, the more likely he is to have the bigger hippocampus and get the girl(s).

What about the natural engineer's specification that sex must confer health and vigor, that sexual rather than other kinds of reproduction persisted after it originated because it increased those options not only for individuals but also entire species, including our own? If sex is somehow "good" for species who keep it around, how—beyond making sure that animals desired and enjoyed it—would it work to confer healthy survival?

Even a high school student with a year of biology under her belt might propose that at the very least, sex, by mixing up new combinations of genes, is a surefire strategy for producing offspring different enough from their parents to allow adaptation to a constant stream of new and unexpected environmental changes such as disease, catastrophe, climate change, and new predators.

August Weismann, in the nineteenth century, was the first to point out the importance of the fact that traits acquired after birth could not be transmitted to offspring. This is because "somatic" cells, those that make up all the various parts of the body *except* so-called germ cells (sperm and egg cells), are the only cells that directly interact with the immediate internal and external environment, and whose genetic material can therefore mutate, or change, for better or worse. Somatic cells die without passing on any copies. Germ cells, on the other hand, hidden and biochemically and physically protected from the world inside ovaries and testes, hold all the stable parental hereditary material. But unless the traits copied and inherited inside reproductive cells had some way to change with the times and needs of individuals and species, nothing would ever change. New species could never evolve and adapt. And new behaviors couldn't either. From thought experiments such as these, and from early manipulations of germ cells and somatic cells, Weismann and others became convinced that only organisms capable of swapping germ line information—what we now call sexual reproduction—would be able to produce new mixes of combinations of genes in each child in every generation. They declared that the reason sex occurred is that only with sex—and here we're not talking about intercourse or wine and roses, but DNA exchange between two forms (or sexes) of a species—can species adapt and change in response to changing conditions of life. Moreover, Weissmann believed, the more complex creatures get, the more they are likely to trigger new species and speed along evolution.

Competing theories exist among evolutionary biologists and zoologists about how "life" as we know it began, and how species are formed. But this notion argues that speedy variation is not only the spice but also the sine qua non of life. In this view of origins, life and sex evolved in a so-called tangled bank of trees, weeds, shrubs, and flowers whose complexity provided the best opportunities for endless and rapid genetic mixing. The problem with this idea is that it emerged before there was any real notion of genetics in scientific play. Thus, proponents of it surmised that on this tangled bank, sex emerged as an optimal way to quickly allow the mixing of new genetic combinations, which enabled species to be nimble and find a "niche" to survive. Unfortunately, this "variation" view of sex, which is almost holy writ in most biology books, flies in the face of what molecular genetics can now tell us, including the fact that each act of sexual procreation, in the words of evolutionary biologist Richard

Michod, "undoes what it creates." Consider, he writes in *Eros and Evolution,* a pair of birds who mate in a forest whose foliage has subtly changed color because of some change, say, in climatic cycles. So their genes sexually mix and by random chance, one of their offspring has feathers that offer good camouflage in the new forest. But when that bird mates, another random genetic shuffle of germ cell genes must altogether destroy or change feather color in her offspring.

As we'll see in more detail in chapter 2, sex in its purest definition *does* achieve genetic mixture by combining the genes both parents inherited from their parents and, just as critical, mixes genes again in each parent when the parent makes eggs and sperm in a process called recombination, which is neither random nor accidental. To understand this, back up to how cell division occurs, since this is the means nature has devised to develop new organisms. During cell division, individual cells split up into two offspring cells, which then do the same thing and create two other cells, and so on. When the most prevalent cells in our bodies do this, the somatic cells, each new daughter cell gets a complete set of gene-carrying chromosomes. And since both of the daughter cells' parents each contributed half of that genetic material, each daughter cell winds up with two copies of every gene and chromosome. During division, the parent cell's chromosomes line up in a special way, where first they split and make new copies of themselves. The split choromsomes create a chemical blueprint from which to copy the new set of genes that will populate the daughter cell. Only after the genes are all copied does the cell itself split up, sending half of a now double set of chromosomes to each daughter cell. The key to this process in somatic cells is that unless there is some accident of nature, each daughter cell is *identical* to the parent cell. This process is called mitosis, and in bacteria and other creatures that have no sperm or egg cells, it is the principal means of reproduction.

In our species, and in all species in which there are male and female, sperm and eggs, the process is seriously different. In humans, our genetic birthright is contained on twenty-three pairs of chromosomes. But unlike what happens in our somatic cells, the rules governing the development of sperm cells in males and egg cells in females are unique.

After fertilization, or mating, these egg and sperm cells jointly form a new embryonic form of the new individual. And that means that each of these gametes must contribute half—but only half—of the offspring's genes. Thus, egg and sperm must themselves be made in a way that limits them to only a half-set, or haploid set, of genes. The distinctive method of assuring this is called meiosis, which in effect *reduces* the genetic complement by half, in one of nature's most delicate operations.

In the case of an egg, or female germ cell, a primordial form of egg appears in a female embryo within a month of conception. It then reproduces through mitosis or normal cell devision that keeps the full set of forty-six chromosomes in each daughter cell as the gonad grows. But by the time this fetus's mother is near the end of the second trimester of her pregnancy, this growth stops. Some specialized DNA activity prepares each premature egg cell in the developing ovaries to await not only the girl's birth but the completion of her puberty before halving the genetic complement and making the egg ready for fertilization, with its half of the blueprint for a new individual. An important feature of this process is that both that specialized DNA activity and the actual halving process, all part of meiosis, are marked by frequent recombinations among and between the forty-six chromosomes in the premature egg. In males, meiosis does not begin until puberty, unlike in females, where the first part of it begins while they are still in their mothers' wombs. But in either case, according to calculations, this phase can produce 8.4 million combinations of genes in an individual's germ cells. Then, when fertilization occurs and conception results, the two haploid sets of genes, from egg and sperm, shuffle again and result in a new individual who gets half of its genes from its mother, the other half from its father. But it also has a set of genes that frequently have been mixed and remixed long before fertilization takes place, shuffled randomly so that each offspring from a given set of parents, unless they are identical twins, is made up of very *different* combinations of its parents' genes. Thus, sexual reproduction allows for not only a mixing of genes that each child inherits from his or her parents, but also for recombination of the genes its parents got from *their* parents, in which whole clusters of genes were exchanged before a final, whole set of 50,000 genes on twenty-three pairs of chromosomes were passed on to the new life. And it is the reason that each of us is a mixture of genes reshuffled from every generation back to the time of our biological Adam and Eve. Again, unless we are identical twins, created from an abnormal split of a single fertilized egg, not even two siblings are genetically alike. (It's also why the genetic variation *among* individual members of a single species, such as humans, is much greater in terms of possible combinations than the genetic variation *between* species.)

Ecologists, sociobiologists, and evolutionary scientists have long proposed that there is something in this mixing and matching of genes that makes animals more "fit" in reproductive terms, probably by making them healthier overall. Some evidence for this idea emerges from the evolutionary record and from experiments that show that many organisms can and do change their means of reproduction in response to environmental change. Although still controversial, some findings in both bacteria and eukaryotes suggest that rates of mutation, which may speed evolutionary adaptation, speed up during times of stress. In harsh habitats, finding a mate is pretty tough, and feeding one—along with the family that comes along—is even tougher. In both plants and animals, when ecosystems evolve into subtle, complex, rich, stable niches, with fragile but steady balances of power, sex wins.

The primary reason this is so probably has to do with what makes ecosystems simultaneously so fragile and so stable (with respect to populations, at least): disease. Pathogens in the form of viruses, parasites, and bacteria make it necessary for organisms to fend off infections that come on suddenly. Pathogens are particularly notorious for their ability to mutate and adapt, reproduce asexually, and create new generations of bugs in minutes, thus enabling them to quickly overwhelm "immunity" in a population of animals that reproduce much more slowly and thus with less variation over time. Those that live within the rich fabric of a plentiful society—enough food, water, air, and a variety of resources to feed whole communities—are also more likely to reproduce faster and meet the demands of a changing environment or an "antibiotic" attack. As it turns out, prominent researchers have plenty of evidence—collected under what is widely known as the Red Queen hypothesis—that most living things are driven to do the things they do just to stay even with the bugs that drive them crazy and drive them to their deaths.

In his 1993 book on this disease-resistance theory of sex's origins, *The Red Queen*, Matt Ridley argues that sex exists because it is the only reproductive design that can constantly mix genes and provide a necessary supply of potentially advantageous, disease-resistance genes that keep us from succumbing to the endless onslaught of bugs determined to kill us off. The process of continually swapping and recombining genes in order to hedge the threats of a constantly changing environment, the theory goes, is like the process experienced by subjects of the Red Queen in *Through the Looking Glass*, namely, as noted in the introduction, that it "takes all the running you can do to keep in the same place."

British scientist William Hamilton of Oxford University, who did the most to popularize this view within the scientific community, helped devise a computer model of sex and parasitic disease that demonstrated, at least theoretically, the major advantage conferred on species by sex. But the advantage would hold only when the "virulent" genes driving parasites were met by equally powerful genes able to resist the disease parasites. In his paradigm, Hamilton created a world in which the most useful "resistance" genes were retained—"conserved" in evolutionary terms. They stuck around and protected the species, or "host," until the "virulence" genes of the parasites got smarter and became more common, destroying most of the resistance genes. Only via sex could new resistance genes emerge and adapt. The checks and balances of virulence and resistance just continued ad infinitum. Thus, for Hamilton, sex is all about disease and death. "The essence of sex in our theory," he says, "is that it stores genes that are currently bad but have some promise for reuse. It continually tries them in combination, waiting for the time when the focus of disadvantage has moved elsewhere."

Further evidence for the Red Queen hypothesis emerges from the patterns of disease and the organisms that cause them, suggesting that we are indeed, like the United States and Soviet Union during the Cold War, in a constant "arms race" with our parasites, bacteria and viruses designed to neutralize each other. The virus that causes AIDS, for example, has mutated, or changed its genetic makeup, more times in a decade than humans have changed in millions of years. The best bet for humans in this arms race is to have better resistance genes, and the best way to get those is—sex.

Indiana University biologist Curtis Lively figured out that if the Red Queen hypothesis were right, there would be an inborn susceptibility for infection in certain snails living in New Zealand lakes from a parasitic worm. In a dazzling series of experiments, Lively took snails and parasites from widely separated lakes. One lake had smooth snails, the other barbed snails, so he could tell them easily apart. He then infected snails from each category with parasitic worms from each lake.

As expected, the worms quickly infected the snails with which they'd coevolved and with whose susceptibilities they were familiar. But each type of worm couldn't infect the foreigner snails. The hometown snails and their familiar parasites had each evolved a system of genetic variability to keep the other in check, the snails via sexual mingling of their genes and the parasites the old-fashioned mutational way. Moreover, Lively also saw that the more threat there was of infection—that is, if parasitic worms overpopulated and got in a position to possibly wipe out the snails—the more sex the snails had. In subsequent experiments, Lively demonstrated the same effect in a small fish in Mexico; and Wayne Potts of the University of Florida at Gainesville found similar support for the Red Queen hypothesis in experiments with ordinary house mice. Through a series of mating experiments coupled with analysis of the principal genes that make up the immune system (called MHC, or major histocompatibility complex, genes) in the offspring, Potts and his team found that females often sneak off from their home nests to mate with males who are more distantly related or total genetic strangers. In fact, half the pups were born to fathers who were not part of the mothers's home nests, resulting in litters with different, varied MHC genes and increasing the probability that the offspring would resist disease genes and parasites.

In 1982, Hamilton and another collaborator, Marlene Zuk, found intriguing arguments for extensions of the Red Queen hypothesis. They proposed that females choosing mates based on the degree of expression of male characteristics—such as size, speed, strength, high coloration, and so on—get heritable parasite resistance for their offspring. In other words, the male traits that females can see and "read" are powerful surrogates for healthy male genes and females read these signals of fitness as "not infested." The degree of expression indirectly indicates a male's lack or relative lack of infestation and thus a lower risk that Mom and the kids will acquire these parasites.

It's reasonable that an animal whose system is healthy enough to be well fed and to afford to spend energy on ornaments is more fit than those less well endowed. Grouse hens who either chose less fit mates or weren't chosen by them would certainly wind up with sons that had fewer ornamental ruffs and tail feathers—in other words, less "sexy" sons, at least those less appealing to females. And that would make their male offspring far less likely to mate with the most successful and fit females and less likely to carry the mother's genes into succeeding generations. None of this search for fitness in birds is achieved at a conscious level, of course. But it goes on, and the same search exists in people, too. Our "markers" of fitness range from classical Greek symmetry of features, beautifully toned and tanned bodies, Armani suits, diamond jewelry, and clear complexions—ornaments and "proof" of resources no less than coloration or tail feathers.

There's more to this story. Consider for a moment a beach in Palm

Springs or Nice, populated by sun-bronzed and oiled bodies of the young, the rich, the famous, strolling along the tide line, posing and posturing, telling stories, making merry, and most of all *watching* other people. Who watches whom? The men surely watch the women. And while the women may (perhaps more covertly) watch the men, they also are watching the other *women* to check out what they're wearing, who are most admired by the most desirable men. If this year's Cannes festival starlet is wearing a gold Gottex thong, it's a sure bet lots more of these will appear on the beach on others—both those who can wear them and those who shouldn't. Among the so-called lekking breeds (where males strut their stuff before females in a kind of male beauty parade known as a "lek"), females also tend to copy each other in their mating preferences, with the most successful females setting the standard for what's appealing in a male. The lesson, from nature's viewpoint, is the same: When females mated indiscriminately with males, those who sought males less big of tail feather were less successful breeders or, worse, might have had sons that other females wouldn't choose. What became important to females in these breeds over time was whether the mate of her choice was also attractive to other *females,* for then and only then could she be sure that she'd have sexy sons, who would in turn be attractive to other females. Is this so far afield, really, from women who find a sexy (read "popular") man more desirable as a mate than one who is shunned by other females? At some level of our physiology, choosing a sexy husband has something to do with having sons other females want. At times, this pattern continues so far that all that sexiness has the potential to be a major burden. The peacock is a case in point; that tail is pretty sexy all right, but he can hardly move!

Following this course through time, it is logical that if this went on forever, and if, as in the case of the sage grouse, only one male of a population gets to breed with all the females, it shouldn't matter which male a female selects. Wouldn't lekking become unnecessary? The disease theorists have an answer. Sage grouse, they explain, have a form of malaria, caused by a parasite that tends to make those who are the most afflicted have trouble strutting their stuff during the lek. Maybe those females are shopping around not just for what's pretty and sexy, but also for disease resistance. Since disease organisms evolve, the strain of grouse that is resistant in one generation might be at risk for disease later on in a never-ending up and down of susceptibility. That will bring new males into the alpha category and inbreeding is avoided. Indeed, scientists at the University of Wyoming have found that grouse males that have no malaria win more females and that if they give males antibiotics, they win more ladies.

Piling up the evidence further, other scientists have found that brightly colored birds of paradise have more parasites (and thus more resistance genes) than duller ones, and that baby barn swallows exchanged with other nests keep the parasite load and tail length of their biological parents, not their foster ones, suggesting that disease-resistance genes are inherited. As it turns out, female barn swallows choose males with longer tails.

Scientists have also shown that when females mate in ways that help them avoid infections, they also strengthen the links between variation in secondary sexual traits and the intensity of transmittable parasites. Even in domesticated mammals and birds, disease-resistance genes have been shown by genetic mapping techniques to be linked with variations in spur length of male birds, male survival, female mate choice, and offspring survival rates. It seems clear that the presence of "good genes" predicts female choice of males, proving that some female birds, at least, evolved to discriminate among males on the basis of secondary sexual characteristics in order to pass on genes for disease resistance that improve fitness in their offspring. And because nature seems to love patterns and conserves what it finds useful, it's likely that other species, including humans, followed a similar course as sex evolved.

Evolutionary biologist Chung-I Wu and his colleagues at the University of Chicago may have followed the story right inside the human genome by exploiting both the DNA sequences being generated by the Human Genome Project and the phenomenon of "sperm competition" among chimpanzees and other primates. Like the house mice families mentioned earlier, chimpanzee societies are marked by males that try to keep a small number of females sexually exclusive to themselves, and females that are, in the parlance of some biologists, "sneaky fuckers." These females sneak copulation with younger males outside the group, where a senior male more or less thinks he rules, and experts on chimps have long believed they do so to keep all the males in the society guessing about their offspring's paternity. Why would they want to keep them guessing? Because male chimps are protective toward their own biological offspring but often hostile, even murderous, toward others'. But since this female behavior means that no male can be absolutely certain of his mate's fidelity, or the exclusion of other males' attempts to mate with his female choices, some sort of "sperm competition" may have emerged to fight off rival sperm in the female reproductive tract. And a growing number of scholars believe that is exactly what did emerge.

In his book *Sperm Wars,* British evolutionary biologist Robin Baker interprets a substantial body of data on human sexual strategies, resource recruitment by females, and infidelity to support his contention that there are clear benefits to males and females from sperm competition within the female reproductive tract. For example, a woman whose life partner has ample resources but is, for her, genetically less desirable than her ex-boyfriend, might have good reason to perpetrate sperm war. She could have sex with her ex and then run home to her husband and work hard at having sex with him. Consciously, argues Baker, she may be trying to cover up her infidelity by quickly following her sexual wanderings with domestic sex. But it's also true, Baker says, that such one-night stands are so rarely detected that this would seem unnecessary unless something else were also going on. And he thinks it is this: that the illicit sex also prepared her subconsciously to perpetrate sperm warfare, to pit one ejaculate against another. To support this, Baker points to a recent survey in England that concluded that one in every twenty-five Britons was conceived as a consequence of sperm warfare. Put another way, he says, "since about 1900, every single one of us will have had an ancestor who was conceived via sperm warfare. Every one of us therefore is the person we are today because one of our recent ancestors produced an ejaculate competitive enough to win a sperm war." Baker also points to studies that show that the less time a man spends with a given sex partner, the greater the likelihood she will have sex with others. Since until recently paternity testing was not available, men who could not or would not want to spend all of their time with their partner—mate guarding or having sex to assure paternity—would benefit greatly from some biological means of addressing the possibility that during sex with his mate, her body already held ejaculate from another man. Thus, to increase his odds of becoming the father of any offspring, he would need to ejaculate more when his mate was *unfaithful* than when she was faithful. Studies have also shown that male ejaculate volume, along with the size, shape, and potency of his sperm, can vary greatly depending on factors such as frequency of sex, or absence from a partner, as well as the hormonal status of the female and the biochemistry of the female reproductive tract. If a female clearly wants, at a conscious or unconscious level, to choose her offspring's genetic father outside of her primary relationship, sperm warfare may well be an adaptive mechanism for both sexes.

Within the relatively promiscuous chimp mating system, a female can be inseminated many times, and males need huge amounts of sperm to get his own to an egg by forcing out other male sperm. (Male chimps, in fact, have huge testicles relative to their body size, unlike male gorillas, whose testicles are tiny and whose access to females is absolute. A single male has a harem and is always sure of paternity.)

Making the case that humans are more like chimps than gorillas, Wu set out to show that sexual selection in the form of sperm competition among males is what drives species apart and creates new ones. But his team's work also fits neatly with some of the Red Queen notion. Delving into the huge computerized files of information of the human genetic blueprint, Wu, along with Gerald Wyckoff and Wen Wang, first looked at genes that make proteins involved in male reproduction. By comparing these genes with similarly functioning genes in our closest primate relatives and other creatures, they learned that they evolved or changed faster than other human genes. Then they narrowed in on three of these genes involved in sperm production, including two that make protamines, which determine the shape and size of sperm. These genes evolved even more rapidly. And when the team compared these human genes with counterparts in chimps, he concluded that the protamine-making genes of chimps evolved as fast as the human genes, but the gorilla protamine genes were slower to evolve.

The team's findings suggest that human societies, at least at some point in their evolutionary past, were more like chimp societies than gorilla societies in that they featured women who had many sex partners, making it necessary for men's sperm to compete with each other. And it's possible that, like the house mice, our female ancestors were buying some kind of insurance against disease. What Wu's findings also suggest is that this kind of sexual selection for a behavioral trait varies from one animal society to another, possibly because outside pressures such as the environment or food scarcity alter the contingencies for survival of a species. Anthropologists and archaeologists have found that in most human primitive societies, and ancestral human groups, the sexes behave most like chimpanzees in that males are polygynous and control numerous females, or at least try to. For example, human social rules in these societies have females strictly sequestered until they marry, at which time they go off to live with the bridegroom's family, where presumably his family can protect their reproductive investment by removing the bride from *her* social support system. But as Wu's DNA data dredging shows, such women found their way around such restraints. The more rapid evolution of human sperm-making genes means they managed to find enough extramarital partners to force sperm competition and, from a practical standpoint, keep the reproductive playing field—DNA-wise—even. Fair competition, of course, is not a "goal" of evolution, nor a permanent consequence. But it may well be a reflection of a useful biological strategy.

A final piece of evidence in the disease theory of sex's origins explains a peculiar thing about sex: that sexual reproduction is far more prevalent, and asexual reproduction downright rare, in species, like our own, that are slow to breed and have long gestations and childhoods. In contrast, asexual reproduction is the norm in water fleas and rotifers that live in here-today-gone-tomorrow environments like puddles or mowed grass; they never have to worry much about parasites or the neighbors because their survival depends on the quick spread of their seeds and spores before their whole environment disappears. They're not around long enough to coevolve with parasites and viruses. The tangled bank theory predicts—mostly wrongly—just the opposite. It would say that creatures like water fleas that live in the face of environmental threat need variety and thus sex the most, while you and I in our stable niches would need it less. Only if the world weren't so full of disease organisms would cloning be the predominant means of reproduction.

Actually, as early as the first quarter of this century, the best biologists found that the Tangled Bank theory of how and why sex originated and persisted made little overall sense for a variety of reasons. For one thing, there is nothing in the evolutionary record to suggest that speed has anything to do with how successfully genes mix, or that animals somehow "knew" they had to mix genes or die. As Ridley put it in *The Red Queen*, "[E]volution is something that happens to organisms. It is a directionless process that sometimes makes an animal's descendants more complicated, sometimes simpler and sometimes changes them not at all. We are so steeped in notions of progress and self-improvement that we find it strangely hard to accept this. But nobody has told the coelacanth, a fish that lives off Madagascar and looks exactly like its ancestors of 300 million years ago, that it has broken some law by not 'evolving'. . . . It has stayed the same—a design that persists without innovations. . . . Evolving is not a goal but a means to solving a problem."

By the mid-1970s, one of the most sophisticated scientists, although not the first, to scratch his head over the nagging question of why sex came to be was John Maynard Smith of the University of Sussex in England. Maynard Smith worried that the most popular explanations for the origin of sex depended on a somewhat tentative but increasingly intriguing evolutionary hypothesis called group selection. Group selection was the label biologists gave to the idea that in groups, some individual animals are altruistic, sacrificing their own futures, and even their lives, to assure that the group or the species will survive. The classic case is that of the vast majority of worker bees in a hive, who give up all rights to send their genes into the next generation in favor of the queen's eggs, providing nutrients to the queen and her offspring in lifelong servitude until they die. Or there's the case of the animal who draws risk to itself by sounding an "alarm" call in order to alert the herd to a threat from a predator.

If indeed the whole idea of life is to create offspring to carry your genes into the next generation, why would you want to do anything altruistic? Or for that matter, why would you buy into any arrangement in which you have to ditch half your genes in the enterprise, which is what happens when we, or any other sexually reproducing animals, make eggs or sperm. Any creature that puts all of its genes into pairs of chromosomes pass on only one of each pair into the sex cells, and must await conjugal union to align with the other parent's single chromosome, form the pairs again, and produce a whole individual. That's a very high price to pay for genetic fitness.

In his book *The Evolution of Sex,* which he published in the mid-1970s, Maynard Smith rained on most existing explanations for the origins and persistence of sex on earth. It's a long way from the Kama Sutra to the pond scum creatures he studied, but in them he uncovered serious flaws and paradoxes in past theories that inform not only the process of evolution, but also those who wrote and practice(d) Kama Sutra.

Bdelloid rotifers live not only in ponds but also in puddles and buckets of freshwater and get about with the aid of little waterwheel-like structures. They can live in suspended animation for long periods of time, even when frozen or heat-dried. It's almost impossible to kill them, even with ice or boiling water. And their sex life is astonishing. Fertile beyond belief, these creatures can colonize huge lakes in a matter of weeks.

What made bdelloid rotifers particularly intriguing to Maynard Smith is that they consist only of females, reproducing asexually for the past 40 million years at least. The very existence of these rotifers makes a lot of conventional wisdom about biology and the origins of sex as clear as, well, pond scum. Here is an organism that has indeed rejected two-sex sex, but managed beautifully to evolve and adapt to change by separating and creating new species of itself. It gives lie to the notion that sex in some essential way is mandatory if offspring are to make it in new ecological niches, places their parents dared not go or could not adapt to. With characteristic charm, Maynard Smith called bdelloids "an evolutionary scandal," happily populating Antarctic ice and hot springs. Some biologists claim there are easily five hundred species of bdelloids, knocking the no-sex, no speciation argument into a cocked genetic cap.

Evolutionary ecologists like George Williams and Maynard Smith first developed and expanded on the idea that perhaps sex was necessary only in those cases in which two genetically different organisms get to a new habitat and need to combine their traits to survive. In this scenario of the purpose of sex, the end game is not to have lots of surviving offspring, but to have a few survivors who are adapted in some special way. As Ridley put it in *The Red Queen*, "If you want your son to become pope, the best way is . . . not to have lots of identical sons, but to have lots of different sons in the hope that one is good, clever and religious enough." Yet upon closer inspection, even this idea didn't quite parse. As Williams first noted, elm trees and oysters reproduce sexually, but these produce bazillions too many seeds or larvae that widely disperse on the wind or in the tides, and find no room in the elm forest or the oyster bed. That means winning the lottery for any given sexually reproduced creature is a really long shot, and a clear case of "why bother with sex?" Under the "traveling offspring" scenario, sex just doesn't pay for enough individuals, and it's wasteful besides. What's more, as the ecologists began to realize, few species ever leave home anyhow.

Enter Montreal's Graham Bell into the debate. Bell began by investigating the Williams–Maynard Smith notion that sex appeared to help animals through new passages in unpredictable environments. He wound up dispatching the central point of this theory of sex's origins to the evolutionary ash heap. Speculating quite reasonably that if Williams and Maynard Smith were correct, one would expect to see more sexually reproducing animals where habitats were harshest and least conducive to just sitting around the homestead. But as far as anyone at that time could tell from field studies, asexual species were more likely to be found in these harsh niches. Even those rotifers, it turned out, could reproduce sexually when environmental threats made food and space scarce.

So where oh where did true sex come from, the kind of sex in which there is a mixing and exchange of genetic material between two "parent" organisms before DNA is transferred to offspring? If life went merrily along for a substantial chunk of its existence without sex, it should have continued on without it unless there was some real advantage to the evolution of sexual reproduction.

In the thirty years since John Maynard Smith asked colleagues to come up with some better answers, scores of investigators have weighed in with theories whose chief common thread is that they seriously challenge the old genetic-variation-produces-faster-evolution-adaption-and-survival-of-species explanation. "The problem in the real world is that selection is likely to be operating on a whole ensemble of traits simultaneously," says University of Chicago biologist Steven Arnold.

Consider something known as conjugation, an ancient form of bacterial sex, sometimes also called "proto-sex," a presexual (as opposed to asexual) form of reproduction. A few members of a bacterial cluster send out a tiny hairlike extension, a pilus, and form a bridge to a neighboring bacterium. Then the empire-building bacterium sends a tiny loop of DNA, called a plasmid, over the bridge for no apparent particular reason and with no particular invitation from its neighbor. Neither the bridge builder nor the new host for this DNA gets anything out of this exchange of material. Only the plasmid wins, who, like every other bit of DNA in the universe, has a drive to make as many copies of itself as possible.

As this theory of the origin of sex goes, the whole conjugation process can sometimes carry pieces of the original bacterium's genes with the plasmid to the new bacterium, giving the new host the opportunity to incorporate the new genetic stuff into its own genome. Michael Rose of the University of California at Irvine and Donald Hickey of the University of Ottawa speculate that somewhere along the line, the parasitic plasmid refused to just ride as a passive passenger on the bridge to the bacterium next door, and instead drove the two bacteria to fuse the way sperm and eggs do during fertilization after sexual union. Then, the theory goes, other bacteria, knowing a good thing when they saw it, committed themselves fully to sexual reproduction.

Well, maybe. But this process begs the whole issue of just why the bacterial civilization would buy in, why any bacterium worth its plasmid would give up any of its genes to another.

One possibility is the foundation of the damage-repair theorists, who embrace the idea advanced by Richard Michod of the University of Arizona, among others, that the fervor to mix genetic material emerged as a means not merely to propel variation, but also to assure means of repairing damage to chromosomes and correcting genetic blueprint errors that could lethally affect protein manufacture and other critical functions.

In the late 1980s, Michod and others went full tilt at the repair theory, for reasons that can partly be explained by keeping in mind what DNA is and what it does, and how rife are the probabilities that something might go wrong when nature transmits its chemical blueprints—in the case of humans, 3 billion units of coded information—to other cells.

Each time one of your cells divides, the genetic material in the nucleus is first replicated, making two versions of the DNA so that one version can be kept in the parent cell while one is given to each "daughter" cell. As we'll see in more detail in chapter 2, because DNA in each cell contains all the information of all the body's cells, careful transcription and transmission of this blueprint is critical. But the process is complicated.

Now here's the problem for evolution: *Any* error in these translations and transmissions of information are potentially threatening if not lethal to a cell and its host. The whole system might collapse unless it has a means of recognizing and repairing mistakes. It turns out that indeed the system has evolved one.

Using a bacterium called *Bacillus subtilis*, Michod and his colleagues demonstrated that these organisms integrated pieces of DNA they found floating around in their neighborhoods, probably remnants of dead *B. subtilis,* and used them like spare parts to repair gaps or breaks in their own genomes caused by environmental catastrophes that he produced in the lab in the form of excessive oxygen or sunlight. Michod's team showed that damaged bacteria, at least lab-damaged bacteria, use more DNA to get their jobs done than do intact bacteria, and that repaired organisms were more successful in replicating DNA than were damaged bacteria.

To Michod, the need for variation is still a critical requirement in sexual reproduction, but he and his supporters consider variation a convenient *effect* of sex, not a cause. Sex, he argues, wouldn't have survived if it didn't first and foremost confer benefit to an individual organism. Variation doesn't necessarily do that at all, but having the tools to repair one's own genes fits the requirement very well. The drive to *incorporate someone else's DNA*—the quintessential characteristic of sex—would, the Arizona group insists, be a compelling reason to start sex. Variation, an effect of this sexual commingling of genes, was a byproduct. But because it, too, conferred benefits—to individuals, nature kept it around.

Beyond the need for variation, perhaps the reason bacteria and subsequently other organisms originated sex—that is, the kidnapping or at least capture of external DNA—was to provide extra sugars (the DNA molecule is made up of strands of phosphates and sugars off of which hang the chemical bases). In time of want, when bacteria crave or need more energy to operate their cellular factories, somebody else's DNA sugars begin to look mighty good, even if the bacterium who's hungry has to swap some of its original genetic fitness in the process. In another variation on this theme, some have suggested that after consuming one of the two strands of "borrowed" DNA (remember, DNA comes in a paired molecule), a savvy bacterium, especially if it has a broken part, might look over the other strand and see if it has a spare part it can hook up with before it goes ahead and consumes the rest of its new meal. Remember that in order to accurately copy itself and replicate, DNA must be able to repair "errors" in its nucleic acid building blocks as they "unzip" along the double helix and lay out the blueprint or genetic code, much the way a bricklayer must "even up" a wall as he goes along to prevent the whole structure from collapsing. Ironically, perhaps, the ability to repair damage and eliminate mutations brought with it another immutable characteristic of sex, along with further clues to its origins: death. As natural historians like Michod pondered further what he called the "dangerous, inefficient and inordinately time-consuming" mechanisms of sex, they reasoned that complex species, like humans, with vast populations of highly specialized cells, must die. They argue that programmed cell death, and ultimately the death of the whole organism, became necessary as sexual reproduction evolved to mix DNA from two individuals. Why? In order to prevent the accumulation of deleterious mutations in those cells that are *outside* of the germ cells (egg and sperm). As we age, we accumulate a huge collection of somatic (non-germ) cells, which happen to be *us,* and we must go before the accumulated bad genes overwhelm the genetic fitness bestowed by sexual recombination in the germ line. As Kate and Douglas Botting put it, "The joy of sex is fun. The point of sex is reproduction. The price of sex is death."

This view of sex's purpose holds that with sex's apparent advantages in helping offspring adapt to new circumstances and eliminate dangerous or weak genes, death *had* to come along if nature wanted to fully exploit those advantages. And indeed, the need for cells, and whole organisms, to die seems to have emerged at the same time cells began their first attempts to reproduce sexually.

To see why this happened, we need to go back to bacteria, the earth's most prolific creatures. Along with a lot of interesting characteristics, bacteria have two that compel us to hear their story: they never have sex—and they never die.

All living things are somewhat arbitrarily divided into five categories (arbitrarily, because in some ways they overlap, and in some organisms it's hard to tell what, for example, is a plant or an animal). But single-cell organisms comprise one of these categories and the earliest of these are the bacteria, making up all of the kingdom of the Monera, which today account for half of all the living mass of this planet.

The genes of these one-celled creatures have always faced frequent threats from the environment, and one response to that threat was clearly to evolve a means to infuse genetic material into each other in order to find a way to repair damage. Scientists have long inferred this, because unless they encountered famine, the death of their host or environment, an accident, or (in modern times) prescription antibiotics, bacteria never face the shadow of death. If death is defined as the extinction of a body, the creation of a corpse, in bacteria there are no corpses. There is no mechanism or "programmed" death built in to their existence and they live on—and on. Their reproduction is *asexual*, achieved by fission, in which they copy their own genetic blueprint and keep splitting themselves into clones that live on forever, as long as nutrients are available.

When Lynn Margulis of the University of Massachusetts at Amherst and other evolutionary biologists look into the world of Monera, in the steamy hot sulfurous tubular vents in the deepest part of the ocean, perhaps, or the nooks and crannies in arctic tundra, what they see is not so much a window on that kingdom as a window to our bedrooms. What they see is how sex—and death—began.

Like anthropologists reconstructing an ancient society from the tales of its descendants, these biologists have revealed the social history of bacteria from the time they first inhabited an Earth whose atmosphere had no oxygen. The Monerans at that point developed the means of living with that problem; and those that did not, died. And when oxygen emerged in our atmosphere, some of their numbers learned to protect themselves from it, and more, to use it. And those that did not, didn't survive. The fact that bacteria are still the most prolific species on Earth demonstrates conclusively that those that did survive had learned many evolutionary tricks.

The surviving generations over subsequent hundreds of millions of years had among them, like Ridley's papal candidates, a few of their kind that experimented with various means of growing, reproducing, and adapting to a changing environment. They lived in clumps, so they could share food. They developed appendages that they could use to store food, and a few pioneers marched out to a new frontier of life known as the kingdom of Protoctista (including some tiny forms) and, later on, Fungi.

Now, most protoctists, which include slime molds and some forms of algae, did something rather revolutionary: they developed nuclei. Back in the kingdom of Monera, Monerans had their DNA hooked onto the inner surface of their single-celled home. But the protoctists arranged theirs into chromosomes and balled them up in a protected cell within a cell, a nucleus. These nucleated cells had two copies of each chromosome, one each from each of the parental cells. In some ways this allowed protoctists to grow vary large, a million times bigger than most bacteria. As they were increasingly able to ward off predator cells and to store more food, the protoctists began to develop more complex internal structures within their single cell. In fact, they swallowed some of their kin bacteria from Monera and the Moneran bacteria—which, remember, had found out how to protect themselves from, and even use, oxygen—became parasites in some respects to the protoctists. But they were also welcome guests because they could produce energy from oxygen. (Our own cells claim the descendants of those bacteria—mitochondria, or energy-producing organelles.)

As protoctists became multicellular, perhaps a billion years ago along the evolutionary trail from Monera, death first arrived. Not from falling rocks, but programmed death, senescence, aging in the biological sense, in which cells and organ systems genetically weaken and die. At the same time, not incidentally, came sex. Nothing steamy just yet, but sexual reproduction as defined as the exchange of genetic material.

As far back as the late nineteenth century, experiments on cells began to suggest that while ordinary single-cell bacteria could undergo fission and go on forever—that this was indeed a fairly efficient way of getting all their genes into the next generation—their bigger single-cell cousins, such as paramecia, displayed a unique (among bacteria, anyway) characteristic. If a single one of these creatures had enough resources, they did not, like the smaller bacteria called protists, go on and on, but over time the *rate* at which they cloned themselves seemed to slow at the behest of some internal clock. And after a certain number of divisions, the offspring just stopped dividing and died. But, the experimenters noticed, if two of these offspring paramecia happened to join forces, to have sex by conjugation, their internal clock seemed to start over and the clones of those cells rapidly divide and create vast new populations again. The "parent" paramecia, however, those that had sex, do eventually die. The conclusion to be drawn, according to William R. Clark, author of *Sex and the Origins of Death*, is that "immortality for bacteria reproducing only by fission is granted automatically; immortality for everyone else depends on having sex." They survive only through their offspring.

Biologist Margulis argues that sex originated when it conferred some real advantage to living organisms, and the real advantage of sex can be traced to a particular population of bacteria or some other single-cell organism that, forced by hunger and thirst, fused "life's two most sensuous thrills, the joys of food and sex," into an act that satisfied "a single desire." She and her coworker—and son—Dorian Sagan, have theorized that eggs and sperm first emerged about 1.5 billion years ago in such creatures as slimes, molds, bacteria, and other protozoa. At some point, Margulis suspects, one of these creatures, facing extinction from starvation, perhaps, ingested one of its fellow creatures—but didn't digest it and survived. If one of these cannibalistic episodes happened to result in a fusion of their nuclei, such fused cells would soon have an edge over non-fused siblings in surviving bad times, and if their fellow cells were even minimally clever, they'd start eating their neighbors, too. Pretty soon they'd *require* each other to reproduce. This process, which Margulis calls symbiogenesis, led to survival of the species. It is as good an explanation of the origin of sex as any.

Some creationists have abused these ideas to advance their notion that Darwin's view of evolution—in which new species arise solely though the long accumulation of random mutations—never happened. But Margulis demurs. "I'm a good Darwinist," she insists, but not a "naive one. . . . What I'm saying is that the accumulation of random mutations is evolution's editor, just that it is not the author of life." In other words, evolutionary change steers things but is not the engine driving the process. "When you ask why we have life, why we have the diversity of life we see, there is enormous evidence that its earliest origins are in symbiogenesis among much smaller creatures than the fossil record can show—one-celled organisms that engulfed and incorporated other small fellows and went on to create an internal community capable of incredibly diverse behaviors." Humans are similar communities of symbiotic cells and organ bits, mergers of organisms that acquire each other's assets. Mutations present *opportunities* for evolution, she says, but they're not enough to explain the evolution of new species. That first accidental ingestion of one ancient organism by another—*that's* the start of it all.

Symbiogenesis in this way is perhaps the earliest *form* of sex and, not incidentally, perhaps prototypical proof of the principle that ultimately, cooperation among and sharing of an animal's or plant's vital elements maintains and enhances life. "Symbiogenesis embraces the evidence that new life forms *and behaviors* [italics mine] are created by the long-term physical association of different species," Margulis says. In this way, arguably, living things that are "social," even at the single-cell level, and who "reach out" to others are the true survivalists, adaptors, and perpetrators of sex and life.

For twenty years, the combative and brilliant Margulis has exasperated a lot of biologists, who charge she has gone "over the top" with her embrace of the Gaia hypothesis. Named for the Greek goddess of the Earth, Gaia holds that the whole planet is but a single giant living symbiogenetic organism and that notions of the origin of any individual living element in it—or its sexual reproduction—may miss the point. Along with the entire Earth, every animal on it is a "walking community," she says. Moving forward from her radical 1965 doctoral thesis, which made the case that mitochondria, plasmids, and other cellular bodies evolved from elements that still exist in the cells of all complex life forms, she has advanced the idea that even cells with nuclei are not necessarily entities unto themselves, but communities of leftover organisms that just happen to live in the same space within a cell's membrane.

"All the steps in evolution are apparent in animals and plants alive today, including humans," says Margulis. "It's just that to uncover the origins of these steps is very hard because the earliest organisms are so obscure and live in remote, hidden, and microscopic environments." But she has found them, in pond scum, oceans, and algal forests and watched these prototypical forms engulf foreign genomes on a regular basis. Some 3 to 4 billion years ago, the earliest nucleated cells emerged by incorporating other species' genomes, and the evidence is as close as our toes and our guts, where bacteria and fungi live in relative peace with our cells.

"Ten percent of our body weight is bacterial in its evolutionary origins," Margulis notes.

Much of what Margulis has to say remains speculation, but the idea that simple life forms originated sex carries logical weight. But was sex, as her work suggests, merely "an accident" waiting for bad times to happen?

Although much of this makes a great deal of sense, nagging issues remain unresolved. If indeed the biggest prize is getting the most copies of their DNA into the next generation, sex as a persistent strategy still misses as a rational approach to efficiency. Asexual reproduction, cloning, which still exists in single-celled creatures; some plants; and about one-tenth of one percent of some higher orders of fish, snakes, and bugs, doesn't need to go through sexual and genetic hoops. That means no mating, no courtship, no searches for Mr. or Miss Right, no time-consuming, energy-hogging efforts that interfere in the search for shelter, food, and protection from predators. Above all, no "war between the sexes" spurred by the obvious burden put on females who have to carry, hatch, birth, protect, feed, nurture, and raise the young while Dad is under no biological obligation to contribute anything beside a onetime gene deposit.

On the other hand, if sex got started, as some argue, to benefit *groups* rather than individuals, why do some whole species become extinct? While group selection possibly explains how sex *persists,* and Margulis's work suggests how sex might have happened, there still leaves the question of why sex was so important.

What else might have induced ancestral life forms to mix up their genes, setting the stage for sexual reproduction the way it has evolved today? After all, mixing up genes from two individually and perfectly fit parents has at least as good a chance of producing something *weaker* or *worse* as it does of producing something better, more fit, and more adaptable. Just because two Nobel laureates mate, there's no guarantee that their offspring won't be a nitwit.

For one answer, we go back again to those paramecia in which, some scientists say, there was the first ever episode of something besides simple genetic exchange between cells. That something was the first ever situation in which the DNA used for sex—that is, for reproductive material such as our own egg and sperm germ cells, or gametes—was separated and protected in the cell from the DNA used for other purposes, such as making energy, getting food, and discharging waste. In short, when more complex life forms began to get a foothold, germ cells were protectively set apart by the nature of the gene-mingling process, while the somatic, or "body," cells reproduced the old-fangled way, by simple fission.

Paramecia did this by essentially forming two nuclei, a macronucleus and a micronucleus. In the macro are the operational genes, the ones responsible for governing movement and breathing, for example. The DNA in these cells *replicates* (but does not reproduce) zillions of times to get enough operational energy to keep going. The smaller micronucleus, however, has only one double set of chromosomes, wrapped up tightly in a biochemical coat of molecules called histones, and it doesn't wear itself out with household chores. It comes to life only when it's time to reproduce the cell.

Experiments have long shown that paramecia can reproduce asexually, in which case the macronuclei and micronuclei divide their DNA in half and give each progeny a brand-new version of its genes from each nucleus. One of the two offspring gets the original stuff and the second a copy. But some paramecia have clearly found each other compatible and are willing to exchange their micronuclear DNA. The organism that two paramecia form now starts out with four sets of chromosomes instead of two. If that went on, the progeny would grow ad infinitum, producing eight, then sixteen, then thirty-two sets, an unsustainable activity.

As it turns out, they undergo instead a process more familiar to human biologists. These cells split in two their chromosomes (which in turn contain the mingled genes of *their* parents), segregating their inherited genes randomly in new patterns. Then, after fertilization they mingle, recombine, and exchange DNA with their partner's germ cell counterpart, fusing and mixing their chromosomes together into a completely genetically novel individual.

The process is, as noted earlier, called meiosis, and its complexity has been known to make biologists-in-training weep. It's a *bit* simpler in paramecia and works something like this: Each micronucleus in each paramecium divides into two micronuclei with a single set of chromosomes in each. They become haploid just like their cousins back in Monera. The haploid set they get is produced by a random shuffle among the "parent" chromosomes, making theses micronuclei genetically distinct from the parent. Then, the two haploid micronuclei divide again, producing four haploid micronuclei (two in each micronucleus), one of which is randomly selected for use in conjugation (sex); the rest are destroyed. These micronuclei divide one final time and then the two sexy cells exchange their haploid micronuclei. Each cell then fuses all that material into a single diploid micronucleus, which becomes the offspring "sex" cell.

In humans, sperm and egg cells also undergo reshuffling and halving in a somewhat similar way, although far more genetic material is shuffled and subject to far more processes. But the result is the same in these bacteria as it is in us: DNA used for reproduction is segregated from all other DNA; and the excess reproductive DNA not sent on to the next generation is destroyed. It is this programmed death of macronuclear material in early nucleated cells that is analogous to our own body's cellular death. All DNA not directly involved in sexual conjugation and meiotic sex can *not* be sent into the next generation because the next generation has no use for it. And it's dangerous besides. Unprotected from the environment because it lacks that histone coat, this other DNA builds up accumulations of potentially harmful mutations and errors that no parent wants offspring to have. Moreover, there is no meiosis in this genetic material to compensate for these bad genes and the new micronucleus is better off building its own new operations center that is free of error. In this view, the reason for sex is that it is a way to make sure that some genes—those in the *germ* line—are, like diamonds, forever. The others are but temporary visitors to the planet.

Without death, there would be no need for meiosis. Without meiosis there can be no DNA exchange. Without DNA exchange there can be no sex. And with sex, there must be death of the bodily self. As a consequence, nature figured out that somatic cells must age and die.

From the tale of the paramecia and the macronucleus, writes Clark, "We see . . . all too clearly why we ourselves must one day die. The only purpose of somatic cells (the non-germ cells) from nature's point of view, is to optimize the survival and function of the true guardians of the DNA, the germ cells." Once nucleated cells became multicellular, he says, reproductive DNA would be kept harmless in a few dedicated cells, which in animals, like in humans, are called germ cells. "Prior to protoctists like paramecia and their relatives, the somatic DNA *was* the germline DNA. Prior to multicellular animals, the somatic cell *was* the germ cell. The haploid germ cells seen in animals like us are in a very real sense the heirs of the micronuclei and the lineal descendants of the early monerans and the asexual protoctists; only the germ cells retain the potential for immortality. At some point in the life cycle of the individual, they may leave the body proper, combine with other germ cells and continue dividing to produce progeny that divide and generate yet another multicellular organism with another set of germ cells. When this happens, the germ cell senescence clock is reset. . . . The rest of the cells of the body . . . are condemned, programmed to . . . die. Once they have carried out their task of ensuring the survival of the germ cells, they and their excess DNA are no longer needed. . . . Sex can save our germ cells, but it cannot save *us*. . . ." In fact, "the advent of sex in reproduction made it necessary to destroy the somatic DNA [us] at the end of each generation."

As perhaps the leading exponent of the sex and death connection, Clark explains that the drive toward ever increasing size and complexity of organisms eventually required the appearance of more and more DNA outside of the germ cells, so that, somewhat ironically, complexity not only required sexual reproduction, but also the generation of some DNA used *not* for a strictly reproductive purpose, but to nurture and guard the sex cells.

Let's suppose that one accepts modern explanations for the origins and evolution of sex. A bacterium or some other single-celled organism came onto the scene and had the planet pretty much to itself for a billion or more years. The first jawed fish appeared about 400 million years ago and hominids, humans' immediate ancestors, a mere 3.5 million years ago. By this time, sex had pretty much decimated asexual reproduction and relegated it to less than one-tenth of one percent of the lowest ranks of life's efforts to reproduce themselves.

So how come at least a few more species, confronted with all that courtship rigmarole, not to mention gestation, birth, and baby-rearing, didn't just chuck it all and go back to the good old days of cloning? Scientists don't know the final answer, although it's commonly assumed that the benefits of sexual reproduction, permitting ever larger and more diverse animals, far outweighed the efficiencies of simple fission. To be sure, some species have hedged their bets. Some worms and a water flea called daphnia can switch from sexual to asexual reproduction, depending on whether there's a mate handy with enough resources to make sexual reproduction worth the effort. A "modern" whiptail lizard called *Cnemidophorus uniparens* is a true parthenogenetic creature, all female, carrying eggs that develop into fully mature animals without the need of any dads at all. But for the most part, sex has won, even though, theoretically and for all practical purposes, every cell in our bodies has all of the genetic material to clone another human being. Assuming we know how to turn on the "development" switch, it should be possible to grow a clone of any of us. And indeed that has happened with Dolly the sheep in Scotland, in which scientists manipulated just one of Dolly's somatic cells to produce a daughter Dolly. Similarly cloned cattle, goats, and mice followed.

Humans are increasingly clever about how they manipulate sex cells as well as stem cells, the progenitors of all cells, germ and somatic. In the first weeks of January 2000, scientists at the Oregon Regional Primate Center, together with reproductive specialists, reported the birth of Tetra, a healthy female rhesus monkey cloned from one-*fourth* of a test-tube-made embryo. The embryo was obtained by first splitting and then reassembling an eight-cell embryo into multiple empty egg "shells," and implanting these split embryos into surrogate wombs. Out of 368 identical "multiples" created this way by splitting 107 monkey embryos, they managed to transfer thirteen of these split embryos into surrogate mothers and achieve four pregnancies. "The birth of Tetra," according to Dr. A. W. S. Chan and his team, "proves that this approach can result in live offspring."

Animals, insects, and bacteria, with their multiple desires, mutinous genders, alternative sex lives, and sometimes violent mating habits, behave in ways that we humans, in our arrogance, consider graceless if not immoral. And yet what we may consider profane in nature is indeed profound.

I thought about this one day while recalling my first look at a peacock's tail at a zoo in Baltimore when I was a small child. I didn't think much of it. Or so my parents tell me. Much later, while reading about the experiment with the "eyes" in the peacock's tail feathers, I was reminded more clearly of a visit to the so-called Peacock Room in the Phillips Gallery in Washington. Created by James McNeill Whistler, this wall-to-wall, floor-to-ceiling, take-no-prisoners chamber is extravagantly painted in vivid hues of Chinese and teal blue, and is excessively decorated with peacock motifs, murals, and brilliant feathers. It was reconstructed at the gallery when previous owners of the room had had enough. Entering the room, I remarked to friends that apart from its resemblance to a cobalt bordello, the room had a certain outrageous charm. The docent overheard and let me know just how outrageous.

Whistler, it seems, had done the room on commission from a British magnate with a collection of Oriental blue ware he wanted to set off and "match" in a room of its own. When the owner went abroad, Whistler went wild, his fellow artists egging him on to ever greater heights of unrestrained peacockery. Predictably, or so the story goes, when the patron returned, he was angry and disdainful of what he considered a bad joke and poor taste. Offended, Whistler persisted, painting a final mural on the far wall, featuring two enormous peacocks in full-feathered splendor.

History did not record whether the owner took a close look at the mural before he paid Whistler—insultingly, in pounds instead of guineas (guineas are a shilling more than a pound and the customary coin of the realm of artists, as opposed to mere housepainters). But without a guide to the profusely oozing opulence of the mural and the room, I would have missed the significance of the unmistakably bourgeois ruffled collar on one of the painted peacocks—and Whistler's trademark white forelock on the other, more dominant bird.

With Whistler's painted peacock, as with real peacocks and with the rest of the planet's living things, what's going on beneath the behavior always is more subtle and more important than preening color and pride, albeit invisible to the naked, untrained, or uninformed eye. With evolutionary biology as our guide, however, we are better able to see what has long been concealed in our nature and nurture, and that the profound is not at all profane.

2

GENES AND GENDER

"DNA neither knows nor cares. DNA just is. And we dance to its music."
—RICHARD DAWKINS, *RIVER OUT OF EDEN*

"Sperm and egg and nature and nurture tell each other what to do. Sex is much more complicated than the question, 'Is it inherited?'"
—JOHN MONEY

The portrait, which hangs in the New-York Historical Society, displays a grande dame indeed. Hair modishly coiffed, set off with a delicate lace tiara. Her fan is held demurely atop a lap swathed in a satin and velvet robe draping across a becoming decolletage. Meet Edward Hyde, Lord Cornbury, governor of New York and New Jersey, later the third earl of Clarendon, who dazzled prerevolutionary Americans with his hoop skirts and elaborate hairstyles. With the normal body of a fully developed male, Hyde considered himself altogether a man, and mostly dressed and acted like one. But he never shed—nor shied from—the feminine side of his personality or his frequent public appearances in women's clothes.

James Morris also dressed his healthy male figure in women's clothes, and in 1974, science gave him an option not open to Lord Cornbury, even if the nobleman had sought it. Morris brought his body in line with his dress when he completed his transformation, medically and surgically, to "Jan." Like American ex-GI George Jorgensen, who underwent a sex change operation twenty years earlier to become Christine, Jan always felt he was a woman, just born in the wrong body. His penis felt like a foreign appendage. He didn't merely get sexual pleasure from donning women's dresses; cross-dressing defined his entire identity.

Anna, a Danish youngster, had no such feelings of being trapped in the wrong body. She was entirely a girl, extraordinarily feminine in her preferences and skills. In school to become a kindergarten teacher, she had a great fondness for and sensitivity to children. A talented linguist who enjoyed cooking, she was shy and sweet and utterly hopeless at arithmetic, geometry, sports, and other "masculine" enterprises. She, however, failed to begin her period or develop breasts or pubic hair. Doctors soon discovered she had no ovaries, only very sparse levels of estrogen and testosterone in her blood; and instead of two X chromosomes, the genetic signature of all women, she had only one X in each of her cells.

The annals of medical and psychological literature hold thousands of such accounts that not only blur simplistic, rigid divisions of humankind into male and female, but also invite attention to what lurks in our molecules to make us masculine or feminine. Long before 1966, when the first gender identity clinic in the world officially opened at Johns Hopkins Hospital in Baltimore, Maryland, transvestites like Lord Cornbury, transsexuals like Jan Morris, homosexuals, hermaphrodites, and others like Anna, a victim of Turner syndrome, were sympathetically understood by a few pioneering doctors and scientists as individuals whose sexual development missed or detoured around some critical turn in the road, whose genes had somehow betrayed their sexual being and identity. Mostly, of course, their sexual ambiguities, their "differences," their inability to fit neatly into one sex or the other, were misunderstood and even feared, striking as they did—and still do—at the heart of our assumption that male and female, the two sexes, are two very distinct and separate roads.

The assumption is understandable. Most of us journey through life at ease with our manhood or womanhood, instantly recognizable in or out of the bedroom as male or female, and endowed with a sexually specific and obvious package of reproductive organs, hormones, and traits as well as an inner "sense" of our gender roles. Somehow, we believe, we are transformed neatly, in just a few weeks, from microscopic blobs of two, then four, then eight identical, dividing cells into embryos and fetuses with an estimated 80 trillion cells, 250 different kinds of living tissue from light sensitive retinas to cuticles, and the potential to become either an unambiguously male or female; and from there in a few months to an individual whose sexual identity and orientation is immediately and eternally "set" in biochemical stone. We successfully navigate the gawky stages of puberty, when bodies and minds seem in permanent disconnect; gracelessly or otherwise, we mature to pursue exclusively the opposite sex with certainty about our own sex and what attracts us to the other; and finally we handle the mechanics of baby making. The game plan is for children to rapidly detach from unisex play to sexual ploy, and to solidify traits that make us a "man's man" (fight yes, cry no) or "all woman" (children yes, CEO no).

Combinations along some lengthy spectrum in between the gender poles are mainly discounted or denied. In fact, the roads to manhood and womanhood defy what one prominent sexologist calls the "bipolar fiction" of sexual development. Theoretical ecologist Dr. Joan Roughgarden of Stanford University, who in 1998 underwent her own transsexual transformation after a half-century as Dr. Jonathan Roughgarden, strongly believes that gender is determined by an instinctive capacity and passion to learn how to behave in a particular way. In nature, she has said, "there are two sexes," one that makes sperm and one that makes eggs. "But how you package those sexes is all over the waterfront." Not only has nature given us hermaphrodites that produce both eggs and sperm, and species that alternate between male and female, it also has produced at least three hundred species with same-sex courtship behaviors, making it likely that such behavior bestowed some evolutionary advantage. Clearly the road to gender identity presents many biological and psychological forks, some masculine and some feminine. To be sure, most of us navigate them easily, making all the critical turns that keep our biological sex aligned with our psychological interests. But largely thanks to those like Anna, Jan, and his lordship who meet detours along the way, scientists now know a great deal about the true nature of the journey to our sexual destiny, the one hidden below any level of consciousness or self-awareness and buried in genes, biochemistry, and our earliest experiences inside and outside the womb. Like astronomers exploring the universe for clues to the origins of matter, sex scientists have begun to discover the collisions and acrobatics of DNA that first determine our sexual identities and behaviors and lay the foundations of sexual orientations.

These primal harbingers of our sexual selves come from two distinct but overlapping sources. The first are certain genes and chromosomes that give us gonadal sex (our testes or ovaries) and genitals (penis or vagina). The other builds on particular biological and genetic influences but also recruits experiences and circumstances that, supported by a brain in constant learning mode and motion, sense and react to events and opportunities.

However obvious it may seem to say that the evolution and origin of new species are intimately aligned with the origins and evolution of genes that fuel sex and our sexual natures, making such a statement concrete has been anything but simple. Only recently, for example, have scientists unearthed any of the genetic circuits that clearly operate specific activities related to differences between the sexes. And very few of such genes have unambiguously shed light on both evolution and direct mating strategies. When they have, the story is compelling, as in one reported in late 2000 in the journal *Nature* by a team at the University of Wisconsin and Washington University in St. Louis. Sean Carroll, professor of genetics at the Howard Hughes Medical Institute at UW, and his group were searching for genes that govern gender-based rear-end coloring in fruit flies. They knew, as Darwin did, that animals assumed bright coloration to promote themselves to a potential mate. In fruit flies, the male's abdomen is heavily pigmented and the female's is not. They discovered that a gene they called "bric-a-brac" accounted for the difference between females and male fruit flies by suppressing color in females. But using genetic analysis, they also affirmed that this abdominal fashion evolved relatively recently in only a small subset of the fruit fly species. They also found that the same gene operates in other closely related flies and the males and females look the same. And in experiments with male flies engineered to have the same coloring as the female, the courted females were lured to sex no less often than when they were courted by a highly pigmented male. For Carroll and his group, the lesson is clear: In flies and undoubtedly other animals, the sex appeal of color for the female wears off over evolutionary time, putting males under constant pressure to evolve some new way of competing for females. "What we found was that the female didn't care and that makes sense in a sexual arms race scenario," says coauthor of the study Artyom Kopp. "The pigmentation . . . is last year's fashion and males are probably forced to evolve new ones all the time." Over time, these changes lead to changes that eventually establish new species and maintain the genetic vigor for which sex was "invented" in the first place.

Whether scientists are filling out the points of interest on our sex-determining gene maps or unraveling our journey to masculinity and femininity, they are finding that the sexes coevolved mutually rewarding strategies. The sexual war dances inevitably end in a pas de deux. Of course, for most of human history, even the brightest minds could barely hear the music.

Aristotle and Hippocrates, like most of their contemporaries, for example, believed that sex determination was a matter of body position during intercourse. Democritus argued that the left testis made girls and the right, boys, and that males developed in the right horn of the womb, girls in the left.

Not until 1827, when ovaries and eggs were discovered and a sperm from a teaspoonful of semen ejaculated with each male orgasm was found to enter eggs, did sex determination become the target of a systematic and rational search. Between 1875 and 1895, the first descriptions of embryos in different phases appeared around the same time that the concept of chromosomes crawled out of Mendelian speculation into factual accounts of human development.

At the turn of the twentieth century, clever experiments that manipulated rabbit embryos disclosed the (then) startling information that the little bunnies inherited different parts of their biologies from their parents.

Finally, between 1900 and the end of the 1930s, a flood of new biological discoveries put some real ingredients in the hands and minds of scientists looking for sex recipes. Among these: sex chromosomes (eventually dubbed X and Y); genes, the basic units of heredity (the name was introduced by Danish biologist Wilhelm Johannson and derived from the Greek for "born" or "produced"); and the sex hormones testosterone and estrogen, produced by the principal sex glands, the ovaries and testes, which also make eggs and sperm.

How these all got blended and cooked, however, required the discovery and development of a sophisticated science of sex in the third quarter of the century. In 1953, James Watson and Francis Crick won the Nobel Prize for their discovery that genes, specific sequences of nucleotides found along segments of huge molecules of heredity called deoxyribonucleic acid (DNA), were coiled up in the nucleus of every cell in the shape of a double helix. The structure they imagined, then demonstrated with the now familiar spiral staircase of colored spheres, was the true shape of sex—and everything else, of course—in the making. It facilitated elegant commingling of genes through a system of chemical attraction, and helped explain how, with each sperm fertilization of an egg, the basic chemicals bonded, lined up, and formed novel individuals.

In higher organisms, cell nuclei contain not only DNA but also a companion molecule, RNA, or ribonucleic acid. Tethered by tenacious formulas to other proteins, the combinations make up the chromosomes responsible for imparting to each new cell the capability of operating the entire complex of biochemical systems that make up life.

By the end of the 1970s, the discovery of restriction enzymes, which won a Nobel for Daniel Nathans and Hamilton Smith at Johns Hopkins, gave gene hunters the chemical knives and splicers needed to cut and paste DNA, to manipulate genes in ways that revealed their function. And finally, in the last decade or so of the twentieth century, scientists discovered a handful of tiny genes that, among the estimated fifty thousand or so genes in the human genome, appear to do the heavy lifting in the making of a man or woman. Arrayed within our chromosomes, sex-determining genes contain the coded instructions to build the proteins and chemicals that express the essential inherited characteristics of a male and female.

Finding those genes, finding out how they evolved, and more importantly figuring out what exactly they do, required platoons of biologists bent on unraveling one of life science's deepest mysteries with only subtle clues and circumstantial evidence. It's easy to forget how little, at first, the pioneer sex gene hunters knew about their targets or about genetic geography.

The entire genetic code lies along double-stranded DNA molecules. And although DNA's familiar double helix looks rather complicated, it might be thought of as beads strung on a string, which are made up of four chemical "bases." Two, adenine (A) and guanine (G), are from a group known as purines; the other two, cytosine (C) and thymine (T), are pyrimidines. The other protagonist in this drama, RNA, is made up of similar collections of bases, with the exception that uracil (U) replaces thymine as the second pyrimidine.

Each of the bases is always partnered with another, according to the chemical rule that a purine must hook onto a pyrimidine. Thus, in DNA molecules, A is bound to T and G to C (the now iconic ATGC acronym). In addition, each base is attached to a phosphate group and a type of sugar molecule called deoxyribose, and this forms what is known as a nucleotide. Of course, it is this double string of nucleotides twisted into a spiral—a double helix—that earned Watson and Crick their fame. Through a process called DNA replication, this double helix is passed from parent cell to offspring cell. This process requires enzymes to first "unzip" the DNA strands, with one strand becoming a template, or blueprint, for creating an exact copy of itself. This assures a facsimile-like transmission of information to the next generation. RNA is the key player in this part of the process, as its nucleotides literally transcribe the DNA blueprint in the nucleus to a smaller, shorter version, then move out of the nucleus and into the cell body, becoming literally a messenger ribonucleic acid, or mRNA for short. Next, tiny protein factors in the cell body called ribosomes cozy up to the messenger and "read" the transcribed code. As they do this, the ribosomes match up to about twenty amino acids, the building blocks of proteins contained like cargo within the RNA's nucleotides. The ribosomal factories recruit what chemical ingredients they need from among other amino acids, and in this way make the proteins or protein parts that carry out our lives.

The early hunters of sex genes also knew about the sex chromosomes, of course, and the fact that a normal female has two X chromosomes (one each from Mom and Dad) while a normal male has one X from Mom and one Y from Dad. They also knew that humans have twenty-two other pairs of chromosomes, wriggly-looking rods called autosomes, and that genes reside—somewhere—on all of them. Thus:

Tightly packed into the twenty-three pairs of chromosomes, which reside in each cell of our body, are two copies of a gene (which, remember, is just DNA's code for a protein) for every heritable trait we have. As first described in some detail in the last chapter, a gene's version of such a trait is called an allele. There is an allele, or gene, say, for eye color, on each chromosome of a matched pair, and within each pair of chromosomes—except for one pair in men, the XY—the same genes are lined up in the same way on both chromosomes. But the alleles aren't always the same on each of the two chromosomes, since although a gene for eye color has to be located in the same place on each rod in a pair of chromosomes, one of the alleles may be for blue eyes and the other for brown.

Further, these scientists knew that chromosomes in our sex cells behave differently during division than chromosomes in the billions of cells that make up most body tissues; that is, they knew about meiosis. In body cells, remember, division is done by a process called mitosis, in which every division produces two identical daughter cells and each chromosome copies itself exactly just before that division takes place.

If you think about it, it becomes obvious that this would not work in the case of sex cells. Parents must produce gametes that contain only half the usual number of chromosomes so that during fertilization, when Mom's and Dad's gametes unite, offspring won't contain double (and lethal or damaging) doses of genes and the biologically active products genes make. Therefore, when a sex cell with forty-six chromosomes (twenty-three identical pairs) divides, it must make either a sperm or an egg that has only one each of the parent cell's twenty-three different chromosomes. To accomplish this, whenever testes or ovaries make a reproductive cell (sperm or egg), only one member from each of the pairs is bustled off into that sperm or egg. This gene scramble alone has, as noted earlier, been calculated to make 8.4 million separate gene combinations in egg and sperm. But in fact there are more than that number, because testes and ovaries—even before they make their sperm and eggs—randomly swap genes between their own pairs of chromosomes in that pair of processes called Meiosis I and II.

There's a running joke among biologists that there actually may be only two or three people in the world who are "fluent" in meiosis and that even they need a program to tell the players apart. But as first noted in chapter 1, what happens during the first step in meiotic sex, or Meiosis I, at its most fundamental level, is the genetic dance that reshuffles and recombines genes from both the parents and their parents represented in those germ cells. This is why each individual is such a random combination of parental and family traits, why Johnny has his father's chin dimple and his mother's flat feet. As an introduction to what this process means to gene hunters, it's worth again, in simplified terms, going through the steps.

First, each of the forty-six individual chromosomes (the twenty-three pairs, one member of each pair from Mom and one member of each pair from Dad) copy themselves, producing twenty-three double pairs. Then, the members of each pair (maternal and paternal chromosomes) join up again, like square dancers responding to a chemical caller, and "outcross," exchanging bits of their DNA and mingling genes from the original maternal and paternal chromosomes. This shuffle means that resulting chromosomes will carry a combination of genes different than that in either parent, a dance that gives no one parent the "edge." When the cell divides, it is these novel combinations of DNA that will go off into separate daughter cells.

Next up, Meiosis II. Remember that each daughter cell now has twenty-three pairs of chromosomes and a mix of maternal and paternal and grandmaternal and grandpaternal genes. In this next step, the chromosomes in each daughter cell line up and divide, resulting in four gametes, each with twenty-three single-stranded chromosomes.

In what will become a male, these four gametes will become spermatids and develop into mature, tadpole-like sperm. Remember, too, that males carry one X and one Y chromosome, so two of the four male gametes carry an X and two a Y after Meiosis II, and there's a fifty-fifty chance that sperm from this future man will carry one or the other, just as there was a fifty-fifty chance that this soon-to-be embryo would become the male that he's about to be. Thus sex ensures a more or less even ratio of male to female at the outset. The female's gamete always has an X because that's the only thing female gametes have to transmit. (Actually, in female sex cell division, what's produced is a large cell, the egg; a smaller cell called a polar body splits off during this process. In Meiosis II, the polar body divides into two and so does the large cell. But the idea is the same.) At the end of the day, Meiosis I and II have produced a means by which sex cells won't result in a lethal double dose of genes.

What complicates the sex cell's task—and went on to complicate the sex gene hunter's—is the fact that meiosis is also the process that assures equal opportunities for Mom's and Dad's genes to mix appropriately, safely, and in a novel way. Every baby is a unique mix of genes, a kind of tossed genetic salad made up of our four grandparents' genes (the recombination part) plus our own mom's and dad's genes (the outcrossing part).

The sex gene hunters knew that each baby's sex was absolutely deter-mined at its conception. But figuring out how, when, and where the "deci-sions" were made; in what genes; and where those genes might be located was the challenge.

The first major clues to inform their search for the roots of gender emerged from the same source that has led to so many other insights into normal development: faults and breaks in the process, so-called experi-ments of nature in which genitals failed to match up with genes and made biological sex an altered destiny.

Among these relatively rare but similar cases of genetic sex scramble or reversal were two particular kinds of individuals. One group, called Turner syndrome individuals, turned out to have had only one X chro-mosome but were clearly female and developed ovarian tissues; the second group had two X chromosomes, like a normal female, but also a Y, and developed testicular tissue, a condition called Klinefelter syndrome.

Physicians and scientists had long speculated about how this could be. At some basic level, it was clear that in the case of the XO (or "X" and "missing" chromosome) "female," or Turner individual, one X was enough. In the case of those with Klinefelter syndrome, it might be that two X's failed to completely overwhelm the Y altogether, or that there was something far beyond a simple Y chromosome that helped turn on testosterone. Microscopic evidence of these individuals' chromosomal makeup, and biochemical evidence of their hormonal makeup, not only confirmed that there had to be something on the Y chromosome that was absolutely essential to producing a male, but also offered powerful circumstantial evidence that the Y is far from the whole story. It suggested that sex determination is a process, not a single event, implying the interaction of genes and their products, and that the process is somewhat plastic. None of these revelations are enough to explain why John became Jan, the earl wore petticoats, or generations of women instantly recognize John Wayne's hip swish as macho. But by following the clues inherent in these observations, scientists could use genetic tools to systematically explore the process of genetic sex expression in animals that they could experiment with directly, without waiting for nature to take some abnormal course.

Mice and other mammals were a frequent choice because they breed and mature rapidly, have genomes with a high degree of similarity to higher-order mammals, including humans, and carry many of the same biochemical and behavioral tools that we do.

Among other things, scientists learned that during the first month and a half of a newly conceived embryo's development, the embryo has the potential to become either male or female, even though all the genes it will ever have are firmly in place. Every normal mammalian embryo at first develops paired but *undifferentiated* gonads and two rudimentary reproductive duct systems, the Muellerian (female) duct and the Wolffian (male) duct. Thus, over the course of decades of study of mouse reproduction at places like the Jackson Laboratory in Bar Harbor, Maine, scientists have learned that an animal's ability to get to male or female depends on whether the gonads fully develop as ovaries or as testes.

Other scientists were to discover that the development of testes or ovaries (at about week seven of a human pregnancy) depends in great part on the action or inaction of a single gene on the Y chromosomes, what some call the "maleness" gene, or SRY, for sex-determining region of the Y. If this gene, inside the cells of the embryo's still sexless gonadal tissue, switches on certain blueprints, then certain hormones are produced, other genes are set in motion, and testes develop. If SRY stays silent, other genetic and hormonal events take place instead and ovaries develop.

Not all of the meaningful operational differences between Adam and

Eve can be found exclusively on this small fraction of the DNA on the Y chromosome, operating in some splendid biologically monastic isolation. For one thing, the Y is relatively sparsely populated with genes— twenty at last count on the main body of the Y, compared to thousands packed into each of the other forty-five chromosomes, including the X, and there is growing evidence that the X is "permissive" in sex determination, not, as once thought, passive. Nevertheless, the big-scale picture so far in mammals, including humans, gives the Y a leadership role in the decision about whether an embryo becomes male or remains female.

The story of Y is, among sex gene hunters and biologists, a long and tortured one, its sex-determining power often obscuring its probable origin as an X with one end cropped, and its complement of genes, at twenty to thirty maximum, in puny contrast to the X's two thousand to three thousand. And although the X and Y were clearly identified a century ago, and the Y's role in determining male sex identified in 1959, the search for the actual genes involved, begun intensively in the 1980s, has raised almost more questions than it's answered about how, exactly, "maleness" gets conferred. Even after 1992, when David C. Page of the Whitehead Institute for Biomedical Research at MIT published the first rough genetic map of the human Y, it's taken another decade for his lab to complete the DNA map, because despite its brevity, it's a mess of multiple genetic copies and inactive gene sequences that Page has likened to "a house of mirrors."

Page and his early partner, Finnish geneticist Albert de la Chapelle, are among the handful of Y chromosome gene hunters who set out on their quest aided by some unusual aspects of ambiguous gender or sex reversal found in individuals with Turner, Klinefelter, and similar sex chromosome– hormone syndromes. For instance, they describe some XX *males* whose testes were smaller than those of usual males and who were sterile.

There also were some XY *females*, who also were sterile and had abnormal female sex traits. Among the Turner subjects they described, all developed as females but with only "streak" ovaries, while individuals with a Y, no matter how many X's (two, three, or even four) always developed as males. Determined to figure out what, exactly, the Y, the X, and their genes have to do with sex and how and when those genes arrived on what we call sex chromosomes, Page, de la Chapelle, and others began to focus on these particular sex-reversed individuals because they revealed clues to where on the Y a sex-determining gene might be.

Specifically, Page and others discovered that in XY women and XX men, a piece of Y was attached to one of the X's on the XX males and a snip of Y was missing in the XY females, offering a critical clue to where on the Y "maleness genes" might indeed reside. Indeed, as early as the 1950s, long before the arrival of the molecular biologist's tool kit, the existence of such a genetic segment was predicted by the great biologist Alfred Jost, who removed the gonads of unborn rabbits in the uterus before other sex organs developed and found that male embryos failed to make any internal or external genitals or organs related to reproduction, while the females did fine, suggesting that the embryonic testes were all that was needed to make a male sex organ.

Page soon was hot on the trail of what was dubbed a "testis determining factor," or TDF gene, and by the late 1980s he thought he had it. Using tiny stretches of DNA taken from normal human Y chromosomes as a template, Page had traipsed up and down the genomes of sex-reversed individuals looking for matches. He was convinced that women with an XY chromosome signature probably had a single copy of one or more genes common to both X and Y chromosomes. He believed this because he noticed that some of these women had anatomic abnormalities—webbed necks, puffy hands and feet—strikingly similar to XO females. And he was indeed able to show that genetically "female" men with two X's actually carried some pieces of the Y chromosome previously invisible in conventional chromosome tests. Page also learned that the small pieces of the Y he found in the XX men was the same piece missing in the XY women! Naturally, this narrowed the search for the genes on the Y responsible for sexual differentiation.

Page published his work in 1987, but then in 1990, Peter Goodfellow, then at the Imperial Cancer Research Fund in London, and Robin Lovell-Badge, of the Medical Research Council in London, won the race to locate what was still being called TDF on the short arm of the Y chromosome. It turned out to be a fragment of DNA. In 1991, working with the mouse version of the Y, the same team reported that an even smaller piece of DNA contained the real business end of the maleness gene, producing a protein that appears to regulate the genes that develop the genital architecture of testes formation in the embryo. Using a fertilized mouse egg destined to be female, they micro-injected this tiny piece of DNA, this functioning mini-gene, into it and altered its gender to male.

Now that its location was certain, Goodfellow renamed TDF the sex-determining region of Y, or SRY, and ongoing studies suggest that the protein made by this gene in turn manipulates and influences the activity of many other genes. SRY is in effect the male "on" switch, but it is also a switch for other genes located on non-sex chromosomes during development. Just how SRY "learns" when to turn its switches on only in males and how it knows exactly when to do so remain a mystery, along with the identity of most of the other genes that are targets of SRY's protein influence.

What else these unusual patients taught Goodfellow, Page, and others was that most of the genetic material on the Y probably had little or nothing to do with sex determination directly, which raised lots more questions about why there was a Y chromosome at all and what else it might or might not have done during evolution. Indeed, the segment of Y DNA identified as SRY comprises only half of one percent of the entire chromosome and it was soon revealed that a tiny region of that segment was common to XX men.

Over the next few years, researchers would repeatedly observe embryonic development in mammals from the moment egg met sperm in efforts to track the action of SRY. They watched how, in early embryos, precursors of tissue that would become ovaries or testes started out looking and acting exactly alike, behaving in a way that molecular biologists, in a fit of cellular anthropomorphizing, called "indifferent." Then in week seven, the SRY gene, if it was available, switched on only in this tissue and testes began to form. These early incarnations of testicles then produced testosterone, which further stimulated male sexual development, along with Muellerian inhibiting substance (MIS), which attacked and broke down any other tissue that had the potential to become female organs. If the SRY gene was not available to turn on, ovaries developed, and without MIS, early forms of the womb, fallopian tubes, and vagina grew while potential male tissue disappeared.

Details of SRY's responsibility for testis making continued to mount, and research, in essence, confirms that it probably alters the very structure of DNA, allowing other factors, such as hormones, to get to work producing males.

Working with cells and genes from mice, Blanche Capel of Duke University has learned that the mouse equivalent of SRY is activated for about one of the twenty-one days it takes for a mouse to develop from a fertilized egg. Capel and her team are studying the small cluster of cells that are the "indifferent" gonads, those that will give rise to ovaries and testes. As in humans, mice develop gonads from a group of cells called the mesonephros. After growing mouse tissue in the laboratory for forty-eight hours, gonadal tissue from one mouse was put near mesonephros from another mouse. If the gonadal tissues come from an XX mouse, the mesonephros cells stay and continue to develop separately from the gonadal tissue. But if the gonadal tissues are from an XY mouse, destined to be a male, cells from the mesonephros migrate into it and form preliminary testes. Something on that Y chromosome had turned on that early precursor of gonadal tissue to develop as a male tissue. Capel continues to search for genes that control such migration, the genes that SRY targets when activated.

Late in 1997, Page and his team at Whitehead answered another fundamental question about the Y chromosome and its single-minded determination of maleness: How did it come to be so thoroughly held on to by evolution and dedicated to male sex determination?

When nature chose two-sex reproduction as the advanced operating system for higher-order mammals it also needed to make sure that whatever gene or genes were dedicated to making males did not "evolve away" during the creation of eggs and sperm, or become exposed to the vagaries and vulnerabilities of non-sex genes that are continuously mutating, recombining with genes on some other chromosome, or even disappearing. Protecting this quintessential determinant of maleness is no mean feat, because of meiosis, when nature creates genetic variety between generations by having pairs of chromosomes swap genetic material.

Page therefore reasoned that in order to keep SRY safe and solely in the male genome, whatever chromosome carried it around (Y, as it turns out) had to fail miserably at what other genes do exceptionally well— recombination—except perhaps at its end, where it shares some genes that also appear on the X. But since recombination is the way our genomes have chosen to meet the challenges of new environments, a failure to recombine would also mean that genes on the Y would mostly become extinct, losing the evolutionary battle for survival. Page's theory predicts what is actually the case: The mammalian Y chromosome is a genetic desert, with relatively few genes and only a few that share characteristics of genes found on its sister sex chromosome, the X.

Does this mean that eventually, the Y chromosome could become extinct? Not anytime soon, although scientists acknowledge that the Y's inability to repair itself or recombine will cause its genes to die off. In this regard, the Y has also emerged as a model of what evolutionary geneticist William Rice of the University of California calls "the adaptive significance of recombination . . . If you stop recombining you have a limited future." Page and other Whitehead biologists have also found that of the relatively small number of genes on the Y still remaining, many have counterparts on the X. These are survivors of the Y chromosome's original full set of genes. It's the other class of the Y's genes that operate, or turn on, in the testis after these gonads develop in the embryo. Unlike the "housekeeping" Y genes shared with the X chromosome, these few are "male fitness" genes exclusively. (Page and other male gene hunters often joke that if ever there will prove to be genes for channel surfing or direction finding, it will be on these "exclusive" regions of the Y, not on those "borrowed" from the X.)

Page has some proof that male-determining genes probably migrated from other chromosomes during the course of evolution and stayed put because they worked so well in a stable chromosomal home. Evidence emerges from his study of one of these genes, called DAZ (short for "deleted in azoospermia," a reference to the fact that the gene is missing in men who make no sperm and are infertile). DAZ seems to have a close relationship with another gene called DAZL, which is found on chromosome 3. He believes DAZL copied itself by accident onto the Y perhaps 40 million years ago, since only in old-world monkeys is DAZ found on the Y. Animals that separated from these monkeys earlier in primate evolution don't carry it. In this way, Page thinks other male-specific genes now found on the Y migrated there, made extra copies of themselves, and found a permanent haven. What's especially intriguing about this work is that it seriously calls into question the idea that the X was the sole or dominant source of Y's genes, and suggests that instead, other outside factors influenced its evolution.

More recently, late in 1999, Page and his colleague Bruce T. Lahn, now at the University of Chicago, reported their discovery of evidence for four separate layers and degrees of separation in the makeup of the human X chromosome, based on their knowledge of genetic evolution. One conclusion they draw from their now fuller knowledge of the evolution of sex chromosomes is that during the emergence of sexual reproduction, the X and Y chromosomes participated in a rivalry for dominance and power that ultimately led to a balance of power—a wary but relatively peaceful truce and coevolution of each sex's biological agendas.

As reported in the journal *Science,* the four layers parallel and reflect increasing levels of genetic "divorce" between the X and Y chromosomes. The researchers say they stumbled onto these layers while investigating the molecular distinctions of nineteen genes shared by both the X and Y. The layers correspond to stages of evolutionary history that can be dated by measuring the actual number of differences between the nineteen pairs of genes found on both the X and Y.

Page and Lahn say that the first of these excluded regions occurred about 320 million years ago when mammals first evolved from reptiles. In some reptilian sex, as it turns out, sex chromosomes as we know them don't determine the sex of the animal; instead, temperature, deprivation, or other outside forces turn on the sex-determining gene or genes. In these animals, this kind of sex determination survived and at the same evolutionary moment that it became dominant in reptiles, the first of the zones excluded from genetic shuffling developed nearby.

Page and Lahn speculate that these zones were created in a process in which a piece of DNA escapes from its home base on a chromosome and gets turned around so that it no longer aligns perfectly with the DNA on its counterpart chromosome. Its square-dance partner never appears and so the exchange of genes never happens. They found three more of these DNA inversions, which greatly enlarged the excluded regions and dated them to 170 million years ago, 130 million years ago, and 50 million years ago.

These recent discoveries, Page says, affirm the dominant role of the Y chromosome as the housing of genes that bestow benefits to males. And, if they ever managed a trip to an X chromosome, they would wreak havoc there. If the DAZ story is right, for example, the Y chromosome may indeed be a hot spot for genes that benefit only males and are potentially detrimental to females. University of California's Rice tested that idea by adding Y-like, man-made chromosomes to fruit flies and breeding the bug. He found that the man-made Y picked up genes that improved male fitness and hurt female reproductive fitness.

But once again, the battle of the sexes played out at the level of the sex chromosomes has been won by both sides. The lack of recombination of the Y chromosome's genes may have given the Y a powerful role in conserving genes that favor males, but it has also made the Y less genetically "fit"; over time it has reduced the number of genes on the Y from a number equal to the X chromosome's to about one hundredth of the X's number. The X chromosome can't reshuffle genes when it's in a male, but it wins in the genetic fitness stakes because when that male transmits the X to his daughter, that X can exchange genetic material with the other X it finds there from the female parent.

For the genes that remain shared territory, with copies on both X and Y, those on the Y appear—when on the Y—to be involved mostly in routine cell functions, although it's quite possible that they activate genes on the other twenty-two pairs of chromosomes that account for important male traits. That remains to be seen, but what Page now believes is that the shrinking of the Y chromosome's genetic storehouse means that the lost Y genes' counterparts on the X have to work twice as smart and as hard in males as they do in females. In females, who have two X chromosomes, one X is automatically turned off in each cell nucleus to avoid double dosing.

Sex gene scientists have now identified a whole array of proteins involved in the activities of entire sex chromosomes and their components, along with mechanisms that do the activating. In some cases, as with the X chromosome, our genetic blueprint carries within it ways of compensating for the "dose" of genetic information contributed by the sex chromosome, and ways to repress or switch off more than is needed. In many organisms, in fact, master control genes regulate sex determination in this way. Scientists at the Howard Hughes Medical Institute at the University of California, Berkeley, for example, recently reported that in mammals, flies, and some worms called nematodes, nature has provided means of balancing the doses of genetic information targeted exclusively to the X chromosomes of one sex. In their study of one of these worms, *Caenorhabditis elegans,* a dosage compensation protein called SDC-2 was found to target only hermaphrodite (XX) worms. What the protein does is repress strongly the transcription (i.e., the copying) of a non-sex gene needed for male development, and at the same time recruits other dosage-compensation regulators known as DPY and MIX proteins to get a 50 percent reduction in transcription of the hermaphrodite X chromosomes. The end result? XX worms that have a total level of X-based gene functions equal to that of male worms. In other words, a great deal of effort to level the sexual playing field.

The Berkeley findings dovetail in some important ways with Page's work. In normal XX women, one of the two X's in each cell is silent, or turned off, in order to avoid "double doses" of the proteins made by genes on the X. Women who are normal have the same concentration of

X-derived genes as do men, who normally have only one X. But an exception to this rule, Page says, is to be found in the X-derived genes that have counterparts to the housekeeping genes on the Y chromosome. The housekeeping genes, remember, are those not directly involved in male sex development and activity. In any case, they stay turned on in both X's, so they equal the male cell's dose of genes found on both the X and Y. Perhaps, then, the short stature, infertility, blood vessel defects, and other problems found in women with an XY chromosome makeup, rather than XX, come from the fact that the single X chromosome makes only half the needed amount of material usually provided by a full complement of housekeeping genes.

Grounded, no doubt, in the evolutionary evidence that both males and females have developed strategies for protecting and reproducing their individual DNA, Page has argued that if both male and female developmental pathways are perfectly evolved and normal, neither one is dominant or superior. Acknowledging feminist angst and anger over millennia of second-class status for females, he insists that nothing in his findings suggest that in nature, female is a "default" pathway, or that the presence of an ovary is testimony to the loss of a testes. Ironically, perhaps, this may disturb feminism's hard-liners who have taken comfort in the thought that female *is* a default pathway and that women are nature's "original" species and men a sometimes useful modification. Nevertheless, evidence suggests that becoming either a male or a female is "a very active process requiring the coordinated activity of many genes and proteins," he has said. "It's like those plays in which a story unfolds, all the characters are in place and all the roles have been learned, but the audience decides how the play should end." In this case, the embryo decides. "Depending on its decision, some of the lines will remain unspoken on any given night."

Indeed, studies of a gene called DAX1 on the X chromosome suggest that X is the active player in the development of ovaries. Because mutations in SRY account for only a fourth of XY sex reversals, it was inevitable that other genes be involved in sex determination. And males have a DAX1 gene, too, since they have both an X and a Y chromosome. Lovell-Badge of the National Institute for Medical Research in London has studied the DAX1 in mice and in 1996 showed that the gene is active in males and females in the same cluster of tissue and at the same time that SRY activates in males.

DAX1 was discovered by Italian scientists looking for the cause of an inherited disease of the adrenal glands. But because it is located on the X chromosome and presented in a double copy in XY females, the scientists suspected the area carried a gene that helps make ovaries. (The double dose meant it was a good candidate for feminization of male gonads in XY females, most probably because it either produced extra amounts of female hormones or increased their uptake by some tissues.) Recent studies in mice also show that DAX1 in the developing embryo is active at the same time and place as SRY. In a nod to what was already known to be the case in fruit flies and roundworms, it appears that it's the ratio of proteins made by the DAX1 and SRY genes—rather than the absolute amount of either protein—that determines whether a male or female will result. In mice and some preliminary human studies, individuals with one SRY gene but two DAX1 genes (instead of the usual one) get a higher ratio of DAX1 to SRY, and thus even when there is a Y chromosome, the male pathway is overridden and ovaries develop. This may be a likely explanation for the development of feminine characteristics that mark Turner syndrome individuals.

Once ratios are part of the picture, it's fairly easy to imagine the continuum of possibilities in the development of our sexual organs and identities. And theories continue to grow over just what the recipe is for making a male or female with ingredients like SRY and DAX1 in the pantry. Lovell-Badge argues that SRY's role is probably to inhibit DAX1 and thus stop ovarian development. But his is by no means the final word on the subject any more than SRY and DAX1 are the only ingredients. For example, a gene known as SOX9 also is a suspect in sex determination. If the level or ratio of SOX9 is upset, sex reversals happen. In mouse embryos with an XY chromosome pattern, therefore genetic males, a mutation in one of the two SOX9 genes (inherited from both Mom and Dad) can trigger their development into females.

The work with DAX1 and SOX9 lends support to the idea that overall, determination of genetic sex is a matter of balance of a small number of genes and the proteins that their blueprints guide to production. More and more, it is clear that this is a dynamic process rather than a purely binary decision that embryos make. A lot of genetic maneuvering goes on in the newly developing embryo, and the development of a boy or a girl is the result of the tipping of a precarious balance in favor of one or the other for just long enough.

The total number of genes influencing sex determination in humans remains unknown at this writing, and although it is likely to remain relatively small, it also has become axiomatic that the harder scientists look, the more they find, especially when gene hunters go looking at sex organ abnormalities. For example, in males with Denys-Drash syndrome (who, sadly, most often die of kidney failure in toddlerhood), gonads and genitals are feminized. Scientists at MIT and the Imperial Cancer Research Fund in London found that DDS is caused by mutations in an inherited kidney cancer gene. Later, Rudolf Jaenisch of the Whitehead Institute in Massachusetts knocked out the gene in mice and produced animals without kidneys or gonads.

Gene hunters also are beginning to look more systematically at all of the X chromosome, although not with the vigor and rigor many women scientists would like. Some feminist biologists blame male chauvinism for the late start on X analysis. It's also true, however, that the story of X is a great deal more complicated than that of Y. X has a huge genome compared to the Y, with many more genes.

The beginning of the hunt for the role of X-derived genes in sex determination had its roots in the fact that both males and females have X and the fact that in most cases, double doses of genes are not healthy. Studies of abnormal development of embryos had made clear that excess X-linked genes are fatal to the developing female. This in turn suggested that in some way, the developing female embryo compensates for or reduces the double dose of X it gets. Somewhere during their development, in short, females selectively turn off or inactivate a lot of X-based genes. Since males don't do this (and don't have to), it's equally clear that females alone have the genes to do this. By inference, then, females have genes that determine their own genetic sex and they can't possibly have anything to do with SRY.

More than thirty years ago, British biologist Mary Lyon suggested that female mammals start with two active X chromosomes, but one shuts down in each cell early in embryonic development in a random way. Thus in any given woman or female animal, some cells will have the maternally inherited X chromosomes shut down and others will turn off the paternal X. She believed this inactivation was permanent and inherited during cell replication, with successive generations of daughter cells having the same X inactivated as the parent cells, making the whole person or animal a mosaic of two cell types. Lyon also suggested that some single genes on an inactivated X might continue to operate and would have the same functioning parts on the Y chromosome as well.

Barbara Migeon, a genetic scientist at Johns Hopkins in Baltimore who pioneered studies on the ways and means of X silencing, says Lyon's hypothesis was so brilliant that although it was completely speculative, science has subsequently proven her right on the mark. "Nothing she proposed has been proved wrong," says Migeon.

It wasn't until 1991 that scientists uncovered, by chance, a human gene that clearly is involved in X inactivation, named Xist (for X-inactive specific transcripts) and found on the long arm of the human X chromosome in a segment called Xq13. While there is no chromosomal equivalent of a feminist perspective, women scientists who work at the level of chromosomal sex can be found smiling a lot, particularly since the discoveries emerged from some women as well as male scientists, including Migeon, Andrea Ballabio at Baylor College, Sohaila Rastan in England, Jeanne Lawrence in Massachusetts, and Carolyn Brown at Case Western.

Migeon, for instance, has found evidence that some genetic elements, maybe genes themselves near the X inactivating center, are necessary for Xist to operate, but no one yet knows if there's some biochemical or single genetic signal that shuts down the whole X, or if each gene on the X makes its own decision to keep quiet one gene at a time.

Inherent in the X inactivation story in mammals is some means of equalizing the proteins and functions of X-linked genes so that males and females get the same amount—except in the sex-determining regions themselves. This notion of "dosage compensation" took serious form before World War II when Hermann Muller began to wonder why genes on the X chromosome of fruit flies made the same amounts of protein in males and females even though the females had two copies of the X. Somehow, he reasoned, the females must equalize the X chromosome gene expression between the sexes, and eventually scientists figured out that flies do this by increasing, doubling actually, the X genes in the male. But different species have done dosage compensation in different ways, and mammals, including humans, correct the "problem" by inactivating almost all X genes on one X chromosome in every single cell.

In recent years scientists have not only isolated several genes that achieve this dosage compensation in fruit flies and worms, but also found that they operate in the same loop of activity that makes the initial decision to create a male or a female. When the dosage correction genes are activated, they regulate expression of all the X chromosome's genes by first binding proteins to the X, forming in turn protein complexes that chemically alter the DNA. That is, they all work by changing the parts of DNA that make chromosomes coil.

Scientists at Baylor have recently found that subunits of these protein complexes are also used by chromosomes undergoing meiosis, that critical early cell division that cuts the number of chromosomes in half to prepare for the doubling of chromosomes during fertilization. In people and other mammals, there is no proof that the very same process inactivates the excess X-linked genes, but Xist remains a strong contender, given that it so far has been found only in the X that becomes inactive.

Uncovering details of dosage compensation genes, including how many are operating, remains a full-time mission for scores of sex-determination investigators. Says Migeon, "attention must turn to what makes a female in an active sense. It's true that SRY has given us insight into what's important in male sex determination, but it's bad science to see the whole story of sex determination as the story of what's important in becoming a male. The X," she acknowledges, "is passive in terms of starting things going to make a male, and maybe genes on the X do help in double doses to make things go wrong, but it's not clear what normal doses of X genes do in females or males," she says. "Remember, there are XX males—but they are not fertile."

Eva Eicher of the famed "mouse house" at the Jackson Laboratory calls the idea that SRY is the master gene for sex determination, or that X is the rest of the story, for that matter, "frankly bullshit. There also are ovarian-determining genes, not just testes-determining genes. This isn't a feminist point of view, just a scientific one. There's too much that can't be explained about human sex determination and all of the abnormalities we see by the presence or absence of the human SRY gene alone."

Eicher's studies of two inherited sex-reversal conditions in mice known as B6-YPOS and B6-Tas, suggest, for instance, the presence of several non-sex chromosome (autosomal) genes without which gonads will not develop normally in these mice, suggesting that these genes may be just as valuable sex-determining switches as SRY. Eicher sees these as tips of a new iceberg in the sea of sex differentiation.

Her convictions have their roots in research that began fifteen years ago on strains of mice she and her team were breeding whose sex ratio was heavily biased in favor of females. "One could almost have thought a virus was present in the mouse room because in one year, in two cases, we saw the presence of more females than chance would predict." The excess numbers of females turned out to be due to XY females, a sort of

Turner syndrome" in mice, and the males were actually hermaphrodites, individuals that had both ovarian and testicular tissues. The strains she was working with were developed by crossbreeding her Black 6, ordinary lab mouse with a Swiss mouse. Only when the Y chromosome from "Turner" mice interacted with the Black 6 strain did ovarian tissues develop in the XY mice.

Because some wild male mice trapped in Italy and Germany produced XY hermaphrodites (essentially females with a male chromosomal pattern that could mate with each other) and also XY females when transferred to the Black 6 strain, Eicher and others speculated that the product of the Y chromosome gene in these mice that parallels the human SRY gene failed to interact normally with products made by autosomal (or non-sex) genes in the Black 6 strain. And it turns out they were right. With genetic mapping techniques, they found evidence for such a gene located on mouse chromosome 4, rather than on a sex chromosome.

David Page has found evidence for another of these autosomal genes on mouse chromosome 2 and some early evidence for a third gene on mouse chromosome 5. What's possibly going on here, Eicher and Page say, is that the product of a gene on one of the autosomes activates SRY or that SRY activates the gene. It's also possible that SRY may keep ovary-determining genes from functioning, making the gene they found an actual ovary-determining gene, just as SRY in humans is a testes-determining gene. Because there is overwhelming evidence that human and mouse genomes have remained essentially intact since they last shared a common ancestor, it is likely that scientists will soon find these genes in humans.

Even for the majority of men and women whose sexual identities have never been ambiguous, the perfect match between what our sex chromosomes and genes say and what we in fact do may be more illusion than reality. As John Money has said, sex is much more complicated than the question of genetics, even when the genetics seem straightforward. Lovers looking through their families' photo albums might have no doubt about each other's sexual orientation, but there is the celluloid evidence of *her* in baseball caps and jeans, kissing girlfriends and there *he* is hugging his boyfriends. He may be a "man's man" and she may be "all girl," both, with an inner sense as well as a public image of their role as man and woman in and out of the bedroom. The illusion, of course, is that the molecules that make us recognizable to ourselves and others as male or female is all there is to it. The truth of the matter, as Havelock Ellis has said, is that the "omnipresent process of sex, as it is woven into the whole texture of our man's or woman's body, is the pattern of all the process of our lives."

Nearly two and a half centuries after Lord Cornbury astonished Colonial America with his cross-dressing, historian Deirdre McCloskey described in her detail-filled memoir *Crossing* the painful physical transformation she undertook in her fifties to move from Donald, an unhappy male heterosexual cross-dresser, to a full-time life as a "complete" woman. Hair removal, breast implants, facial reconstruction, hairline repositioning, eyebrow-ridge grinding, voice operations, penis removal, neovagina construction. But the real sex change, she writes, was accomplished only when she was accepted as a woman among women, to "pass" as a member of the female sex so that passersby no longer paused to wonder if she were male or female. Writing of herself in the third person, she says "she had started to forget what it was like to be a man. . . . what it felt like to not understand relationships because you find them boring. Or to feel that you are by rights the local hero. Or to feel that people should serve you. Or most superficially and most fundamentally to think of men as 'we' and women as 'they.' "

More recent was the headline-grabbing account of a normal male infant whose circumcision was so badly botched by a Canadian surgeon decades ago that his parents, in understandable despair, had him castrated after they were told that little could be done to confer a "normal" male life on their son. A short time later, the parents heard an interview with the world's leading theorist and thinker on gender identity and consulted this psychologist, John Money of Johns Hopkins. According to Money, he counseled the family and suggested a number of options, one of them to attempt to "reassign" their son's gender, by rearing him in every possible way as a girl. The parents, divided by some accounts in their commitment to this effort, say they tried and ultimately failed to rear "Bruce" as "Brenda," and after years of reinforcing feminine behavior and punishing male traits, and of female hormone treatments, their son subsequently decided to live as a man. Publicizing the book about him by John Colapinto, "Brenda" told reporters that he "never felt like a girl, no matter what was done for and to him." While it is easy to sympathize with this individual's travail, the book contains distortions of both the known facts in the case and the theories and research of John Money. Contrary to Colapinto's interpretation of events, Money's position has never been that gender can be randomly assigned or that biology has nothing to do with gender. Money had theorized in the 1960s that because gender is in substantial ways a psychological and cultural construct, there are those situations in which reassignment early in life can lead to successful gender identity that is different than biological gender. Indeed, at least one of Money's cases, another boy reassigned as a girl after a circumcision injury, is living successfully, and from all accounts happily, as a woman. Although gender assignment "experiments" of this kind—because they can never be controlled—can never be proven to work or not work, in Bruce's case, at least, nature may have beaten nurture by an evolutionary mile. Bruce, now renamed "David," is alive and living as a married man. But all that such cases can really suggest is the complex interaction of nature and nurture in the determination of sexual identity.

What Bruce and Dierdre are reporting, of course, is that it's not just our obvious physical or biological traits (hair, breasts, penises, ovaries, testes, and genes) that define our sex for ourselves or others, but also our moods, inflections, gestures, walk, and attitudes, our inner sense of where we fit in the world our species inhabits. It's nature *and* nurture ever at play with each other and ourselves. Watching Jean (formerly Jonathan) Roughgarden undergo her gender transformation, friends and colleagues reportedly were struck by how her emotional and behavioral traits tracked her new physical emergence as a female. She became, they said, softer spoken, less assertive and aggressive, more nurturing and friendlier. In a new book, *Evolution's Rainbow,* she makes the case that a binary paradigm for gender as well as sexual mating is woefully outdated.

Just a few centuries ago, progressive thinkers believed there was essentially just one sex, and that sex organs were simply mirror images. The womb and vagina were the penis and scrotum "inside out." These days, all manner of biologists, psychologists, and sociobiologists, or at least their interpreters, have sometimes moved to the other end of the reductionist spectrum, seeing male and female as so different—men from Mars and women from Venus—that they might be different species altogether.

The truth lies someplace in the more complicated middle. Menstruation, lactation, and gestation are the marks of women, but not every menstruating, lactating, or pregnant woman is what we call "feminine." Some women never give birth or nurse an infant, but they are no less feminine. Some men never throw a ball side arm, or walk like John Wayne, but feminine women still are attracted to those who have no history of boyhood boisterousness or aggressiveness, bravery, physical and technical competence, or love of math. Not all men are sperm producers; women

may be brave, while men may be physical cowards. Women can be war-
riors and men tender nurturers of infants. The way men and women eat,
sit, stand, walk, make love, talk, and think are both biologically assigned
and socially achieved. We may think we *know* that a man who swishes
his hips is not quite "manly" and a woman who wants to play pro football
not quite "womanly." But do we really? Tomboys and effeminate males,
dogs and snails and puppy dogs' tails, sugar and spice and everything
nice—they're *part* of what boys and girls are made of. But *what* part,
exactly?

John Money learned to his despair, perhaps, that attempts at decon-
structing gender can be devastating at worst and at best, only partly possible with even the most advanced technological methods of tracing
behavior to its biological and psychological roots. Even the sequencing of
the genome is never likely to reveal any single human genes that entirely
explain coquetry or swagger. "No one," says sociobiologist Harmon R.
Holcomb III, "has identified any particular stretch of DNA as a hereditary
factor specific to any human behavior, psychological mechanism or behavioral predisposition." Including gender.

So what, exactly, do we know about the biology of sexual behavior
and orientation? How might that biology operate—and be operated on—
in the context of culture and environment, factors not hardwired in our
genetic blueprints? Do we get to choose any of our sex behaviors? Are
we "assigned" our gender by others merely on the basis of obvious sex
characteristics? How much of a role—and at what point in development—do the environments in which we're conceived and the cultures in
which we grow up play?

One way of thinking about this is to picture chromosomal sex and
gender as processes that go on both inside the genetic blueprint but also
in non-chromosomal parts of our biology and in our external cultural
environments as well. One example of this comes from the work of scientists at Harvard who found that an unfertilized egg gets ready for its
long developmental journey by first sorting outs its own RNA (the genetic
material that translates DNA's blueprint into an organism) and putting
that RNA into very precise spots within the egg. With this sort of "housekeeping," the egg in some sense gives its own DNA an "edge" after fertilization, when the first cell divisions that halve the new embryo are more
likely to produce "daughter" cells with different distributions of RNA,
which in turn sends them on different paths. This environmental maneuvering and "positioning" of RNA seems to help a developing embryo's

other genes "know" which end is up and how to orient for head, tail, back, front, left, and right. But such asymmetry in the distribution of an egg's RNA also suggests that far from being a passive recipient of fertilization, the egg has an "interest" in directing mom's genetic heritage through the fertilization process—and has the genes to carry out that interest. John Eppig of Jackson Labs, an authority on egg development, who likes to start interviews with the observation in Latin *Omne Vivum Ex Ovo*, "all life from eggs," calls eggs "the molecular and physiological bridges between generations and drivers of the strategies of gamete development." He adds, "in the earliest stages, the developing embryo's whole physiology is most like that of the egg alone, before fertilization, with lots of signaling going on between itself and body cells around it. Like sperm, eggs need to get signals from their surrounding environment to become what they are and nature has highly conserved this signaling mechanism, even though gametes are different. Eggs secrete substances that *can change* [italics mine] the default pathways for development and there is two-way communications, with eggs sending out signals to body cells and marking them and vice versa. By doing so, eggs create the means of their own survival."

Having testes, ovaries, eggs, and sperm in no way accounts for our full definition of sexual identity. Nevertheless, sex-determining genes and the proteins they make do provide blueprints for some secondary sexual traits, and some of the traits we identify as masculinity and femininity, as well as gonads and genitals. As noted earlier, many fully masculine and feminine individuals are infertile, and sex hormones don't have to be procreative in order to be creative in our bodies. Moreover, hormones march to many drummers, those whose rhythms are biologically predetermined, and those whose rhythms are determined more incidentally, by such things as stress or famine or a good feast. There is more than one story in literature and lore of men who managed in dire female-less emergencies to produce at least colostrum, the pre-milk protein that has formed the first breast-expressed nourishment for eons. And there are more than a few stories of hundred-pound females managing Sylvester Stallone–style physical strength to lift life-crushing weights off loved ones.

It should be clear by now from studies of such transgendered individuals as Deirdre McClosky that our gender characteristics are not binary at all. Men have the cultural reputation of the sexual predator, but women are notably sexually aggressive. Jezebel was no aberration and behind closed doors, some studies suggest, women are as likely to ask for, or even demand, sex as are men. Likewise, men as well as women have heart-bursting emotional attachments to their children, even though women are considered to be the "nurturing" sex. "Real men really do know all the words to the little teapot song," my son, Adam, told me, "but what we also have are the tears and the tenderness that come without plan or artifice. When I watch Ellie play and laugh, the joy is an almost over-whelming feeling in me, but so is the poignancy of knowing that all too soon, her innocence will have to end. Sometimes the pain of that is so intense, I almost can't bear it." This is from a sensitive male, perhaps, but also one whose male physicality and personality are unmistakable.

The search for sex-determining genes that help explain more than the bedrock physical requirements for male or female reproductive organs has, like the search for the business end of the Y chromosome, depended heavily on subtle clues from nature's "mistakes." A natural population in which to go hunting for the biology and sociology of gender, according to MIT's David Page and his colleagues, are the estimated 2 percent of all men in whom no apparent abnormality exists to explain their defective sperm-making ability and infertility. These men have all the right anatomy and make all the right hormones, but still can't make decent sperm. Was something wrong with their hormone-receptor apparatus? With their psy-chological makeup? With some genetic or biochemical "downstream" event that had so far eluded detection as part of the gender-developing process?

So far, results of Page's work suggest that there is nothing much psy-chological about fertility or infertility in this category of men. To reach that conclusion, Page and his team conducted a search for what they reasoned would have to be a gene for sperm production alone. Using a nearly complete gene map of the human Y chromosome, they found strong evidence that such a gene exists by tediously tracing missing or abnormal sequences of DNA in men with sperm-making problems, then comparing these Y fragments to the entire Y chromosome. In this way, they pinned down a particular portion of the Y that is missing in 13 percent of all men with severe sperm-producing defects. By comparing the Y DNA of the *fathers* of these men, all of whom had a perfect, intact Y, they could conclude that the missing pieces somehow "disappeared" or were missing in the infertile sons.

Holding that thought, Page and his team then cleverly turned to Turner syndrome females, who also have inherited somehow an incom-plete Y chromosome. The investigators found that all lacked a portion of the Y chromosome in an area that could be also involved in the infertile males he and his coworkers had studied. After a string of experiments, they identified at least two genes, known as RPS4Y and RPS4X, that appear to be codes for a piece of a protein needed by *all* cells that eventually become sperm.

Another set of genes, it turns out, named "zinc finger" genes for their chemical characteristics and shape, have been found to have similar properties, and may be responsible for loss of eggs and ovarian failure, also found in Turner female patients.

What's really intriguing about all this may come from some preliminary work by Elizabeth Simpson, a researcher focused on sex-determination genes and behavior. While working briefly with hybrid mice at Maine's Jackson Laboratory, she had been looking at other zinc finger genes and the proteins they make. Using a genetic-engineering tool, she "knocked out" a zinc finger gene on one of these animals' non-sex chromosomes, number 10. The protein this gene makes is only expressed in spermatids and spermatocytes, the precursors of sperm in males, so obviously it has something to do with sperm making in mice; it is speculated, though not proven, that it also has a close counterpart in humans.

When Simpson knocked out this gene in her lab mice, she found some startling sex-related behaviors. The males became overly aggressive, frantic, and completely unwilling to give up a fight. "They bit the hell out of the females and in the weeks before we realized what was happening, the lab technicians would routinely find females dead in the breeding cages. It was very upsetting," she reported. The same gene knockout in the females produced very different behavior, however. Although their sex hormones were at different levels and all the females got pregnant, they seemed to lose their maternal instincts. They abandoned their infants and routinely failed to feed or nurture the pups.

Simpson, interviewed in her laboratory, was reluctant to speculate too much about the behavioral impact of the knocked out gene. She has found brain damage in both males and females with this zinc finger protein missing and the behavioral genes are almost certainly tied to the damage there. But as part of the gender identity story, her preliminary findings are enticing because the zinc finger gene involved is only known to be turned "on" in male germ cells. "Clearly, however, it's expressing in the brain of females, too." Whether or not this story reveals a process that might help explain the connections between genes and behavior, and our perceptions of masculine and feminine behavior, remains to be seen, of course. But it's clear at least that the process is complex and likely to produce numerous variations on the gender behavior theme.

Evolutionary biologist Brian Charlesworth, in an essay on the evolution of sex chromosomes, has argued that genetic sex determination in higher-order animals, including humans, may sometimes have evolved from "environmental" sex determination in which circumstances drove the need for each sex to have genes that protected its own sexual development. In this sense, the "battle of the sexes," along with the means of its resolution, may begin very early, playing out in each sex's efforts to get its share of the DNA to hold sway over the other. Like a couple out on a first date, sex-determining molecules flirt, play hard to get, dance, tease, and consider carefully the final decision to embrace and lock together their futures and their fates. Moreover, the emergence and expression of such molecules are likely to run along a finite but flexible continuum, with "doses" determined in each individual by a variety of influences, internal and external. In short, no one size and amount fits all. It's possible that this process explains at least some of the gender differences *among* individuals of a single sex, as well as differences *between* the sexes, although at present there are no clear examples.

Among the more creative experiments designed to answer fundamental questions related to how male and female genders are fabricated are those that attempt to eliminate one sex or the other in the reproductive process. If reproduction can be achieved this way, it might be argued, then it is strong evidence that males and females have within them, despite the sexual demarcations we all recognize, the capacity to do it all by themselves by somehow "willfully" turning on dormant biological processes—or at least passively doing so in the absence of any other option. Given that a man's sperm or a woman's egg already carries a full genetic complement, it's logical to ask whether we really need both a mom and dad to create another. Remember, Dolly, the famously cloned ewe, is like all cloned specimens, not a "sexually reproduced" animal. In a tour de force of manipulation of genetic switches, scientists controlled the genes that regulate development. No mingling of genes occurred to meet the demands of evolution for variation and adaptation to new challenges.

In contrast, there is the work of Davor Solter, a mouse biologist who began decades ago searching for the broad mechanisms by which mammals regulate their sexual genes and development. He has conducted the insightful experiments that more definitively answer the question of whether normal males and females could be made if both of the fused nuclei of a new embryo came from dads or if both came from moms.

Beginning in the 1980s, Solter used micro-tweezers and a steady hand to remove the dad's contribution to the nucleus of a newly formed fertilized embryo and inserted a second female "pronucleus," the nuclear material in either a sperm or egg that unites during fertilization with its counterpart. In one of his experiments, for example, the all-male-derived embryo failed to develop embryonic structure but the all-female-derived embryo failed to develop the placenta and they both died. Once there is commitment to sexual reproduction, Solter concludes, "for normal development you need one male genome and one female genome." Inherent in his conclusions, he says, is that normal development requires genetic factors that go far beyond a few single genes.

What kinds of factors? Well, consider that if, at the level of sperm and egg fusion we cannot consciously "choose" our sex or even our gender preferences, can the reproductive environment be manipulated to do so? Kingdoms have been won and lost on the determination of couples to bear sons, or sons who were vigorous and masculine. Alas, unlike female queen bees and wasps that can "willfully" produce different ratios of sons and daughters depending on the fortunes or misfortunes of the hive and the swarm, we humans shouldn't count on choosing what sex we will have anytime soon. In truth, it's not the insects' sex chromosomes that do the job for them, anyway. All do-it-yourself attempts to alter the sex ratio have met with no more success than efforts to influence the outcome of an honest coin toss. In frustrating testimony to nature's preference for tried and true strategies, there is not a shred of scientific evidence that timing of sex, baking soda douches, climate, food preferences, position during intercourse, or any other popular gender selection "folk" formulas ever altered or out-maneuvered nature's plan for a fifty-fifty sex ratio, or increased or decreased the prevalence of transgendered individuals.

Unconsciously, on the other hand, such manipulation may indeed go on. J. G. Vandenbergh and C. L. Huggett at North Carolina State University have studied mammals in which sex ratio alterations are known to be related to environmental factors, and speculate that in house mice and gerbils at least, the position of a fetus in the uterus and who he or she is next to can influence anatomy, physiology, and behavior. For instance, the distance between anus and genitals of females who developed in utero between two males is longer than of females not between two males. Similarly, when females were located in the uterus between two males, litters contained 58 percent males; and when the females were not between two males, the ratio was 42 percent males. Nature may be too practical to have evolved any other approach. Based on observations and studies of sex ratios by Sir Ronald Fisher, William D. Hamilton, and Richard Dawkins, it appears that only if the sex ratio is even can male and female both enjoy the same advantage when it comes to sending their genes into the future via a system in which each child has one mother and one father. This may seem painfully obvious, but in fact it could have been otherwise; that is, in an evolutionary sense, it might "pay" for women to have more sons or fathers to have more daughters—as it turns out, it doesn't.

In a serious "mind" game, Dawkins has demonstrated this by asking what would happen if Mom's or Dad's genes, conserved over evolutionary time, had the power to "choose" the sex of their offspring by, say, Mom creating chemicals in her vagina that slightly favored her mate's son-making sperm, over daughter-making sperm? Which would Mom choose—to make daughters or sons? Stability and reproductive fitness of the species, on average, would predict that you would choose to have a child of whatever sex was scarcer. If males were being killed off in wars or females dying of disease or early childbirth, you would want to be sure that *your child* would find a healthy *mate* of the opposite sex. "The well-designed parent is strictly indifferent about whether a son or a daughter will be born," he says, because Mom and Dad can't benefit by favoring one sex over the other.

As for the seeming economic hogship of bigger males, our physiology works that out, too. All of that extra growth in males comes generally *after* junior achieves puberty, at least in animals and in our Stone Age ancestors, so the *cost* of keeping males fed and happy is no longer a parental expenditure requirement from nature's standpoint. Even when male babies are more likely to die than females, and women outnumber men when courting time comes around, it still would make no sense for Mom or Dad to favor a larger number of sons, thinking that the sons would then be more assured of a mate. But what the parent's genes might also ask is whether to have a girl who will survive childhood or a son who may die in infancy, requiring the parent to spend more resources to have "backup" sons. "The total amount of goods and energy invested in sons (including feeding infant sons up to the point where they died) will equal the total amount invested in daughters," Dawkins notes. Even if more men than women die as adults because of wars or disease, parents' genes still wouldn't win the DNA survival test if they chose to favor sons over daughters, thinking at first blush that males will be scarce and females more likely to seek their sons, whatever they look like or however much they earn. Why? Dawkins again: "The decision facing a parent is . . . 'should I have a son, who will likely be killed in battle after I've finished rearing him but who if he survives will give me [extra] numbers of grandchildren? Or shall I have a daughter who is fairly certain to give me an average number of grandchildren?' The number of grandchildren you can expect through a son is still the same as the average number you can expect through a daughter. And the cost of making a son is still the cost of feeding and protecting him up to the moment when he leaves the nest. The fact that he is likely to get killed soon after . . . does not change the calculation."

Dawkins goes on to hold the same mirror up to species where sex selection is "decided" by the offspring themselves (fish called blue-headed wrasses, whose largest female members can and do change sexes if alpha males in a harem of females die, and whose bachelor males can spend their lives as productive females). The sex ratio is still fifty-fifty, based on function.

And finally, Dawkins asks, if you were the parent, would you have a son who has a 90 percent chance of ending up a bachelor, giving you no grandchildren, or a daughter who will be in a harem and probably give you an average number of grandchildren? The answer rests in the indifference of DNA to any one individual it is housed in and its focus on keeping its sequences going in the species. If you have a son, he may be the one to have a harem and could produce more grandchildren than dozens of daughters could. "The average number of grandchildren you can expect through your sons is the same as the average number you can expect through your daughters." Nature still stacks the deck at fifty-fifty. "What is really going on," Dawkins concludes, "is that genes for maximizing grandchildren become more numerous in the gene pool. The world becomes full of genes that have successfully come down the ages. How should a gene be successful in coming down the ages other than by influencing the decisions of individuals so as to maximize their numbers of descendants."

One conclusion from all of this is that evolution has not only given us sexy sons and sexy daughters for the purpose of reproducing our species, but also sexy sons and sexy daughters who can't or won't, or who have just a little too much or a little too little of certain traits and interests, to find and appropriately use Mr. or Ms. Right. Harvard MBAs might think nature a wastrel, squandering precious resources on bachelor seals, peacock tails, elaborate bowers, and extravagant bird-song. Similarly, we might speculate at how "wastefully" the whole sexual repertoire of convoluted genetics and physiological processes chews up most of our reproductive years with hairdos, fancy clothes, romantic weekends, cosmetics, courtship, and endless cycles of involuntary and voluntary sexy folderol that make us attractive to the opposite sex. Why would nature build Rolls Royces for the masses, instead of Chevrolets? Why do we need breasts and penises and ovaries and testes and hair and eyes and arms and legs—and minds, for that matter—just to reproduce our DNA?

The answer again rests in the overarching need for species flexibility and plasticity, the ability to stay the course whatever may come. Nature's sex- and gender-determining machinery favors long-term survival of the species DNA, not yours or mine alone; and it cares little if parents run out of resources and steam and retirement income as long as junior can grow up to shoot Grandmom's and Grandad's DNA into the next generation.

The need is for genes to switch on and off when needed to accomplish particular ends. Within each of our cells, the goal is the same—keep the house in good shape for those eggs or sperm. Every new roll of the genetic dice, every new mutation of a gene influences every new conception in probably several ways. One of these is generally dominant for survival. But as a whole, there appears to be purpose and value for a species to hedge its gender bets, making some individuals more fertile than others; some alpha males and females, and others betas, gammas, or deltas; some feminine males, and some masculine females.

"Whenever natural selection favors a gene because of its beneficial effects in youth, say on sexual attractiveness in a young male," says Dawkins, "there is likely to be a downside: some particular disease in middle or old age, for example. Nature would not want to select for disease genes that turn on in the young, of course, but might favor modifier genes that switched on after you had children to eliminate you and your wasteful use of resources as soon as your children reproduced." (Cancer, as famed cancer gene hunter Bert Vogelstein of Johns Hopkins has noted, is principally a disease of aging, and thus a "side effect of having evolved to the point where we live long.")

In the end, humans are programmed to "spend" their genetic resources early, not late. "Everybody," says Dawkins, "is descended from an unbroken line of ancestors all of whom were at some time in their lives young, but many of whom were never old. . . . We inherit whatever it takes to be young, but not necessarily whatever it takes to be old. . . . genes don't care about suffering because they don't care about anything. Both bad fortune and good, when 'decisions' about determining sex or its regulating genes are at stake, are expected and meaningless *on the average.*"

As a consequence of this kind of thinking, the hard-line genetic differences in the two basic human sexes—marked by the division of labor in gonads, egg and sperm—should not obscure what else is going on in the process of becoming male and female. And scientists have begun to tease out more details of how the process unfolds, and how uncertain, fragile, and tenuous it often is.

There are genes and there is gender. There are sexes and sexual roles assigned by nature and society and there are the roles we acquire and draw about ourselves. And like the sexual stars they are, both our genes and our gender identities require top billing if a couple are to perform their most important roles well.

Long before a couple comes to enjoy and appreciate the way their bodies "fit" together, they begin to become familiar with their bodies and their minds' interest in and willingness to participate in sex. Watch a fetus in the uterus on ultrasound play with its penis. Whether or not this gives the unborn infant pleasure is impossible, of course, to know for certain, but it may be a rehearsal to perpetuate reflexes, to fill a penis with blood. It may be DNA's way of making sure this boy will someday know what else to do with it. Long before a male has his first intercourse, he has wet dreams, sex dreams full of fantasies and imagery of copulation and orgasm. There is rehearsal going on, driven by genes, experience, and emotions. We are huge collections of sexual molecules, but also full of good old-fashioned animal sexuality.

Socially and emotionally, people fall in love as well as in lust, a part of sex that requires not only the right gonads and gonadal sex, but also the playing out of our masculine and feminine roles. Whether or not we accept that sex is a "drive" or that we are "motivated" for sex or have instincts or needs for sex, men and women all have the capacity to move ahead or not with sexual behavior, to have bodies and minds that get exhibited or inhibited. Our sexual lives are moving all the time along a continuum of do and not do, mate and not mate, arouse and not arouse, what sexologist John Money calls "thresholds" for the release or inhibition of behavior. Argues evolutionary biologist David Crews, "Gonadal sex is a discrete characteristic that categorizes the individual, whereas sexuality is a suite of continuously variable traits that is unique to each individual." There are ovaries, testes, sperm, and eggs, but also non-egg and non-sperm biological influences (so-called *epigametic* factors) that trigger the "separate, distinctive physical and behavioral characteristics that mark human sexual behavior. . . ."

All of the evidence so far collected by gene hunters, psychobiologists, and anthropologists seems to be telling us that while genes are at the foundation of sex and sex behavior, they don't operate in a vacuum. They are influenced not only by other genes but also by events outside the control of individual animal genes, or even whole human populations of genes.

In their landmark book *Man and Woman, Boy and Girl,* John Money and Anke Ehrhardt concluded that "your gender identity is more than your gender role; it includes . . . imagery and unspoken text that may be known to you alone. The two are joined . . . [together] its determinants are not polarized into nature versus nurture, prenatal versus postnatal, biological versus social . . . [but are] the outcome of nature and nurture supporting one another at each stage and phase of development." It is certainly possible to carry that conclusion too far—as in the sad case of Bruce/Brenda, which ironically formed one of the foundations of their controversial work. But it is likely that scientists have yet to see the boundaries of the continuum, or really test how far it can be pushed one way or the other.

Using extensive histories of hermaphrodites and intersexuals, people born with ambiguous genetic and gonadal sex and genitals, Money and his colleagues discovered various combinations of ovaries and testes, hormones and hair, gonads and sex chromosomes, gender role and orientation. These intersexuals made a compelling case for the plasticity and flexibility of our male and female "sides," and for the sex role you are assigned by family and society having something to do with the outcome of your masculine or feminine behavior.

A few celebrated patients who were "assigned" a gender identity contrary to their genetic sex at birth (because of ambiguous genitalia at birth and parental wishes to offer normal lives to these children) have countered that they could not really ever overcome their genetic and hormonal sex. But in a study of seventy-six patients (none of whom were, like Bruce, a normal XY male born with normal genitals), Money demonstrated that twenty had a contradiction between gonadal sex and the sex of assignment and rearing and all but three of the twenty considered themselves "male or female" based not on what their gonads and hormones "said," but what their family and friends perceived them to be.

Money also compared internal reproductive organs and external genital appearance with ultimate declarations of male and female, masculine and feminine. "Since the uterus is the organ of menstruation and the prostate the major organ of seminal fluid secretion, it is necessary to compare maleness or femaleness of gender role with internal reproductive equipment," he says. And since it goes without saying that the external genitals are the sign from which parents and others take their cue in assigning a sexual status to a neonate and in rearing thereafter, "it is also necessary to look at gender identity and external genital appearance." In seventeen cases in whom an assigned sex and rearing were inconsistent with uterus and prostate existence, gender role agreed with *rearing* in fourteen. "So far as the evidence goes, there is no reason to suspect a correlation between internal accessory organs and maleness or femaleness of gender role," Money concluded, although it also is true that in intersexuals neither the uterus nor prostate reaches full functional maturity with taking hormones. As for external organs, again in the study of seventy-six intersexuals, most of the twenty-three who had lived most of their lives assigned to one sex or the other did so with a "contradiction between external genital morphology and assigned sex." While these people clearly had distress over the inconsistencies, they had mostly succeeded "in coming to terms with his or her anomaly and had a gender role and orientation wholly consistent with assigned sex and rearing," rather than determined by whether they had a penis or vagina.

Regardless of how history assesses Money's work, or that of other gender experts, it can be fairly said that strict "social constructionists" or strict "essentialists" arguing it's all nurture or all nature have little credibility. Even if by some stretch of the imagination, we could envision a set of genes that absolutely determined gender and sexual behavior, the strict essentialists would have a hard time excluding the unique human mind and its capacity for learning and social engineering as part of the gender-acquiring process.

In 1993, Janet Shirley Hyde and Mary Beth Oliver conducted an analysis of gender differences in sexuality, in which they collated studies of 239 different groups of men and women comprised of 128,000 in total, and looked at gender differences on twenty-one different measures of sexual attitudes and behaviors. These included attitudes about premarital sex, casual sex, intercourse in a committed relationship, intercourse during engagement, homosexuality, civil liberties for homosexuals, extramarital sex, sexual permissiveness, sex guilt and anxiety, satisfaction, masturbation, kissing and petting, intercourse, oral sex, age at first intercourse, sex partners, and frequency. The scientists found that the largest gender difference was in the incidence of masturbation, and the second was in attitudes about casual sex. These were enormous, they report, compared with gender differences in math ability and verbal ability. But surprisingly, perhaps, there were many variables that showed *no* gender differences, such as civil liberties for homosexuals and attitudes toward sexual satisfaction. Other gender differences were moderate, such as attitudes toward premarital sex and guilt.

The large differences in attitude about casual sex fit well with mating strategies that have proven their evolutionary mettle. Short-term mating or casual sex will logically be more important to men's sexual strategies than women's, since women historically have needed reliable signs of commitment as a requirement for intercourse. But it also is possible that this "sexual double standard" is what is taught by society in which men have organized the learning. What once was a reasonable biological/survival strategy in the Stone Age became entrenched socially, despite the altered imperatives in modern society to eliminate male-female conflict for which there is no longer any need. After all, in evolutionary terms, women, at least in the developed world, have just as many resources or access to resources as men nowadays. What distinguishes humans from other animals are our brains and the capacity to learn from experience. Stone Age biological imperatives are retained in the power structure, but there is nothing in our "genes" to argue that this must always be the case. In the same vein, we can speculate that in the Stone Age, women did not "disapprove" of casual sex, because the strategy men used was deemed necessary to survival of his genes—and her son's. Today, as the survey showed, women far more than men disapprove of casual sex, but still participate in it, a double standard that belies the biological imperative of our Stone Age ancestors and is perhaps a frustrated bow to social convention and male-dominated social structures.

"Evolutionary psychology," says David Buss, "predicts that males and females will be the same or similar in those domains in which the sexes have faced the same or similar adaptive problems. Both sexes have sweat glands because both sexes have faced the adaptive problem of thermal regulation. Both sexes have similar taste preferences for fat, sugar, salt, and particular amino acids because both sexes have faced similar food consumption problems. Both sexes develop calluses (to protect against skin friction). . . . both sexes appear to have cognitive and emotional mechanisms designed to detect cheaters and foster reciprocal relations."

"In other domains," however, Buss notes that "men and women have faced substantially different adaptive problems over human evolutionary history. In the physical realm, for example, women have faced the problem of childbirth; men have not. Women therefore have evolved particular adaptations that are lacking in men such as a cervix that dilates to ten centimeters just prior to giving birth, mechanisms for producing labor contractions and their release of oxytocin into the bloodstream during the delivery." They had faced different information-processing domains as well. "Because fertilization occurs within the woman . . . men have faced the adaptive problem of uncertainty of paternity in putative offspring. Men who failed to solve this problem risked investing resources in children who were not their own."

There is no "grand design" to any of this in the biological record, just happenstance. DNA has no "purpose," no plan for one sex's domination over the other. Whatever set evolution in motion took advantage of mutation and adaptation, those biological "editors" of which Lynn Margulis speaks, just as the human mind and human society take advantage of resources and experiences. Sex-determination genes may be selfish, but biologically influenced behavior need not be, and in fact often is not. The "fundamental desires of women and men differ in some domains, perhaps most notably in the sexual domain," says Buss, but "fulfilling each others' evolved desires may be one of the most powerful keys to harmony between a man and a woman."

Our genes and our experiences would seem to operate best of all—at least for the purposes of sex—when we seek *cooperation and mutual satisfaction.* "Those who fulfill each others' desires have more harmonious relations that are marked by kindness, affection, and commitment," he says. "Our evolved desires . . . in short, provide the essential ingredients for solving the elusive mystery of harmony between the sexes." In this way, evolution also has given humans the "capacity to control our own destiny," he concludes. We do not control our sex genes, at least not yet, but alone among the species, we have a lot to say about how we use them.

3

WHERE "IT" HAPPENS

"There is more wisdom in your body than in your deepest philosophy."
—FRIEDRICH NIETZSCHE

Bodies perform lots of tasks. They propel us, keep us upright, eat, think, learn, paint landscapes, earn a living, and sleep. Sexual bodies, on the other hand, have only two, closely linked goals—to be attractive and to have sex in order to reproduce. But how do our sexual bodies do that? More to the point, perhaps, how is it that the feeling is mutual—what his body signals, her body "reads"? And vice versa. How is it that attraction signals are pretty much universal and frequently hijack rational thought or common sense? Playwright James Goldman captured the process just right in *The Lion in Winter*, in which an aging, imprisoned Eleanor of Aquitaine describes and defends her scandalous emotional and physical abandonment of a royal fiancé, and ultimately ruinous seduction, within hours of meeting Henry II: "He [Henry] rode into my view," she says wistfully, "with the head of Aristotle and a body that was sin itself."

As direct descendants of Stone Age man and woman, modern humans, clearly including Goldman's king and queen, represent ancestors who bequeathed us leonine heads, long, thick penises, fleshy lips, pendulous breasts, smooth skin, beards, muscled male chests, ticklish ears, jutting jaws, fleshy buttocks, rosy nipples, silky hair, smelly secretions. It's the sexual body's "wisdom" that guides our sexual behavior.

Ethologist Desmond Morris, who coined the phrase "naked ape" to describe humans, notes that for whatever functional reasons of cooling or heating they may have evolved, hairy armpits and crotches are sexually erogenous traits, possibly because incidentally they could also hold pheromones and scents that trigger such primal urges as sexual excitement and kinship. Similarly, the tongue may not be a faux penis in everyone's romantic suite, but its texture, musculature, and (for some) taste make it difficult to dismiss as an important part of the sexual body. It seems our prehistoric relatives transmitted not only their genes, but also their shift at puberty from a childhood aversion to "yucky" raw, salty seafood and saliva to adult enjoyment of tasty French kisses, vaginal secretions, semen, and oyster bars. By the time of the Neanderthals, five thousand to ten thousand years ago, the sexual bodies that we admire today were in place, and our minds already were "wired" to let them shape our longings and our fashions for eons to come. "No individual can think his way around his own attractors," note pyschiatrists Thomas Lewis, Fari Amini, and Richard Lannon in *A General Theory of Love.* Blaise Pascal put it better more than three hundred years ago when he wrote that "the heart has reasons that reason cannot comprehend."

During the months a fetus swims in the womb, floating, bouncing, and responding to movement and sound, it is responding also to the touch of its own body. Experiments show that what the fetus touches of its own self stimulates nerve cells to sprout and grow in particular directions, feeding the pattern that will become the circuitry of the brain, the neural pathways that later will govern behavior, including sex behavior. Speculative but intriguing is the notion that ears and nipples may be erogenous zones in part because formative brain circuits connected to the genitals are also stimulated, coincidentally or because of proximity, by nerve pathways feeding to and from the penis and clitoris. Foot fetishes, found almost exclusively in men, may derive in part from intrauterine contacts between the unborn's feet and penis, as the developing brain is rapidly making nerve cell connections in those areas responsible for erection and arousal. In fact, boys have erections and girls secrete vaginal fluids while still in the womb, and as anyone who has had children knows, even babies masturbate, rehearsing the physiology that will generate adult sexual response.

If everything goes right, from infancy on, our limbic brains—the seat of feelings, instincts, and mood chemicals—connect our thinking higher brain to the pleasurable feelings, comfort, and sense of security we get when we stimulate and enhance our bodies. We respond eagerly to touch, massage, and certain kinds of faces and we never forget the *sense* of the sensation. "Children as young as two months show aesthetic preferences and judgments of faces similar to those of adults," says University of New Mexico biologist Randy Thornhill, an authority on the links between physical attractiveness and reproductive fitness. Infants don't "learn" these preferences; they're built in because they serve important signals of genetic and reproductive health and fitness. As children mature and reach puberty, body features under the influence of estrogen, testosterone, and other sex hormones become fixed and recognizable messengers of sexual maturity. They are so-called secondary sexual traits that operate as further advertisements of genetic quality. Deep voices, beards, armpit hair, thickening male jaws, cheeks, and brow ridges distinguish young men's faces as prepubertal estrogen declines. In teenage girls, the ratio of testosterone and estrogen shapes smaller lower jaws; less protruding noses, chins, and brows; smoother, softer skin; fuller lips; and a face whose lower portion is smaller than a boy's. By young adulthood, when sex organs are mature and fertility established, our preferences for certain body shapes and sizes are entrenched.

An important corollary to Thornhill's conclusions is that these preferences arose in males and females in parallel and in concert. Indeed, common sense as well as scientific investigation tell us that bodily attractants and preferences did not emerge in a simple process of cause and effect (he wants it, therefore she develops it). Instead, the bodies and looks we prefer got to be that way because members of the opposite sex not only found them, over evolutionary time, to be pretty good signals— clues to sexual desire, availability, and fertility—but also developed their own ongoing responses, even trumping the others, to set in motion sexual desire for these traits and more responses. In love as well as in coevolution, minds and hearts influence and change not only ourselves but our partners.

Sociobiologist Michael T. McGuire of UCLA sums it up succinctly if a bit formally: "Research findings over the past decade have clearly established that others' behavior and communications have physiological effects on persons and animals that encode such information." Pretty soon in this process, McGuire suggests, individuals learned to seek out situations that altered their responses in ways that felt good or protected them. This is a pretty good way to think about the concept of coevolution, and in his studies with some of our closest animal relatives, vervet monkeys, McGuire has found strong evidence for this theory. When the animals moved the body in submissive ways, for instance, such as lowering the head and "hiding" the sex organs from more dominant animals, the blood levels of the brain's "calm-down" and depression chemicals rose in the submissive animals. Body and brain work together to facilitate survival and opportunities for sex that aren't likely to be dangerous.

Darwin, and later British biologist Ronald Fisher, also supposed that in Greek tycoons, sexy film stars, and male peacocks, certain body parts stuck around and became "sexy" because females prefer them that way. Darwin called this sexual selection, because it was the sexual imperative that drove the selection and survival of a particular trait. After watching endless bouts of animal courtship and (in Fisher's case) studying the biological processes that attended these bouts, they argued that at some time in history, some random gene produced a male peacock with more splendid, colorful tails than other male peacocks, and men with relatively larger penises. If, for their own genetic or opportunistic reasons, peahens or females chose bigger tails and bigger penises, then any male offspring these females had would tend to have the same traits. Soon they would become the platinum standard for sexy. The pressure works on the other end, too. Fewer and fewer females will want the ninety-seven-pound-weakling equivalent of peacocks or men, and pretty soon, Greek islands, Hollywood, and zoos are filled with the bodies that the females desire. Any rebellious female who thwarted the new standards, either because she was no prize or somehow unable to respond to them like the others, would produce male offspring who would have a harder time getting a girl. So, if you were a peahen or a woman who didn't give a damn about what all the other girls thought was important in a mate, you probably lost the descendant lottery.

The body's naked truth in advertising, ultimately, is a matter of survival. If male vervet monkey scrotums are bright blue and their testicle size is linked to the amount of testosterone he secretes—a marker for fertility—it's only because over time, females learned that if they preferred male vervets with lots of testosterone who also happened to have blue balls, they'd get more offspring and more resources. "Physical attractiveness is a health certification. It tells us about the deep health of an individual all the way to the immune system and the capacity to provide protection," says Thornhill. And as with vervets, some parts of the human body, more than others, have become sexually sensitive, responsive, and desirable.

Testing and measuring this kind of phenomenon in people poses its own challenges, but none that have stopped Thornhill and his colleague, psychologist Steve W. Gangestad. They intensively investigated muscularity, health, vigor, social dominance, and wealth in 203 romantically involved couples and calculated the ratio between the number of sex partners each member of a couple had had and the amount of facial symmetry or evenness of their features. The more even the men (by the scientists' careful measures and the "handsome" ratings of their partners), the more sex the men got, even when women complained about the time or attention they were given. "Men with symmetrical faces *are judged by women* to provide better protection, to be healthier, stronger and potentially wealthier," Thornhill says. "So they get more sex. Men and women," he believes, "coevolved genes and body traits that helped them control fertility and choose mates wisely."

Other studies affirm that men tend to equate even features in a woman—beauty—with sexual availability, and women to equate handsomeness with mens' success and power. Psychologist R. Glenn Hass of the City University of New York asked subjects to rate forty photographs of opposite sex strangers on campus walks. Half were asked to judge the "success" or "sexual availability" of those pictured just from their appearance. After shuffling the photos, the subjects were asked a second time to rate "attractiveness." Hass secretly *timed* the ratings cumulatively to investigate what he *really* wanted to know: how men and women "instinctively" translate a body's "look." He assumed that the faster a man or woman rated a body trait, the more likely that judgment was "spontaneous" and "built in" to our genetic makeup. His results were that men are more likely to spontaneously translate a woman's physical appearance in terms of her sexuality, while women are more likely to translate physical good looks in terms of "success." That's because if the men in his study *first* rated a photo of a woman for sexual availability, they more rapidly rated her for attractiveness the next time. Women who first rated a photo of a man for sexual availability more rapidly rated him for success and resources the second time.

Given the central role of sex in the lives of animals, and the fundamental efficiency of nature, it should be no surprise that many body features—breasts and penises, for instance—not only perform ordinary reproductive jobs but also coevolved to do double duty as a sex shopper's guide. Whether or not his sperm are potent or her nipples will ever squirt milk, his penis rises in response to her breasts in a Wonderbra. The sexual body is a combination of reproductive organs and physical preferences. We can't use the former if we fail to negotiate the latter.

To many, it might seem that our sexual bodies—and the enhancements we make to it in the gym or the salon—coevolved to wind up the way they are along some rational or even predetermined route. In other words, we might be tempted to believe that there was some intrinsic *reason* for long, curved necks or big penises. Or, for that matter, for Manolo Blahnik spike heels, barbell biceps, and waist cinches. The truth is otherwise. Alexandra Basolo, a University of California at Santa Barbara biologist, for example, has evidence that some body parts that we now consider sexually attractive evolved for no other reason than a whimsical preference for some male adornment or ornament somewhere in females' evolutionary past. The preference may not coevolve at all, but in fact precede it. The girls want something for no apparent reason, and establish that fact in no uncertain terms to the guys in the neighborhood, in effect setting the standard of the handsome sexual body without so much as a by-your-leave.

Basolo's evidence is based on experiments she conducted with a small freshwater fish found only in Central America. She found that originally, the fish only had males with short tails, until one species in the genus developed long swordtails. What grew up were two distinct species whose only difference was the length of the tail. Female swordtails clearly preferred their males with long swords—the longer the better—and the sword seems to have no other fishy purpose than to attract the girls. Basolo then took the swordless fish, sewed colorful artificial swords to them, and let them swim around females of their swordless species. As Basolo suspected, the females went for the cosmetic surgery, and the longer the better, though they had never seen such tails in their own species before.

Later, scientists from the University of Texas repeated the experiment with transparent plastic swords the females presumably couldn't see and demonstrated that it was the color, not the tail itself, that was attractive. It wasn't that the swords helped propel the fish better, either; the females just liked the way they looked.

Why the color preference exists remains open to speculation; it may have to do with some resemblance to the foods the females like, Basolo says. What seems clear is that the human taste for big, decorative parts of the body, enhanced with ornaments or cosmetics or left alone, may be as natural as it is in fish and an important part of getting sex done. More than twenty separate areas of the human and primate brain help process colors and shapes. That much redundancy in our evolutionary and biological makeup suggests nature wanted to make sure we could see the body beautiful, whatever it looked like, although an alternative possibility is that we developed these color processors for some non-sex purposes like food gathering or defense, but then recruited them for sex as well.

Whichever scenario ultimately proves true, enhancements to the sexual body also seem as universal and natural as fish tails. Imagine the woman of your dreams dressing for a date, shaving her legs and underarms, and perhaps waxing away some pubic hair as well. Strapping herself into the Wonderbra to create an uplifted (read, "young") bosom, she uses lipstick and rouge to "flush" the face, eyeliners to emphasize the pupil, making it look larger and mimicking the widening of the pupil that occurs during sexual arousal. (Belladonna, a drug eye doctors have used to dilate the pupil, was part of women's cosmetic kits for decades. The word itself means "beautiful woman.") Tans mimic the healthy glow of skin in the throes of passion when chemicals dilate blood vessels, wet the palms with sweat, swell the sex organs, excite the heart rate, sensitize nerve endings, and leave us breathless.

Voltaire wrote that if you asked a toad what it found beautiful, the answers would be the female toad's body "with two huge round black eyes coming out of her tiny head, large flat mouth, yellow belly, and brown back." Humans, too, have developed bodies that in the eyes of the beholder become the objects of desire: thick, long (when erect) penises, long hair and beards (until very recently in evolutionary time), rippling pecs and biceps, *Baywatch* breasts, red, fleshy lips. When any of these are a little lacking, we improvise with whatever the multibillion-dollar cosmetics, fitness, and plastic surgery industries have to offer. And although men more than women value looks in a sex partner, women notice the way a man walks and talks, along with his buttocks, chin, jawline, hands, and muscles. If men today do not go to the same lengths that women do to enhance their physical traits, that hasn't always been the case: Codpieces are a centuries-old invention. So, in contrast to the purely reproductive body comprised of ovaries, testes, prostates, wombs, and an assortment of glandular hormones and brain chemicals, the sexual body is dual purpose, made up of these plus erogenous and responsive zones, the parts of us that tingle and sizzle with pleasure, attract the opposite sex, and dance to the music of our biochemistry.

Reproduction is the business end of sex. On the surface, it seems simple enough: During intercourse, a man ejaculates millions of sperm cells through his penis into the upper reaches of the female vagina. If a woman's ovaries have recently released an egg cell, or ovum, one sperm will make its way there first, penetrate the cell, trigger a series of minute but cataclysmic chemical reactions, and nine months hence, a 200-billion-celled new human is born. That's the plan. Getting that feat accomplished, however, is anything but simple and requires a collection of organs working in complementary fashion.

His collection consists of a system of muscles, sphincters, valves, tubes, sacs, glands, and containers that make and deliver sperm and produce orgasm. The human penis, when limp, is about the length of a man's middle finger on average, much larger, both in absolute terms and in relation to body size, than a gorilla's or other apes'. Again, on average, an erect penis is about six inches in length. William H. Masters and Virginia E. Johnson, the renowned sex physiologists, called erection the "great equalizer" of penises, in that smaller penises grow relatively longer than bigger ones during erection. Moreover, while it may put a damper on *Penthouse* columnists and pornographers' fantasies, size has nothing to do with satisfaction, either reproductively or sensually. The one and a half inches of a gorilla and orangutan and the five to six inches of a human are functional equivalents. Notes zoologist/biologist Jared Diamond, "[T]he 1.5-inch penis of the male orangutan permits it to perform in a variety of positions rivaling ours and to outperform us by executing all those positions while hanging from a tree. As for the possible use of a large penis in sustaining prolonged intercourse, orangutans top us in that regard too (mean duration fifteen minutes versus a mere four minutes for the average American man)."

Inside, the penis is a composite of sponge-like tissues laced with tiny blood vessels and nerves, but without the bones that some other male mammals have. The urethra runs through it to empty urine from the bladder and to move semen and sperm. When ejaculation occurs, a valve shuts down the urinary pathway to the bladder and two small glands deep inside at the base of the prostate release a chemical that neutralizes any leftover urinary acids that might kill sperm on contact.

At the tip of the penis rests the glans, whose nerves make it exquisitely sensitive to touch and tension. In the uncircumcised man, a foreskin sheath covers the glans until erection, during which the foreskin pulls back. Why there is a human male foreskin is a puzzle, although some have suggested (without too much evidence) that because it tends to accumulate secretions that "glue" it to the penis, and because such adhesions may make first intercourse a bit uncomfortable, very young male virgins will delay first intercourse until they are presumably more prepared to provide for offspring and have had longer experience with masturbation. The female hymen may have the same evolutionary function.

Under the penis rests the scrotum, a bilateral sac. In each chamber lies a testicle, or testis, an epididymis, which is a duct, and muscles that pull or relax to keep the testicles closer or farther away from the body and thus maintain the temperature of the testicles somewhat lower than 98.6 degrees Fahrenheit—the optimal temperature for making sperm. Both egg- and ball-shaped testicles are the same size, although the left testicle tends to hang lower than the right. The egg-shaped testicle contains several football-field lengths of ringed tubes that make sperm, pass them into the epididymis, where they undergo maturation to become the familiar tadpole-shaped swimmers, and then pass through the vas deferens into two small containers, the seminal vesicles. The sperm sit there until ejaculation, when they first mix with the semen, which nourishes sperm and keeps them alive in the vagina and through their journey into the neck of the womb or cervix and onward to the egg. Rounding out the reproductive end for men is the prostate gland, palpable from the rectum and sitting just beneath the bladder. It makes the opaque, slightly milky part of the semen. When orgasm (usually but not always a part of ejaculation) is imminent, the muscles of the nerve-rich prostate and other muscles in the vicinity contract in spasms that shoot the semen down the urethra and create the ejaculate. What "comes" is the sperm-containing semen and the orgasm.

A woman's reproductive system is less obvious than a man's and more complicated, as well. Of course, nothing remotely sticks out the way the penis does. And the clitoris, which produces the same pleasurable sensations when stimulated as the nerve-rich glans, or tip of the penis, has no role whatsoever in urination, conception, fertility, or birth.

A woman's principal reproductive organs are her ovaries. Corresponding to a man's testicles as producers of gametes, each of the two resembles a one-and-a-half- to two-inch-long squashed oval resting deep inside the lower abdomen, on each side. Unlike the sperm, which are continually made anew, all of the eggs a woman will ever have in her ovaries—about four hundred thousand—are there at her birth, in the immature stage. The ovaries store them until at puberty, biochemicals trigger their maturity and a fully ripe ovum is released at the rate of about one each month—at mid–menstrual cycle. Unlike other female mammals, women have no periods of estrus or "heat," and can and will mate anytime of the month or year, but sexual interest and intercourse do seem to fluctuate according to phases of the menstrual cycle.

In any case, with each menstrual cycle an egg matures in a little sac on the surface of the ovary, then bursts through a small opening in the fluid-filled sac, and travels, along with its own fluid and some nurturing cells, into the opening at the end of the top of the fallopian tube, which has moving "fingers" of thin tissue to pluck the egg down the tube toward the womb. There, hairlike cilia continue to herd the egg along. The egg stays in the tube for about three days before it gets to the uterus. If it stays unfertilized by sperm, it disintegrates and becomes part of the "menses," which are made of the blood-filled tissues along the lining of the uterus or endometrium that prepares each month for implantation and pregnancy by thickening and growing. If conception occurs, the fertilized egg continues down the tube, dividing as it goes. At about the sixteen-cell stage, it implants itself along the wall of the uterus, a pear-shaped muscle about three inches long in a woman who has never been pregnant. The uterus will stretch to many times its unpregnant size to accommodate the growing fetus and, with the help of abdominal muscles, expel it during birth through the vagina. The vagina is also a very pliable muscle, a tube about four inches long that has been likened to a balloon without the air.

The vagina's inner tissues are moist, much like the mucus membrane that lines the mouth. This area is also without nerve endings, and thus insensitive. But the outer reaches are highly arousable. In a virgin, the vagina is partly covered by a thin membrane called the hymen, named for the Greek god of marriage and the word for membrane, a tribute to the belief that the tissue's presence was considered a sign of virginity and purity. In fact, it is not an intact membrane—all have at least a small opening through which menses can flow—and most break or stretch long before a woman's first intercourse.

What shows on a woman from the outside is known collectively as the vulva. The vulva begins with the fatty hillocks of the mons veneris, or mounds of Venus. Some studies of desirability suggest that the plumper or more pronounced the vulva, the more appealing to men because it signals health, and the ability to grasp and hold the penis more closely during intercourse, increasing tensions, friction, stimulation, and pleasure. The mons overlays the pubic bone and is covered with hair post-puberty. Beneath it lie the outer lips, or labia majora, which act as a protective cushion around the genitals. When these are separated, the inner lips, or labia minora, show a second set of skin-covered tissues, bare of hair and exceptionally sensitive to touch. These come together at the abdominal side to form a small flap or hood of skin that covers a small, hard, sensitive bump, about the size of a small pea. This is the clitoris, the erogenous center of a woman's body and the site and trigger of a woman's orgasm. There also is a clitoral shaft, like a penile shaft, containing the same kind of sponge-like tissue that makes a penis erect. When a woman is aroused, blood fills the sponges and enlarges this tiny organ.

The evolutionary presence and purpose of the clitoris remains a subject of intense debate. In males, of course, orgasm is absolutely critical to ejaculation and impregnation of females, because the spasms literally push sperm into the vagina. But some, like anthropologist Donald Symons, among others, don't think the female orgasm—or the clitoris that makes it possible—is adaptive at all because it is of no specific value to reproduction.

Some in the debate maintain that both the clitoris and the orgasm are merely "fellow" travelers with something else nature needed for survival. In a famous essay, evolutionary biologists Stephen Jay Gould and Richard Lewontin liken the clitoris and its pleasurable sensations to the triangular "spandrels," or wall spaces, in some cathedrals such as San Marco in Venice. These spaces became favored places for frescoes and paintings, but were merely architectural leftovers that came about by virtue of the angles at which Gothic vaulted ceilings were constructed to support roofs. Just as cathedrals don't need spandrels, women don't "need" orgasms to reproduce; they are happy leftovers.

Others argue just as convincingly that it's unlikely the clitoris survived the eons unless it had some purpose. For one thing, all female primates and higher mammals, not just humans, have a clitoris, and a chimp's is actually larger than a woman's. And recordings of the copulatory screams of female chimps strongly suggest that they have multiple orgasms. For another, an orgasm, with all of the extraordinary physical and emotional sensations it evokes, is something that men *like* women to have (as well as themselves) because it would seem to tell a man that the woman is happy and less likely to look elsewhere for satisfaction. If orgasm isn't necessary to get a girl, it may indeed be important in retaining her and guarding her to make sure any offspring is his alone. It's likely that women fake orgasms precisely because they keep men emotionally seduced.

Finally, as Timothy Taylor points out in his *Prehistory of Sex,* the clitoris is "not a small bud of underdeveloped cells," or some leftover, but a logical part of the human body plan no less than the glans penis. The clitoris, he says, "is actually no smaller than the penis although in humans much of it is hidden. When flaccid, the glans clitoris is partly or wholly covered by a fold formed by the inner lips of the vagina, congruent with the fact that all humans' original body plan was female. After the first inch or two the clitoris divides and the remainder of its erectile tissue runs down for six or seven inches to either side of the vaginal opening," a size some men would kill for. In female lemurs, primates with whom we shared a common biological ancestry 20 million years ago, the urethra, or urine tube, was contained in the clitoris, so that urination in females is the same as in males—through the penis. In fact, says Taylor, it's just possible that Gould's idea of an orgasm as spandrel may more appropriately apply to the male orgasm. "While a penis," Taylor writes, "seems necessary for normal insemination, it does not need to spasm to ejaculate; it could just produce a directed flow, as with urine." What is just as likely, he says, is that male orgasm is a vestigial form of the vaginal contractions used for inhaling sperm. The battle of the sexes aside, Taylor notes, there's a lot of "overlap" between male and female sex organs, which in turn is reflected in a lot of overlap between sexual behavior and sexual thoughts and feelings.

Continuing our anatomical journey takes us just below the clitoris to a small opening through which women urinate and below that to a much larger vaginal opening through which menstrual blood, vaginal discharges, and babies pass. Anthropologist Helen Fisher and others have noted that a modern woman's vagina tilts down away from her anus, a position that invites face-to-face mating and the missionary position, which in turn lets a man's pelvic bone rub against her clitoris, enhancing orgasm for her. Says Fisher, "The downward tilting human vaginal canal could have evolved via sexual selection. If [our earliest human ancestors] had a tipped vagina and encouraged face-to-face coitus [almost universal in human populations], her partners could see her face, whisper, gaze and pick up nuances of her expressions. Face-to-face copulation fostered intimacy, communication and understanding. So, like those ancestral females with pendulous sensitive breasts, those with tipped vaginas perhaps forged strong bonds with their special friends and bore disproportionately more young—passing this trait to us."

Through the vagina, about a finger length deep, are moist mucus surfaces that expand during sexual arousal to prepare for penetration by an erect penis. The outer third of the vagina is very sensitive, containing many nerve endings. If a woman contracts and relaxes the muscles of the pelvic floor, the so-called Kegel exercises, the action stimulates the nerve endings in this part of the vagina and rewards her with intense feelings of pleasure during intercourse. The vagina continues at an angle toward the back, and if you extend a middle finger length as far as possible, you'll hit the rubbery lump of tissue that is the lower part of the muscular uterus, the cervix. The cervix has a small depression, called the os, through which sperm and menses flow, and it is the os that dilates during labor to allow a baby to pass through it.

Although all of this is pretty much common knowledge, at least among those in the know, it was not until 1999 that a Dutch team published in the *British Medical Journal* one of the very few ongoing efforts to bring the latest imaging technology to the anatomical study of human sexual arousal and intercourse. One of few, because of the nearly universal ethical, moral, and scholarly squeamishness of doing so. Indeed, the team of investigators at the University Hospital in the Netherlands, including a gynecologist, physiologist, anthropologist, and radiologist, noted that the "hypothesized anatomy of human coitus as drawn by Leonardo da Vinci in about 1493 and by R. L. Dickinson in 1933" has been almost impossible to prove or disprove even though MRI has been around for more than two decades. (In his drawing "The Copulation," da Vinci imagined the erect penis during intercourse as perfectly straight in the missionary position, distorted the sexual bodies, and showed that semen was supposed to come down from the brain through a channel in the male spine. Dickinson did a little better by placing a glass test tube as big as a penis in erection into the vagina of the female subject, aroused by clitoral stimulation with a vibrator, to guide him in constructing his drawing.)

Determined to do better, the Dutch team recruited eight couples and three single women who participated in a total of thirteen experiments over eight years. Six of the couples succeeded in partial though not complete penetration while performing coitus in an MRI tube—until the advent of Viagra, which serendipitously became available during the experimental year. Two couples were then invited to repeat their efforts after the man took one 25 mg tablet and completed penetration that lasted long enough (twelve seconds) for images to be taken of completed coitus. The experiments did indeed show that MRI imaging of male and female genitals during coitus is feasible; and they suggest a number of intriguing results. According to gynecologist Willibrord Weijmar Schultz and his group, during missionary position intercourse, the penis has the shape of a boomerang (their word), and a full third of its twenty-two-centimeter (eight-and-a-half-inch) length consists of the root of the penis. "Scanning the position of the human genitals during coitus gives a convincing impression of the enormous size of the average penis in erection and of the volume of vaginal and pelvic space required by the pendulous part of the penis." They also said their pictures challenge the conclusion of Masters and Johnson that during female sexual arousal without intercourse the female uterus increases. "In contrast to Masters and Johnson, our images did not show an increase. . . . Their interpretation may have been caused by the raising of the uterus or filling of the bladder during their experiments." MRI further showed that during female sexual arousal, there were striking changes in the anterior, or back part, of the vaginal wall and that this part of the vagina is most sensitive to stimuli, along with the urinary bladder.

To say that their experiments have created controversy is a vast understatement, as the authors themselves acknowledge. One Canadian radiologist, for example, sarcastically congratulated Schultz and others for their "millennium paper," calling it a "Dutch treat." Wrote Leonardo Martin of McMaster University, the experimental design may be wanting with respect to overturning Leonardo da Vinci's drawing. "Leonardo's volunteer was not subject to a 1.5 Tesla magnetic field, a force which perhaps could boomerang anyone's '22 cms.' I am concerned the conclusions drawn may be premature and 'phallatious.' " But he congratulated them again for their "creative presentation." More sober critics noted that the restricted space within the MRI may have forced the copulating couple into unnatural manipulations; and that the resolution of the images was too poor to clearly see female anatomical structures such as the clitoris, vagina, and urethra during arousal, opening the possibility that the women faked it. Others noted that the reason so few men could sustain an erection and achieve complete penetration in the MRI was due to "prying eyes, instruments, strange partners"—and that masturbating men would have had as much success as the masturbating women imaged during their arousal without a partner. "The mind," this writer noted, "is a wonderful and very successful sex organ."

Responding to the critics and commentators, the scientists acknowledged the difficulty of doing observational studies of humans during sex but were steadfast about their quest. "The main purpose of the . . . paper . . . was to show the scientific world that it is possible to perform sexological research in a MRI. Further research is necessary to increase our knowledge." Whether or when their experiments will be affirmed, or new research done at all, is open to question.

For all the wonder they inspire, the sex organs and their function are much more than the joining of genitals and the union of gametes. They are to the sexual body what Vesailles's extraordinary, ahead-of-its-time plumbing is to the palace—a marvel, perhaps, but not what we think of when we think "sex." Indeed, the total experience of the sexual body is not just the sum of its parts. Masters and Johnson named their masterwork *Human Sexual Response*, rather than, say, *Human Sex*, or *Sex and Reproduction*, because while some of the instruments of reproduction are involved in the body's sexual response, it's clear to anyone past puberty that sexual response can and does occur without reproduction and that we evolved our sexual bodies and physiology to respond pleasurably and sexually without any thought whatever of having a baby. The body's "advertising" campaign, therefore, has been honed over millions of years with the help of generations of "focus groups" voting "yes" or "no" to the messages we carry on and inside of us.

One way to understand the nature of the sexual body as a highly evolved, integrated reproductive and erotic "package" is to peer into people's sexual body preferences across cultures. The more widespread a preference or a trait, the more likely it is a trait that meant business. A 1995 poll in Germany, for instance, suggests that as we shed Victorian notions of modesty, the size and shape of genitalia are moving up in the ratings as overtly sexy. But with respect to what shows at the beach, for men the top honors go to full mouths, large breasts, slim bodies and waists overall, rounded hips, long legs, and trim thighs. Curvaceous, heavier bodies of the kind depicted in Boticellean Venuses and Madonnas are less popular. Long hair and tight buns are relatively more popular in the West, while in the East, clear skin, dark eyes, straight noses, and some fleshiness around the middle are considered sexy. What women seem to value most in a man's sexual body are tallness, muscularity, physical dominance, and strength. Surveys have suggested that for men, the ideal female face is oval, with a straight, diamond-shaped nose, blue eyes, fine brows, well-proportioned ears and earlobes, and a middle-sized to larger mouth, while male faces ideal to a woman are square with more bony, somewhat humped noses and heavier brows.

Overall in these surveys, women prefer tall men (six feet or over) and muscular bodies, with a high ratio of lean muscle to fat all over the body. Women want men with small bottoms, large penises, and narrow hips and waists. A widely quoted study conducted by researchers at Loyola University in Chicago shows that only a small number of women want muscle men, and most want a man who is V-shaped, with slim legs, thin trunks, and broad upper bodies. The ideal male sexual body belongs to Greg Louganis, not Arnold Schwarzenegger, to fine-boned men, not Hulk Hogan. And while tallness is highly rated among women, talent and money can take down NBA heroes, as witnessed by the reproductive success of Mickey Rooney (we've lost count of his wives and kids), Charlie Chaplin, and Pablo Picasso (five feet four inches tall).

Research also suggests that the best sexual body is a quickly apparent *combination* of shapes, colors, and odors, everything we collectively call "sex appeal." Madison Avenue pays high dollar to photograph perfect hands and feet that sell diamond rings and toenail polish, but in the real world, it's the gestalt that counts. In search of sex, we look for the "look" we seem to automatically love and lust for; not surprisingly, our choices of cosmetics, clothes, and other enhancements match our notions of what's erotic. In sum, sexual bodies are the impressionistic catalogs of reproductive wares. Long before a couple gets to sex, they get to shopping; and with a few primitive societal exceptions, they don't, at least at first, get to view directly, much less try out, the reproductive product. It's the rare fellow who gets to see vagina, womb, or clitoris. What they *do* see is what sells sex. And that starts with the body's largest organ, the skin.

For European men, and men of European descent, the sexiest women have skin that is youthful, clear, and childlike—peaches and cream, dusty, rosy, dewy, and firm. Their women, however, want men whose skin is unscarred, yet ruddy or tanned. "Over time," says Peter Frost of the Universite Laval Department of Anthropology, "hardwired mental linkages may have formed between certain aspects of human skin color, the type of encounter it called to mind, and the appropriate state of [sexual] readiness."

Male skin, Frost explains, has more melanin and hemoglobin than female skin, making men browner and ruddier and women paler. Ancient Greek and Egyptian urns are famous for the pigments that painted males almost red and brown and women beige and white. The sexes, it seems, differ in both their "constitutional" pigmentation and tanning potential. "This sex difference begins at puberty when girls lighten in color and may widen in adulthood as male pigmentation darkens in response to repeated tanning," Frost says. As our ancestors evolved into families, the differences in the complexions of men and women became a main source of skin color variation, with darker skin signaling "man" and paler skin "woman" or "infant." In this way, they armed themselves for courtship and mating by having strong visual, quick clues to what another animal intended. If a male saw another male in the distance who was, like himself, relatively dark, that could mean impending conflict or competition for a female. If he saw a pale face, it more likely meant a child, and no threat at all—or a female with whom he could mate.

Texture and odor feed our universal interest in complexions as well as color. Think back to your first kiss and the feel of his lips or your first touch of a man's penis. One like velvet, the other more like satin. The soft skin of a baby's cheek has been poetically and accurately compared to the soft skin of a woman's thighs. Soft skin, therefore, may have evolved as a sexual body preference because it signals "youth" and perhaps helplessness. The combination of softness and "fluffiness"—as in fur—has evolved as a highly erogenous set of circumstances as well. A woman's pubic hair, the stuff of legend in "skin" magazines, may not only hold hormonally driven "sexy" odors, but also be a visual cue to fertility and sexual receptiveness. Animals ruff their fur as part of the mating game, and as Blackgama advertisers know, there is something erotic and erogenous about a pale woman in a dark ranch mink coat, or a dark-skinned woman in white fox.

Frost notes that in modern Western societies, particularly, skin tone differences have been submerged by intermarriage, protection from the elements, cosmetics, and tanning salons; but from the standpoint of our hardwired preferences, skin color is incorporated into our search for mates. Our complexions are tip-offs to our hormonal and reproductive status. Studies in college women have even shown that when women are most fertile, as measured by their sex hormones, they react differently to men with different skin colors and different-sized faces, preferring the darker skins and larger faces that signal "testosterone" and "fertility" in these men. Tanning may also be preferred because in certain circumstances it hides imperfections.

Along with texture, shape, and color, the sexual body varies in size.

But size doesn't always count, and while men make a big deal about the size of their penises, the fact is that few women will ever get to measure or compare them. Nor do women need to care, despite many women's insistence that they do, because studies incontrovertibly confirm that female pleasure is independent of penis length or circumference. In sum, adult male penises, in all their size variations, are *relatively* about the same from a woman's standpoint. She cares how big a man's penis is about as much as men care about the size of the female vagina. (Gynecologists at Cleveland Clinic Foundation actually measured the vaginas of 104 women and found that the vagina's breadth, length, or atrophy had little to do with how sexually fit or fertile a woman is.)

On the other hand, it is true that the size of men's testicles are a rough estimate of sex drive, thanks to correlations with certain male sex hormones. But there is growing evidence that if size matters in any way, it is *women's* sex habits that drove evolution of both the size of male genitals and *preference* for them.

The principal reasoning behind this line of thinking is that women can and do have more than one mate, and can and will mate with any man they choose, so men must compete for their favors. In gorillas and walruses, where one male serves an entire harem and has no competition from females, the size of his penis and his testicles are not as big an issue. He needn't compete for her favors. In the case of chimpanzees and some whales, the better hung they are the better chance they have of getting their sperm to the females of their choice before some other animal gets there first. Back in the pygmy chimp (bonobo) colony, the females will mate with everything that isn't nailed down or dead, constantly, without fear or favor. The males need all the testosterone (and testicles to make it) they can get just to stay alive under the females' sexual demands. Male chimps clearly evolved their conspicuous genitals for the same reason that human males did: A prominent, distinctive penis helps proclaim sexual readiness and fitness. It's very likely that when apes began to walk upright and Lucy's ancestors became fully two-footed, they did so in part because they could parade their genitals. And succinctly if rudely put, females selected the men with biggest balls because that meant they would get relatively, albeit probably unnecessarily, more sexual stimulation and pleasure. Throughout the animal kingdom, quantity counts. Some insects have large penises, snakes and lizards have two that they use alternately, chimps have the ability to display an erect penis and flick it with their fingers as they stare open-eyed at the object of their desire. While it would feed the macho mentality to suggest that sexual bodies evolved so that males could fight each other for favored females, the truth is that they probably evolved to charm the ladies, at least in the case of animals in which the females have lots of choices and exercise them.

By Timothy Taylor's reckoning, the large relative size of the human male's penis, even in its flaccid state, has nothing to do whatsoever with sexual prowess or the capacity to deliver large quantities of sperm (primates deliver more, better, faster, and with "vanishingly small penises") and everything to do with the fact that women liked this advertisement. He points out that having an "unruly" member that got erect even if the man didn't want to advertise himself to enemies was a liability. Moreover, unlike gorillas and other animals, from a distance, prehistoric man and woman may not have been able to easily tell each other apart, Taylor suggests, given that their total body sizes were very similar and probably covered with hair. "Once the penis became a visual criterion of manhood, its evolutionary growth was guaranteed," according to Taylor. "In any generation there will always be some women who favor bigger penises and few or none who positively favor smaller ones, so a trend toward greater length was established."

If the question of size seems paramount when discussing male members, it pales in comparison to women's breasts. There may be sexual body parts more subject to nonsense, but at times it's hard to think of any. The object of infant nourishment, childhood hilarity ("look, you have boobies!"), preadolescent despair (flat-chested girls), "second base" campaigns (horny boys), surgeons' artistry, and fashion's whims, an adult woman's breasts draw a relentless stream of male dichotomous Madonna-whore thinking. They are functional milk factories and they are erogenous zones—for him and for her—simultaneously.

In his groundbreaking book *Phantoms in the Brain*, University of California neuroscientist V. S. Ramachandran describes a unique set of nature's experiments—amputations of body parts—to illuminate the way the brain, and we, get to have and experience sensations, particularly erogenous ones. Quoting an Italian neurologist, Dr. Salvatore Aglioti, Ramachandran says that a substantial number of women who have radical mastectomies for cancer experience "vivid phantom breasts." The Italian wondered what body parts are mapped in the brain near the breasts and by "stimulating adjacent regions on the chest he found that parts of the sternum and clavicle, when touched, produce sensations in the phantom *nipple.*" The neurologist further learned that a third of women with radical mastectomies reported "tingling, erotic sensations in their phantom nipples when their earlobes were stimulated. But this happend only in the phantom breast, not the real one on the other side." One possible explanation, according to Ramachandran, is that the nipple and ear are next to each other on the brain-body sensation map. It may also explain why at least some women are highly aroused when their ears are nibbled or licked during foreplay. Almost half a century ago, a Canadian neurosurgeon, Wilder Penfield, more infamously known as a psychosurgeon, exploited the availability of patients' exposed brains prior to lobotomy to apply electrode stimulation and "map" where the brain perceives sensations from parts of the body. As it happens, Penfield's map shows the female genital brain area adjacent to the nipples area.

Clearly, there is physiology involved in what makes breasts erogenous zones. But there also is evolutionary biology and culture at play as well. In their extensive studies of facial and body symmetries as powerful clues to sexual and reproductive health, Randy Thornhill and his colleagues have amply demonstrated the links between physical attractiveness and mating success. What they have done also along the way is explain why certain body shapes and traits *are* so attractive to us. It's not just that wrists, ankles, knees, feet, hands, eyes, facial bones, and breasts are symmetrical and therefore attractive; people have come to find those parts of our anatomy attractive in and of themselves.

Especially breasts. Over millions of years, humans have come to associate symmetry and the parts that are symmetrical with sexual pleasure as well as with reproduction, and it's hard to say whether it's the symmetry or the breasts themselves that are the more powerful reinforcer of sexual behavior and the incentive to keep producing offspring.

Thornhill, along with Anders Pape Moller in Denmark and Manuel Soler in Spain, began a famous study of breasts and their sexual functions with the observation that those of human females are large compared to those of our close primate cousins, and their speculation that large breasts are much more likely to be uneven, or asymmetrical, than small breasts.

In one of their investigations, the team obtained breast measurements from Spanish and U.S. women. In the Spanish group, average fertility was higher and breast-feeding ubiquitous, while the U.S. group had lower fertility and breast-feeding was less common. (Both pregnancy and nursing can affect the appearance of breasts, and the researchers wanted to be sure to include women in both categories.) The American subjects, fifty women, were all patients of an Albuquerque, New Mexico, plastic surgeon, Dr. Patrick Hudson. The women had come to Hudson for breast reduction or augmentation during 1993, meaning these women believed their breasts were either "too big" or "too small." First, the scientists estimated in each woman the symmetry of her breasts by measuring the distance from the suprasternal notch (approximately the center of the breast bone) to the center of each nipple down to one millimeter. Breast size was defined as the mean of the distance from the notch to the left and the right nipple respectively. While there was considerable variation in breast size, asymmetry, and other variables, women who had children had an average of 19 percent less asymmetry than women without children in the Spanish group, and a whopping 54 percent less in the New Mexico group.

Thornhill and his colleagues are convinced that breast asymmetry may play an important role in sexual selection and mate choice in humans because it is a reliable—and very visible—indicator of fertility, independent of age. The results also suggest that this is why nature makes so heavy an investment in spending resources on building large breasts during puberty, long before a girl needs to nurse.

It was Desmond Morris, the noted ethologist, who first popularized the idea that as our ancestors began to walk upright, they began to perceive the most erogenous parts of the body as those that stuck out in front, the easier to be seen and paraded. For example, a woman's use of lipstick is painting the lily of her already everted lips to imitate the labia of her vagina, and her breasts may well be look-alikes for fleshy buttocks. Not everyone agrees, to be sure. Psychologists at York University in Canada, Ontario, evaluating the genetic basis for male love of breasts, suggest that Morris's notion that big breasts are a "genital echo" of large, heart-shaped buttocks or fatty vulvas is probably incorrect. Their study of college men showed a particular interest in the least buttock-like parts of a woman's breasts, and their interest depended more on tactile enjoyment than just visual stimulation.

In any case, Stone Age males began to be turned on more to women with the largest breasts and reddest lips undoubtedly because they stood for signs of fertility, so ever bigger breasts and red lips persisted in both modern woman and our ideas of what is a sexy body. That accomplished, large breasts also could have been a deceptive come-on when women needed to find provisions even if she wasn't interested in or capable of pregnancy and lactation. In other words, breasts as we know and prefer them today may be, like large penises, *secondary* sexual characteristics whose central if not sole value is to attract the opposite sex, not feed babies. "Choosy males prefer females with symmetrical breasts," not principally big breasts, they say, and such males "may experience a direct fitness benefit in terms of increased fecundity and an indirect benefit in terms of attractive or fecund daughters." Indeed, some studies have shown evidence that breast asymmetry is inherited, so that males who mate with women with symmetrical breasts are likely to sire daughters with even breasts as well.

Why would this be so? The answer begins with understanding why women have breasts in the first place. Or, more precisely, why they have big breasts.

Nursing babies is not the answer, because even animals with small flat breasts, such as apes, can do so, and most breast tissue is fat, not milk producing. This fact leads to another theory, that breasts may have been evidence of abundant fat reserves. Or they may have been evidence of ovulatory potential, or to provide a way for infants to hang on. Evolutionary economist Edward Miller of the University of New Orleans finds that women probably evolved to have an optimal amount of milk-producing glandular tissues. All of the fat is actually, as anthropologist Helen Fisher notes, a pain in the chest and in the way, as the increased rate of reduction mammoplasties in the well-endowed attests.

In theory, men might gain an edge for their genes by selecting women with bigger fat deposits—even if the cost to women for having this extra fat was increased death rates due to reduced ability to run or gather food—because the benefit of the extra milk going to his offspring was worth it, to him, anyway. But in humans and most other primates, the survival of a man's offspring is highly dependent on Mom's well-being, too. Human infants are notoriously dependent for years, and how is Dad going to slay dragons and bring home the venison if he has to milk goats for the brood? A better argument is that larger fatty breasts are there to attract men because, well, they like soft fleshy things they can fondle. In this case, flat-chested women would have their genes die out, but this idea only explains how a man's preference for females with breasts was maintained once men came to be introduced to them and like them. It's fairly easy for Darwin's theory of sexual selection to explain how a trait is retained once it is established, but explaining how it got to be preferred in the first place is trickier.

Noting that female primate breasts enlarge only during pregnancy and lactation after puberty, Miller points out that males may have evolved to be more generous providers for women with bigger breasts, in order to care for what may be their offspring in utero. But it's also been documented by such things as rates of births of girls who have small rather than large breasts that in situations where males can have a go at any female they want—harems, for example—smaller breasts are preferred and major mammaries are considered unattractive. Dominant males may have rejected big-breasted females in this situation because they may have believed them to be already pregnant or nursing and they rationally chose the smaller-breasted females to impregnate.

Because nursing is one of nature's contraceptives, although not a perfect one, nursing women are at least temporarily infertile. Thus, pendulous breasts might initially have been a turn-off to a successful male because he wouldn't want to waste his bank account and his beef on someone who already is carrying a child, even his own. A pregnant woman would not be ovulating and of no immediate reproductive value. So in this case, dominant males would have rejected big-breasted females.

But this seeming contradiction really isn't, Miller says, once dominant and nondominant males alike became able to kill meat on the hoof and bring it home. In this scenario, larger-breasted women were now free to mate with nondominant males, who presumably were happy to provision the larger-breasted females and their offspring.

"It is certainly possible," Miller also argues, "that if a female's mate regarded her as less attractive because of enlarged breasts, she would find it easier to mate with others. Her mate might, unfortunately, provision her less well, given his lack of alpha male status and prowess. However, if large breasts came to be viewed not as a deceptive signal of pregnancy and lactation (or even more sinister, as a device to cheat on a husband), and came to be viewed perhaps as a useful way of avoiding unwelcome attention from a dominant male, including rape, that's good news for her."

The bottom line, for an economist anyway, is that early on, at least, women with enlarged breasts were better provisioned. In nonhuman primates, like chimps and gorillas, where male provisioning was unimportant, preferential fat deposits in the breasts did not emerge as part of the sexual body. Jane Goodall, in her field studies with primates, saw that female chimps appear to reward any male with sex if the males feed them. In fact, females in heat were better at begging for meat from males than those females who were no longer or not yet able to reproduce, and bonobos copulated with males who had fruit they wanted. Hunter-gatherer human ancestors had the meat to give the females, and they of course gave it to the girls who offered sex. But what kind of economic resources would females most reward with sexual access? Probably those that came from males who, the females guessed, were more likely to stick around and not leave them, at least during the critical nursing period. If, in fact, females consciously or unconsciously understood their vulnerability during nursing, then females paired with males who provision them as they nurse would *over time* leave more descendants, and females would then prefer such guys as mates. From an evolutionary standpoint, these females learned to reject men likely to leave them at this time, and to increase the pool, perhaps, of males who stick around.

Clearly, this kind of female strategy would confer great *disadvantage* to any males who didn't like big breasts. For their part, women may have begun to intelligently deduce and remember and teach to their young that certain men would stick around and others would leave. Even today, Miller notes, women seem to be close observers of the reliability of *other* women's mates and to socially denounce the scoundrels who walk out of relationships and divorce or abandon them. In any case, once males preferred lactating women with big breasts and women learned to form bonds with male provisioners during lactation, the stage was set for the conservation of males and females with these genes and with genes for big breasts. "In this account," he says, "disappearance of a dislike for enlarged breasts and attraction to enlarged breasts is partially a device for extending male provisioning beyond a short-run purchase of sexual access. It may even have been a necessary step, since without it males would continually replace their pregnant or lactating mates, with ones whose flat chests indicate they were capable of being impregnated. Long-term pair bonding or marriages, required that a male's attraction to a female continue past her pregnancy and lactational period."

By now in our evolutionary history, breasts are taking on Promethean value. Extra resources in women's reproductive interests were important all the time, not just when she was nursing or pregnant. When she has such resources and is not lactating, the extra food still helps increase the probability, by adding to her body fat, of her ability to have the next baby at some point and gives her a real advantage over her flat-chested sisters, whose breasts only contain gland, or milk-making, tissue.

Suppose that food is so abundant that lactating females can feed themselves. She can not only bring home the bacon, but fry it up in a pan and let the devil take the man, except to impregnate her. The population would grow, Miller says, and food scarcity would eventually stop growth when females without long-term mates were unable to get enough food to support lactation—if they needed to carry infants along while gathering food. Such females would need to wait a long time to build up enough fat stores to have another baby *unless* males provisioned them. They might become infertile due to lack of nutrition, and the population would decline. Assume that this resulted in some kind of population stability: When food was scarce, and the female needed lactational provisioning, she needed males; when food was abundant, she didn't. In these circumstances, male provisioning is likely "to make a real difference to a lactating female." This benefit to the genes of the male who gives her food and whose child would benefit is likely to exceed the benefit from provisioning a non-nursing woman. In this case, males are likely to be selected for attraction to large breasts and females are selected for visible breasts. Since human males are the only male primates who regularly give women lots of food, they are the only species whose males are likely to have evolved to provide provisioning when it's most needed, and the only species whose females increase male gifts of this kind by *simulating* or faking lactation through appearance: large breasts.

Breasts might have evolved this way, but, notes anthropologist Helen Fisher, they might have evolved also as honest signals of the ability to ovulate, because women in their prime tend to have firmer, larger breasts than prepubertal or postmenopausal women. And if humans are genuinely the "naked apes" that Desmond Morris claims we are, then a love of voluptuous mammary glands may indeed have evolved only in us because among other primates the only time breasts grow large is when females are nursing their young. Of course, if Miller is right, today's cosmetically enhanced or naturally wondrous *Playboy* magazine breasts are dishonest signals, and one could conclude that it's men's baloney detectors that need some fine tuning.

Whatever the reason that big breasts developed and became so attractive to men, according to Fisher they remain a "bad design" that stuck around because they also evolved to be erogenous zones. "These protuberances around the mammary glands seem poorly placed. They bobble painfully when a woman runs. They flop forward to block vision when she leans over to collect food. And they can suffocate a suckling child. Moreover, breasts (of any size) are sensitive to touch. Why? A woman's nipples harden at the slightest touch. And for many, fondling the breast stimulates sexual desire. . . . for whatever genetically adaptive reasons . . . ancestral males liked females with these sensitive pillowy appendages and bred more often with sexually responsive, big-busted women—selecting for this decor."

No matter their ultimate origin or purpose, there is no doubt that breasts, along with faces and limbs, carry many anatomical signals. The signals may be read on many levels by the opposite sex, but one of the quickest—even in a glance—is how even, or symmetrical, they are. Humans have a visceral (and mostly negative) reaction to unevenness and there doesn't remain much doubt that balanced body proportions and symmetries are "programmed" to be sexually attractive because they signal health and reproductive fitness. Biologists are so certain of this that many biologists have gone searching for the genetic waters that feed it.

In recent years there has been an explosion in understanding the genetic control of the development of the body and particularly of the genes that are directly connected to the symmetrical lineup of our bodies. Scientists theorized that if they could first identify genes involved in obvious developmental defects, then try to link those defects to defects in genitals, they'd have some powerful clues to the role that symmetrical bodies play in signaling genetic health and reproductive fitness.

What has emerged from their investigations is the story of a cluster of genes, called the homeobox, or HOX, that occur in humans and all other animals, from mammals to insects. These genes function collectively as a master set of instructions, blueprints for the development of the body's plan. (As it turns out, a similar set of genes that controls the development of the forebrain is also being tracked in the genome.)

In the body, HOX genes are responsible for an organism's directional development. They tell a growing embryo where the head belongs relative to the feet, which things go up and which down, which set of cells must produce arms and which legs, which fingers and which toes. Originally discovered in that master of all lab creatures, the fruit fly, they were found to lay down the bug's entire head-to-tail pattern. They appear to do the same thing in every other creature, from worms to people. Humans, it appears, each carry thirty-nine HOX genes arranged in four sets. In February 1997, scientists at the University of Michigan, led by geneticist Douglas Mortlock, uncovered a HOX gene mutation that causes genital as well as feet and hand asymmetries, or abnormalities. The finding specifically affirmed that similar inactivation of a single gene in mice causes the same problem.

This link between limbs and genitals could show why HOX genes and the parts of the body's plan they code for are so highly conserved and valued by nature. The Michigan group studied a family with a hereditary abnormality of thumbs and big toes that were stunted and shifted toward the elbow and knee. The entire problem is caused by twenty missing amino acids that alter the transcription of some target genes and send the wrong messages from blueprint to body development. The family calls their defects "foxy feet" and "butterfly fingers" and they cause no apparent disability in their skeletal strength. But three women who have these deformities also have womb defects and one is infertile. All the rest have fertility problems as well, and all carried the mutation in the HOX genes that was previously found in the analogous gene in mice and fruit flies, which caused similar problems. The male mice with the same defect have malformed penises.

One possible conclusion is that when mutations in the genes we need for normal genital development occur, a means would be needed by nature to telegraph that news to prospective mates because they reduce reproductive fitness. By linking such genital defects to defects in the body that are highly visible, nature has evolved a means of alerting animals to potentially defective mates. Too bad for the individuals, but good for the survival of the species.

HOX gene expert Denis Duboule of Geneva even speculates that the developmental gene link between limb and genital may explain a great biological mystery: Why do so many vertebrate animals have a balanced five fingers and toes on each hand and foot, a number conserved for 300 million years? The digits, it seems, like San Marco's spandrels, may just be along for the reproductive ride, because the HOX genes happen to link both genital and digitary development. It's not that five fingers on a hand are so vital, but the right lineup and number of genitals certainly are, and nature made sure that the genital genes stayed put, even if it meant carrying the pentagonal plan of fingers and toes along with it. As Michael Coates, a London paleontologist, put it in an article in *Science*, "stability in the distal part of the limb is favored because otherwise you mess up the genitalia." Could it be, then, that long, even fingers as a signal of genital fitness is something men and women evolved to care about? Why not?

Obviously, the shapes that our sexual bodies take are critical to the signals we send. Genes play a large role in this as do hormones, which soften figures in women and outline muscles in men. However, along with symmetries, other aspects of shape have become important as sexual signals as well, but some of these may be more culturally determined. Think Marilyn and Twiggy. Or, consider photographs of the Rockettes and Hollywood chorus girls of thirty years ago. *Vogue* supermodels today wouldn't be caught dead on a runway with *those* thighs, or hips that failed to wrinkle a sable coat. This contrast in what is considered chic offers convincing evidence that in the Western world, at least, curves and fatty hips once considered alluring are now considered ugly. The oversized pasha, sumo wrestler, and nineteenth-century squire are not likely to find themselves modeling Ralph Lauren.

In women curviness is a good signal of the ability to have a pelvis spread for pregnancy and enough fat distribution to nourish an infant if times get tough. Painters like Renoir, Raphael, and Rubens reflected society's appreciation of fat-hipped women, with thickened waists and folds of abdominal fat that would send them to personal trainers today.

But a closer look at even the "heroin chic" ultra-thin bodies of mannequins makes clear that the sexual bodies we love have to have the right *proportions.* The trimmings are as extraneous as the excess fat. In fact, evolutionary psychologist and anthropologist Devandra Singh of the University of Texas in Austin believes proportion, in women at least, is the whole story. A charming and scholarly man, he can't quite stop himself from scanning a woman's—*any* woman's—body from top to toe immediately upon introduction. In this he is not unlike most or many men, but what *he* is doing is not so much admiring as measuring. As he sees it, humans are captives of Stone Age minds as well as bodies, acting in ways that evolved hundreds of thousands of years ago as a response to evolutionary pressures and unifying romantic love, sexual attraction, sex behavior, and gender orientation, along with parenting, fidelity, infidelity, courtship, and marital and social kinship bonding.

Singh has a lot of company. Computer studies suggest that it's not so much that the eyes, or ears or brows or noses or chins or jaws are perfect in the sexual bodies we love and lust for, but that they are proportional as well as symmetrical. And evolution was pretty precise about it all. In one survey, faces that are considered sexy have eyes one-fourteenth as high as the face and three-tenths its width, with the distance from the middle of the eye to the brow one-tenth the height of the face, and the nose taking up no more than 5 percent of the entirety. In another study, men and women used a computer graphics program to make composites of their ideal faces. The ideal female faces had a shorter lower face, with a nose-to-chin proportion typical of a prepubertal girl, the lips of a young teen, and a small mouth. Scottish and Japanese researchers did studies suggesting that such disproportions as higher cheekbones, slimmer jaw-lines, and shortened mouth-to-chin and nose-to-mouth distances were preferred. And every man seems to love oversized eyes, probably because they evoke "youth." Baby animals and humans alike are considered "ador-able" in large part because of the relative size of the eyes to the faces in infancy. Eyes don't grow; heads and faces do.

When Singh went searching for the "ideal" sexual body, he considered the scope of the territory. How would he find common ground between Bambi Tucker's ivory muscular legs delicately posed astride an urban cow-boy and a kimono-clad geisha's tiny bit of exposed nape? He narrowed his search for what men found appealing in women's sexual bodies by first checking out *Playboy* centerfolds and beauty contests—an admittedly limited cultural view, but one that had the strength of staying power. With precise measurements done over three to four decades of these magazines and flesh markets, Singh learned that indeed the ideal measurements of a woman's figure had changed. Weight went down substantially and figures got taller and leaner. But the key finding was that the *ratio* of waist size to hip size was essentially the same. To get this ratio, he divided the waist in inches by the hip in inches, incorporating the buttocks with the hip. According to Singh, *healthy* women always have a waist-hip ratio below 0.8 no matter how much they weigh or how tall or short they might be or how big or small their breasts are. For women in their prime of repro-ductive life, the ideal range is somewhere between 0.65 and 0.8. This translates to something hovering around the Barbie doll or hourglass fig-ures of twenty-four- to twenty-eight-inch waists and thirty-six-inch hips or twenty-seven- to thirty-one-inch waists and forty-inch hips. The lower the ratio, the more sexually attractive the woman's body appears to be to a man. When Scarlett O'Hara asked to get her eighteen-inch waist back with breath-defying corsetry, she was adhering to a prehistoric call to advertise her pre-partum fertility. Whether the call was heard by Rhett Butler or not is up to literary whimsy. Kate Moss's flat-chested, disap-pearing body act may seem unattractive in the extreme if fertility signaling is at stake, but her waist-hip ratio is probably in the 0.7 range or better, putting her right in Singh's statistical ideal.

In order to rule out variables such as pretty faces or breast size in his determination of what Singh considers *the* component of the sexual body, he also asked hundreds of men and women of different cultures and ages, from teenagers to octogenarians, to rate pictures of twelve female figures on the basis of their health, looks, sexiness, beauty, and apparent desire and ability to bear children. The faces and breasts were kept constant, only weights and hip-waist ratios changed. Four of the women were normal weight, four too fat, four too thin, and waist-hip ratios went from high to low.

Results show that both the men and women surveyed preferred the women of normal weight with the low waist-hip ratio. Narrow waists won each time as well. The men neither preferred fat or thin, nor equated them with fertility or beauty, respectively. In his search for bodily icons of beauty in the United States, Singh created Barbie dolls, twenty-six of them with a range of waist-to-hip ratios from 0.4 (so nipped in "she looks sick") to 1, which is essentially a straight, stick figure. He also arrayed the dolls in every possible way, including placement near a model brick house. He tried it with line drawings that captured the same effect as the Barbie dolls. (Barbie, incidentally, is not anatomically correct on a number of scales, either genitally or limb-wise; her legs are far too long for her trunk, but her waist-to-hip ratio is right on the national mean. Recently, Barbie's corporate manufacturers introduced a more "natural" version and it remains to be seen whether she'll hold the same appeal.)

The results again showed that men don't like extremes; for example, seventeen-inch corseted waists. Studies found waist-to-hip ratios of 0.5 or 0.6 to be considered equally attractive and 0.7 average and most common. Singh also found that having a pretty face (like Barbie, ever youthful) can help overcome a bigger waist. "There's no questions in my mind that the .7 waist-to-hip ratio is the best marker for fertility and a good proxy for absence of diseases and reproductive fitness. A .8 ratio is linked to high health risks, as is .4 and below," he states. On a regular Barbie doll, the ratio is 0.58, or just a shade under six.

The catch-22 for American women, as aerobics instructors can attest and Scarlett clearly knew, is that the waist is the better marker than the hips. If you increase the waist to meet the hips, even if you keep the 0.58 ratio, it is less attractive. Since American men seem to want skinnier hips, women need difficult-to-get-and-keep waist circumferences. In other research, Singh has learned that if a man can't get 0.58, he'll take up to 0.7; that physicians can sometimes detect ovarian diseases from rear photos based on a too-low waist-to-hip ratio; and similarly men with prostate cancer taking steroids have 0.7 and lower ratios, and women are "freaked out" seeing them. "Men have a normally high waist-to-hip ratio, and testosterone restricts fat on the buttocks so that in men .8 to .95 is normal. If he drops below .8, it demonstrates a distinct lack of testosterone. Examples of the ideal male abound, but the best is probably Michelangelo's *David* in Florence."

What about fat elsewhere on the body? When the major sex hormones kick in during puberty, fat is significantly and distinctively redistributed in the male and female body. In him, "baby fat" and pudginess disappear pretty much evenly along the face, torso, and limbs as muscle mass bulks. In her, the "baby" fat around the face and middle go away and fat deposits on the hips give her more of an hourglass figure until after menopause, when her waist thickens again to more of a male's waist-to-hip ratio. (No wonder liposuction is so popular.) Obstetricians also know that women with more of a man's waist-to-hip ratio have more difficult childbirths—and may tend to be more athletic as well.

Walking a bit along economist Miller's path, Judith Anderson has studied the amount of fat perceived as attractive in different cultures and found that women's economic and political power correlates with thin and fat. Where women are powerful, as in the United States and the West in general, thin, or at least slender, is the standard of attractiveness; where there is less power, the standard is for fatter women. Powerless women say "her fat is her fortune," a position consistent with other mammals because fat is a reproductive asset. Where subsistence economies are the fate of most women, energy resources are absolute constraints on a woman's reproduction. Fat is her advertisement of capacity to bring nutrients to Dad's genetic offspring.

Significantly, those who study fat pathways in animals as well as humans have complementary evidence for this effect. One is Arthur Campfield, a neurogeneticist who has found that a protein called leptin, produced by fat, probably helps regulate satiety because it signals how much fat is in the body to the brain's hypothalamus, which coordinates basic body functions such as eating. Some obese people may make leptin at a greater rate to compensate for a faulty signaling process or action. Leptin's signaling ability may also help explain the high rates of regaining weight found among dieters. In some studies, women have been shown to have significantly more leptin than men because they have a higher body-fat content than men. Now, leptin presumably did not stick around over the eons in order to regulate fat in people who routinely eat too much and want to look better in bathing suits. Instead, it did just the opposite: For our Stone Age ancestors, Campfield points out, the real threat was never overweight but starvation, so leptin was conserved as a sensitive indicator of low energy source. Thus, it lets an animal know to do something important to get food. Why? Because when there is impending starvation, animals should not spend energy on sex and mating. Nature fixed it so that our ability to conceive, carry a pregnancy to term, and even impregnate a female diminish in conditions of starvation, because as a long-term strategy, having babies you can't live long enough to feed and raise to reproductive age doesn't make a lot of sense—and presumably there won't be much food to give to the newborn, anyway. Ten thousand years ago, getting fat on a kill would be useful, or at least OK, because in leaner times our ancestors would "feed" off of their own body fat. Today, when few Americans truly go through periods of extreme starvation, leptin appears to control fundamental hormones and calories that get preferentially sucked up into fat cells. It appears that we evolved to store fat, but not *be* fat.

Marilyn Monroe curves are clearly, if not solely, related to the amount of fat in the hips, buttocks, and breasts. Most women over twenty-five can attest that, as women fatten, the fat goes preferentially to those areas. This, as the leptin studies affirm, is because of the relative distribution of fat cells in the body and the action of female sex hormones. Men tend to distribute fat more around the middle.

In a series of studies, Nigel Barber, of Birmingham-Southern College in Birmingham, Alabama, has managed to integrate a lot of what Singh, Miller, and others have reported. He suggests that preferences for fat in the opposite sex varies historically, as well as biologically and culturally. Like Singh, he has traced the curves of Miss Americas and *Vogue* and *Playboy* models over time, but considers *bust*-to-waist ratios the changing icons of the sexual body. At the turn of the twentieth century, the typical *Vogue* model had a bust measurement twice the size of the waist—the well-known "Gibson girl" hourglass look. Then things really changed. *Playboy* and Miss America stayed in tune, but *Vogue,* which is read strictly by women, took a different route, each "standard" going its way over time.

To get an idea of the divergent paths, consider Jayne Mansfield, the classic *Playboy* centerfold, with a bust-to-waist ratio of 1.5, or what Barber calls "moderately" (?) curvaceous. (Assuming a classic twenty-four-inch waist, this would mean the equally classic thirty-six-inch bust.) Julia Roberts, at 1.2, is not curvy at all. Rubens painted women at 1.54 and 1.56, very curvy on the Barber scale, not to mention hippy. But still not the extreme of the *Vogue* hourglass look. They were very fat but had narrow waists, confirming Singh's reliance on waist-to-hip proportions as a measure of the sexual body beautiful. Over time, even *Vogue* provided variations ranging from hourglass to straight-hipped. Barber speculates that perhaps these fluctuations were due to prejudice against women in the workplace, where women perceived a need to hide their curviness—their femininity—down to about 1.2. "But I have my doubts," he says. "My alternative centers on reproduction and it's more complicated. I say women are in two different modes. One is the work mode, which advertises delayed reproduction and in which the standard of selection is slender. When they're in the reproductive mode, they get married early and don't tend to have careers and are more curvy and show it off."

Barber set out to test this model by measuring the bust-to-waist ratio in *Vogue* models and other U.S. women. Using information from a long-term survey of such ratios, he and his colleagues examined samples of measurements every four years from 1900 on. Barber then lined up his findings with historical advances in women's activities and fashion-model preferences. He concludes that fashions of bodily attractiveness are influenced by an "evolved psychology of mate selection." He predicted that as women begin to go to college and seek careers, the bust-to-waist ratio should go down; that it should increase when there were excessive males in a population and decrease as young women opt out of marriage markets and stay single to pursue careers. And that is what he found by tracking the Standard and Poor index, per capita GNP, college enrollment, marriage, the percentage of women working outside the home, and the birthrate. "The curvaceousness of *Vogue* models dropped like a stone as more women got Ph.D.s and went into careers. *Vogue* is clearly sensitive to its readership. The birth rate correlated too. As bust-to-waist ratios dropped, the birth rate would drop 12 years later."

The unmistakable lesson to be drawn from these studies is that *whichever* bodies we're born with, our minds came along for the ride and developed to the point where we can extend the duration of an alluring sexual body. Wigs and toupees aren't *really* mistaken for the real thing, but they tell the world we are still vigorous. Breast augmentations or reductions confer and convey our appreciation for symmetry and proportion. Liposuction restores ideal waist-to-hip ratios so that fashions proclaim our sexual interest long after childbearing is possible. Penis extenders hint at testosterone and sexual pleasure for partners even when parenthood is no longer uppermost. In all cultures, the decorations, cosmetics, and plastic surgery may have been designed in some part to fool or trick the opposite sex, but probably more to give each other more of a natural good thing.

Regardless of society's motives or agendas, chastity belts and heavy veils—even successful promoters of "just say no" campaigns—have lost the evolutionary war that gave us our sexual bodies and minds. If social or religious convention only allows the eyes to be visible, the eyes become erogenous, no doubt the reason veiled orbs are frequently outlined with heavy kohl and highlighted with mascara. If napes of necks and a flash of ankle are all that show out of Oriental kimonos or Victorian hoop skirts, then napes and ankles become irresistibly sexy. Though codes of sexual honor and chastity defined the age of chivalry, Lancelot and Guinevere still got together.

It's very likely that when apes began to walk upright and our ancestors became fully two-footed, they did so in great part because they could now parade their genitals for females to see and choose. We will always find a way to advertise our wares—honestly or dishonestly—even if it takes tight jeans and panty hose, because once we knew that certain signals promised sexual rewards, the bodies that enabled those signals became our billboards.

What directs all of those signals, of course, are those most distinctive features of Homo sapiens, our minds and brains. Everything we consider unique to our species—our sense of right and wrong, our conscious awareness of our existence, our creative adaptation to the world around us—have been attributed by many evolutionary biologists to the pressure of natural selection. But, as evolutionary psychologist Geoffrey Miller, author of *The Mating Mind*, told *The New York Times,* "natural selection is about living long enough to reproduce." It is sexual selection, he says, that "is about convincing others to mate with you. It's about animal minds, our ancestors' minds, influencing other minds." In sum, it is the brain that is the principal organ of sex.

4

WHERE "IT" REALLY HAPPENS

"From the brain, and from the brain only, arise our pleasures, joys, laughter and jests, as well as our sorrow, pains, griefs and tears."

—HIPPOCRATES, FOURTH CENTURY B.C.

Have you fallen in love "at first sight"? Found Bach a grand prelude to sex, but not Beethoven or the Sex Pistols? Wondered why the man of the hour wants action thrillers, filled with sexual violence, and why she prefers romantic chick flicks? Know women who can't find their way anywhere without "turn left at the red barn" instructions and men who navigate never-before-seen cities in foreign countries with nothing more than his "hunches" and a map? Welcome to the wonders of the sexual brain.

Far more than penises, vaginas, ovaries, and testicles, the brain and mind, our wondrous capacity to learn from and adapt to experience, are responsible for obvious and subtle differences in how the sexes play out their biological and psychological sex roles. Men and women—males and females of all kinds of animals—are different not only in the way they look or in their sex chromosomes and hormones, but also in the way they think and make love. Decades of research boil down to this: The brains of males and females, while comprised of the same general anatomy and chemistry, in some ways develop differently and thus, in some ways, react to and process the world and its sexual challenges differently.

Both he and she have a right and left brain; both he and she have the means of expressing love and lust, perceiving it in the other, feeling it, and responding to it. The overall organization of male and female brains is much more alike than different. But to deny the differences is to deny evolution and some universal differences in male and female behaviors and response to many stimuli. For example, the neural highway that carries information across the two hemispheres is smaller in his brain than hers, which some have interpreted as a possible explanation for why his means of expressing terms of endearment is more isolated than hers from the "intuitive," perceptive, and emotional right brain. (The speech center is predominantly on the left side of the brain in both sexes.) Moreover, women tend to have a more diffuse brain organization, in which some parts of language capability reside on the right side as well. The result: Her feelings flow into words; his are clogged in the left brain. He *feels* the same feelings; he just can't get them out so easily.

"Although every adult is essentially familiar with sex differences in the genitalia and other physical attributes and the changes that occur during a lifetime, the fact that sex differences in brain function and structure also exist is less well appreciated," says Roger Gorski, one of neuroscience's preeminent brain anatomists and a pioneer in the discovery of sex differences in the mammalian brain. Gorski was the first to show that a small region of the hypothalamus called the preoptic area, or POA, is considerably bigger in male rats than in females, a size difference generated by sex hormones before and right after birth. His colleague Laura Allen found an almost identical difference between the genders in humans. "Why we have such a hard time with this is hard to say," Gorski says with genuine frustration. "It should be a no-brainer," so to speak. Some attribute it to an overdeveloped sense of political correctness, scientific sexism, or just plain ignorance. But like it or not, says Gorski, men and women differ in their brain biology, and therefore in their sexual *and* nonsexual behavior.

Some of these differences involve learning and ability, some reproduction, and some emotions and motivations. Mark Jude Tramo, a neurobiologist at Harvard University specializing in the part of the higher brain that processes sound and music, has studied the physics and molecular biology of sounds made and received by macaque monkeys during sex. Tramo has identified a uniquely male "copulatory scream" that is identical across this species, impossible for females to make, and impossible for normal females and other males not to hear and instantly understand as a sexual message. Given that macaques, humans, and many animals all have some sense of combinations of sound that seem "good" or pleasant and others that are bad, we can assume that our brain's capacity to sense these sounds are vital to sex and therefore to survival. And that the brain's musical capacity is different between the sexes should come as no surprise given the different sexual agendas of each sex in courtship, mating, and reproduction. Male and female brains even differ in their capacity to develop and express mental diseases such as schizophrenia (in men, it's later to show up than in women). Scientists at Johns Hopkins have discovered striking differences between men and women in a part of the brain—prominent in Einstein—linked with ability to estimate time, judge speed, visualize things three-dimensionally, and solve mathematical problems. The differences, the researchers say, may underlie well-known trends that vary by sex, such as the fact that more men than women are architects, mathematicians, and race-car drivers. The key is in a brain region called the inferior parietal lobule (IPL), which appears significantly larger overall in men than in women. The area is part of the cerebral cortex and appears on both sides of the brain, just above ear level. Also, there's a symmetry difference, with men having a larger left IPL than right. In women it's the right IPL that's somewhat larger, though the difference between the two sides of the brain is less obvious than in men. Godfrey Pearlson, the Hopkins scientist in charge of the project, cautions that "there are plenty of exceptions, but there's also a grain of truth, revealed through the brain structure, that we think underlies some of the ways people characterize the sexes." Earlier research by Pearlson also showed that two crucial language areas in the frontal and temporal lobes of the brain were significantly larger in women, perhaps explaining their advantage in language-associated thought.

Presumably as well, the sexual neurochemistry of our brains operates and catalyzes the activities of the mind and body in sync with brain tissue already feminized or masculinized by prenatal and pubertal hormones. Dr. Martin Kafka (no relation, he emphasizes, to Franz), senior attending psychiatrist at McLean Hospital in Belmont, Massachusetts, and a Harvard faculty member and expert on sexual addiction and paraphilias, believes he has found a neurochemical link between men with hyperlibidos—sexual predators and compulsive seekers of sex—and women with severe eating disorders such as bulimia and anorexia. Kafka was aware that males account for 95 percent of all sexual predators and females 95 percent of all anorexics and bulemics. In the early 1980s, he began to focus on the similarities of behavior in both groups, and came to believe they all suffer from similarly brain-directed inability to regulate appetites and satiety. One thing his initial observations about all this told him is that if brain chemicals—particularly the so-called monoamines dopamine, norepinephrine, and serotinin—were involved, then sex hormones, especially testosterone, and such influences as a history of sexual abuse would not account alone for these dysfunctional syndromes and dangerous syndromes. Moreover, if he were right, then paraphilias and perversions could be more widespread than society knew. And indeed this turned out to be the case in his analysis. For one thing, investigations showed that only a fourth to a third of sexual predators had any history of sexual abuse that might account for their behavior. And a variety of studies have found that a consistent one-third or more of men have rape fantasies, and a fourth are aroused by pedophiliac pictures.

In his search for a role for a monoamine in his patients, he took for his first text a 1969 paper in the journal *Science* in which researchers gave male and female rats a drug, parachlorophenylalanine, that lowers serotonin levels in the brain. Among other things, the rats soon became compulsively aroused, mounting each other without letup. Given some serotonin, their sexual appetites disappeared. Testosterone levels stayed the same throughout.

Kafka has given Prozac, Paxil, and Celebrex, which alter serotonin transmission or uptake in the brain, and dramatically altered the course of sexual behavior in some of his patients with severe paraphilias and out-of-control libidos, making a case for his monoamine theory of sexual impulse disorders. Intriguingly, his studies suggest at least anecdotally that treated patients lose their deviant natures but function well with what society considers normal sexual urges. That is, they report they are now able to fall in love and feel in control of their urges. Kafka's model might seem contradictory if not ironic, since serotonin is generally an inhibitor of mood, and decreased serotonin levels are a key player in serious depression. But in this case serotonin works like an aphrodisiac, or at least its ratio to other neurotransmitters makes it work that way. If it's true that drugs like Prozac can stop "bad" sex but encourage "good" sex, it might also be true that sex and lust and love are played out in separate brain tissues in some complex calculus as individual as the stimuli that turn us on. Studies of women with epilepsy have shown that lesions in a particular part of the temporal lobe can lead to promiscuous behavior. In a rare disorder called Kluver-Bucy syndrome, damage to a part of the emotional or limbic brain called the amygdala may lead patients to have lust for everyday objects, as do fetishists. Drugs like Prozac, generically known as selective serotonin reuptake inhibitors, may work in Kafka's patients by gently reshaping rather than eliminating areas of the brain that use the neurochemical.

Perhaps like the unusual memory losses—the inability to remember names of animals, for instance—experienced by people with particular kinds of brain damage, Kafka's work suggests that prenatal influences, injuries such as a brain infarct that selectively damages small clumps of tissue, or seizure disorders might well be responsible for creating separate kinds of sexual desires or appetites.

Studies support a monoamine hypothesis for other aspects of human sexual activity as well, notably falling in love. Psychiatrist Donatella Marazziti of the University of Pisa last year published an article in *New Scientist* magazine noting how lovesick youth's obsessive thoughts about the object of their affections resembled the thought patterns of people with obsessive compulsive disorder. OCD patients experience anxious thoughts and high levels of alertness. They feel compelled to repeat certain tasks over and over until they are rendered all but incapable of normal everyday functioning. The newly lovestruck? They praise their beloved's characteristics for days on end and experience anxieties and sensitivities on a similar scale.

When Marazziti and her coworkers studied serotonin levels in twenty "lovesick" Italian students and twenty OCD patients, they found similarly low levels of serotonin, prompting her comment that "when you're in love, maybe you're a little bit crazy. That may be true." She and a few other neuroscientists now speculate that obsessive behavior is adaptive in the context of sex and mating. Without such intense "anxiety" and attention to a desirable mate, bonding, falling in love, and caring for offspring might happen too infrequently. Says Marazziti, "The evolutionary consequences of love are so important that there must be some long established biological process regulating it. [Our work suggests] that the serotonin 5-HT transporter might be linked to both neuroticism and sexual behavior as well as to OCD."

Given the structural and chemical differences in the sexual brain and between male and female brain, there is another, more compelling reason to be interested in the brain's role in sex. Remember the seduction sequence in Woody Allen's cosmic and comic 1972 version of David Reuben's *Everything You Always Wanted to Know About Sex but Were Afraid to Ask*? While Allen works madly to reach home base with a girl in his car, the film switches back and forth between the car interior's sweaty human enterprise and the filmmaker's metaphorical control central, where Tony Randall as the operator calls a "code red" when Allen begins to lose his erection. "What the hell's going on down there!" yells the operator. "We need an erection!" The answer appears shortly when control central security guards bring the operator a just-discovered intruder—a priest!— with a monkey wrench that the cleric had used to sabotage the control room winch used to raise Allen's organ. In this cinematic farce, conscience hilariously made a "coward" of Allen, but the scene cleverly demonstrates the serious scientific fact that beyond its role in the differences between male and female behavior, the brain is where sex and all of its accompaniments fundamentally "happens." Our conscious and unconscious minds are *always* in control of sex.

A two-plus pound convoluted archive of everything in human history important enough to keep our species alive and reproducing, the brain is the overarching originator and regulator of all human behavior. Nothing that goes on in our bodies and minds goes on in isolation from the brain's sensory and intellectual cells, circuitry, and chemistry. In addition, what "happens" is both anticipatory (looking for Mr. Right, being in the "mood") and more reflexive, automatic, and stereotypical (erections, lubrication, ejaculation, orgasm). There may be infinitely variable forms of the former and fairly limited aspects of the latter, but the brain is the key to both.

Our brains are why our perceptions are as potent as our preferences, why the one seems to tune in to everything the other does. Our brains are what get lured to her perfume, his cologne, her figure, his face, her walk, and his touch. The means to make a young man's fancy more likely to turn to love in springtime resides in the brain, genetically photocopied from animals that prefer to mate in spring because in conditions of more light, brains produce more of a protein that serves nerve cells needed for mating. What gets our couples sexual radar beeping and keeps the electrochemical sexual system of hormones, glands, nerves, and organs humming, from first flirtation to gamete exchange, isn't our genitals.

Instead, we feel the pain of rejection and the ecstasy of orgasm well above the gut or waist, in the *brain*. Not surprisingly, many studies confirm that the brain's capacity for coalescing thought and action makes it possible for sexual stimulation, arousal, and orgasm to blot out pain, like an erotic anesthetic. And, when at the point of temptation wedding ring morality keeps a penis in the pants, or when a wife is your most sought after lover long after the wedding presents have worn out, it's merely the brain in action.

Similarly, if any of us have fallen in "love at first sight," it's because well before we first spotted each other across a crowded room, our "mind's eye"—not our baby blues and browns—"saw." And what the brain "saw" was a vast complex of signals, triggered by a literal room full of stimuli that all needed processing. In that room were faces, breasts, sounds, tongues, hands, pheromones, perfumes, foods, songs, sighs. For *his* brain to appeal to *her* brain, and vice versa, each had to calculate the rational and emotional total of all of the "others" each had met, dreamed of, or fantasized, and decide the price was right. In their minds they were not really meeting each other for the first time, but "recognizing" each other with what John Money has called a lovemap, "an idealized and highly idiosyncratic image" that has formed in their brains by dint of what they have learned. They have fallen in love with a Rorschach brain blot, says Money, a love-print as unique to them as their DNA-driven finger-print.

Lovemaps are found on no neuroanatomical chart, of course, but their existence is a logical outgrowth of what the 1990s, the decade of the brain, has taught about the central nature of sex: that the brain is the great integrator of sexual thoughts, instincts, and actions. Only the brain is capable of organizing the many reactions and actions that make a woman's nipples tingle and her milk let down the moment she hears her infant cry, but not when she hears her neighbor's. Only the brain can make a woman in love climax just imagining her lover's tongue on her clitoris or her breasts. Our mind-brain connections, male or female, have evolved to work in concert for all things sensual and sexual, as demonstrated in the power of fantasies to arouse and odors to evoke memories. Our senses recruit tastes, touches, smells, sights, and sounds all over our bodies, but only in the brain do these march in patterns of "love." Both romance and orgasm happen in the chemistry and architecture of the brain molded by molecules that alter and enhance mood. It's the brain that governs, per-mits, or represses the intense emotions triggered by love, infidelity, com-mitment. "The only thing that arises in the gonads is sexual *reproduction*, and you don't even *need* two sexes for that," says behavioral endocrinol-ogist David Crews.

A central characteristic of a lovemap, it's worth emphasizing, is that it is not some prenatal fait accompli, present and cemented at birth. It is instead formed by the interaction of what nature provided and nurture piles on. "Like a native language, it differentiates. . . . sometime between ages five and eight or even earlier in response to family, friends, experiences and serendipity. It is a developmental representation, or template in your mind/brain and is dependent on input through the special senses. It depicts your idealized lover and what, as a pair, you do together in the idealized, romantic, erotic and sexualized relationship," says Money. So, well before we physically meet the man or woman of our dreams, we "meet" that person in our brains, the proto-dream face, race, color, odor, features of our ideal mate. And for each of us, our brain forms the specific lovemap that may have nothing to do with some generalized concept of "beautiful" or "handsome."

Brain science has confirmed these broad statements by laying an increasingly detailed knowledge of the brain's anatomy and biochemical makeup over knowledge of sexual behavior. Every new piece of evidence affirms the connections between our oldest neural components, the so-called "reptilian" brain responsible for such automatic functions as blood flow, and the new brain, or neocortex, the thinking brain, with its sidekick hippocampus which organizes and holds on to memories. The link between the two is what's unique only to mammals, the so-called limbic brain. With its potent chemistry set of dopamine, opiates, oxytocin, serotonin, and more, the limbic brain is where love, learning, memory, jealousy, lust, tenderness, and emotional attachments come together.

Experiments on primates, our closest nonhuman relatives, show that impulses activated by sexual arousal and orgasm are relayed via electric impulses and brain chemicals to the pleasure and emotional centers of the brain, reinforcing in the strongest possible way the desire for sex. If these areas are directly stimulated by experimenters, the animals become sexually aroused even in the absence of the opposite sex and every other known social "cue" for sex. Direct electrical or chemical stimulation of the pleasure zones in the limbic system, moreover, can rival the lure of erotic arousal and even orgasm, and some people, through intense concentration alone, are able to "think" themselves to orgasm by imagining acts of sex or pleasure. (Surely pornography's power derives from the brain's ability to make such leaps from imagination to ejaculation.)

Inescapably, it can only be our brains that account for the unique place of humans in the sexual spectrum. In the calculus of the human brain, it's no accident, scientists say, that the large size, volume, and weight of our gray matter relative to our body mass evolved to stand out among animals at the same time that we evolved to become the planet's sexiest creatures.

It's worth noting the latest piece of evidence for this in the completed rough draft of the human genome. Along with the finding that the human genome may have only about thirty thousand genes, it turns out that these genes are distributed not evenly along the genome, but in clusters. And one of the lumpiest areas codes for activity in parts of the brain that account for our comparatively much higher order thinking and learning ability. Dependent on seasonal sex, like every other mammal? Not we. We are in an eternal pre-arousal state, primed and ready. If the neurons say yes, the word "no" never has a chance. In its awake state, the brain gets information from its own "remembered" neural circuitry and from light and sound frequencies, biochemical message carriers, touch, taste, smell, and hormone baths. It has chemical, electrical, and physical properties that act and react in sensitive feedback loops to govern and regulate what goes in and what comes out. It has adrenergic chemicals that keep us awake and alert and cholinergic chemicals that dominate when the brain is asleep, ready for the neurological bursts we call dreams and fantasies.

Brain scientists argue ferociously over just how the tissues and cells in the brain concoct awareness, joy, seduction, arousal, or next week's shopping list. Some, like MIT linguist Steven Pinker, believe that genes and the sculpting process they control are so completely shaped by evolution that they are essentially "hardwired" during prenatal life to give rise to a limited set of activities—sexy or otherwise—by the mature mind. Others, such as Dr. Terry Sejnowski at the Salk Institute in San Diego, counter that the brain's circuitry is so flexible and its functions so plastic and widely dispersed, that whatever parts are prewired merely form guide wires over which the limitless bounds of experience construct the mind as we go through life. For evidence, they point to hemispherectomies, in which removal of an entire half of a child's brain (for intractable seizures or tumors) do little to stop such a child from learning just about anything. One camp says there are specific, exclusive parts of the brain for courtship, arousal, perception of beauty, and orgasm; the other that our minds are not cemented in the Stone Age, but evolving still.

The evidence in support of any extreme position is scant. But there is little doubt that even if genes, ultimately, control behavior, they must do it by controlling the development and operation of the brain and the products that genes make, including hormones, tissues, receptors, and neurochemicals that integrate the influences of our environment into our brain's perception, building bridges between psychology, emotions, learning, memory, and action.

So, while we're busy diddling around with body parts down *there*, the lights are flashing up *here*, blinking, glowing, pulsing, crackling like rhythmic Christmas tree lights on a house roof. There is simply no exaggerating the brain's role, and if the lexicon of sex were truly accurate, the first thing to pop into our minds when someone says "sex," or "orgasm" or "foreplay," would be the 50-billion-celled hunk of jellylike matter encased in our bony skulls. And while there are parts of the brain that regulate what is "automatic" in us (gamete formation, sperm transport) and others what is deliberate (flirtation and intercourse), to ask what part is the sexual brain is like asking what part of a pyramid is responsible for its shape and stability.

With the exception of the sex chromosomes, men and women have the same genetic composition, so how the brain gets "sexed" has a lot to do with what those sex chromosomes do. Many brain scientists now accept that primary differences most likely are a matter of hormonal influences on our brains early in fetal, newborn, and childhood life. Once testes are formed, they make a testosterone "environment" in the womb to masculinize and organize "male" behaviors.

In the uterus from prenatal development and in infancy, a brain is sexist, sculpted by hormone-promoted growth and experience: Fetal neural circuitry depends in part on what it feels and touches, on its movements and position within the womb, and even on contact with its own genitals. Recall that foot fetishes, almost exclusively a male sexual behavior, may result from brain maps formed before birth by the juxtaposition of a fetus's feet and penis in the womb. Ears and nipples in females may also be erogenous zones because of prenatal sensitization of this kind. Studies of girls exposed prenatally to excess androgens, due to a genetic disorder called congenital adrenal hyperplasia, grew to be "tomboys," while other studies demonstrate that "sissy" boys have too little or the wrong ratio of other brain-shaping hormones during development. Male rodents deprived of androgens after birth (by castration) never learn how to mount females and instead learn such female behavior as lordosis, the arching of the back in a sexually receptive posture. Other studies in rodents show that if there are more males than females in a litter of unborn pups in the womb, the females will be more masculinized in their brains and behavior and vice versa. Clearly the uterine environment of the mother is having an impact on the brains of the pups, and the maternal uterus is of course in turn influenced by the mother's environment. "There is ample room for early experiences, including the very different social stimuli presented to boys and girls, to cause the sex difference found in the adult [brain]," says Marc Breedlove, a brain scientist interested in sex differences. "No one yet knows precisely how these differences develop, so it is not clear whether this sexual dimorphism is a cause or an effect of human sexual differentiation." But the smart money is on some combination of predisposition—to masculinity or femininity or transsexualism—while social experiences enhance or diminish the predisposition and sculpt the brain.

As a child grows, all parents recognize the emergence of dogs and snails or sugar and spice. Even the dimmest observer can see that boys and girls play differently, act differently, think differently, and desire differently, from their play to the way they form relationships. He wants to play doctor, she wants to play nurse. He wants to fight. She wants to pretend she is a mommy. He wants to drive around with his buddies. She wants to have pajama parties and share "secrets" with her girlfriends. By the time puberty kicks in, boys and girls consider the opposite sex a species of alien. Hormones exert their second level of control, changing their bodies, their reproductive capabilities, and their brain's perceptions and controls—making and using hormones and neurotransmitters differently and selectively, converting these chemicals in different ways in different doses. As the hormones cycle with menstruation, she undergoes mood changes along with the physical ones. As his ratio of male to female hormones changes, he gains self-confidence, competitiveness, and aggression.

With some evidence to support them, many serious psychologists will say that this is all due to learned responses to social cues, to conditioning that starts very, very early in a child's life. But it's hard to ascribe all of the effect to such a mechanism, given what is known about the way the brain is shaped prenatally by sex hormones and other biochemicals. It's difficult to imagine a developmental scenario that relies fully on conditioning until puberty, when hormones then suddenly take over brains and bodies. More likely, there is a continuum of effects by both conditioning and chemistry, experience and biology. Whatever the extent of either influence, by the time we come of sexual age, sometime in middle school, our brains and behavior become as sexually distinct as our genitals. For example, studies show that because of differences in a flat region of the temporal lobe called the planum temporale, a female is more likely to hear well out of both ears, while a male is better able to detect written words with his right visual field than his left. A girl is "suddenly" more sensitive and a boy more decisive. A woman in the making begins to flirt with her eyes and hips, and a man-to-be to show off and strut his stuff. A high school boy doesn't understand *why* his girlfriend is pissed off and she doesn't understand how he could possibly miss it. *"What did I do?!"* he yelps. *"If you don't know, I can't tell you!"* she retorts. A boy forgets male faces and a girl never forgets a name. A girl becomes intuitive while the boy next door is literal. Under the influence of an increasingly feminine mind and body, an adolescent girl heads for romantic novels and her opposite sex heads for auto mechanics. What were merely sex-specific *tendencies* have now become more obvious.

Consider the issue of spatial relations and tasks. Several years ago, my husband and I were headed for an Argentinean restaurant in Los Angeles for dinner. "Did you get directions?" I asked. "I've been there before, once, I can find it." It turned out that "been there" was four years earlier. But he drove down a main road, took a series of rights and lefts, and as we approached yet another stretch of commercial businesses, said "I think it's on the left-hand side of the road a few minutes from the last intersection." "Sure, sure," I thought quietly, up until moments later when he pulled up and parked in front of the restaurant.

How did he do it? Research suggests it was his brain at work in a way that mine isn't. Recall that some higher thinking functions of the brain seem to be relatively more controlled in one half of the brain or the other, so that speech is processed more in the left brain and spatial reasoning in the right. In rat studies, Barnard College's Christina Williams has found that females navigate by using "landmarks" to get through given spaces, while the males used geometric angles and shapes. If no landmarks were around, the females eventually learned to use the geometric "cues," but the males could not adapt to the use of landmarks. This appears to be the same sex difference in brain organization that makes it easier for men and harder for women to use maps, and for men to disdain women's use of "the red barn" or "lamppost" when asking for, giving, or following directions. (Women, however, may have the last laugh in their capacity to use both sides of their brains with more flexibility, or, in scientific terms, to be less "lateralized." They are, for one thing, less prone to permanent damage from strokes to their speech centers.)

Evolutionary biologists and anthropologists looking at such sex differences in the brain speculate that during the Stone Age, when our modern brains were evolving, and our ancestors emerged from isolated wandering to form societies, circumstances in all likelihood required them to live in societies that were much more sex-divided than ours are today. Men, because of their strength and size, ranged far from home to hunt and bring home the bacon, while women, vulnerable as a result of pregnancy, birth, and lactation, probably remained back at the cave to cook what men brought home. Women were probably also responsible for maintaining the hearth, gathering fruits or nuts nearby, and watching the kids' every move. It stands to reason, then, that natural selection pressures favored men with the ability to navigate without landmarks, and favored women who were tuned into smaller forays marked by familiar sites and minor but potentially life-threatening changes in their immediate surroundings.

Another intriguing piece of evidence for this process has come from Steven Gaulin and Randall Fitzgerald of the University of Pittsburgh, who studied a species of prairie rodent in which a male mates with several rather than one female. To keep himself supplied with female companionship, such a polygynous (promiscuous) critter needs to roam farther and farther afield and still get back home successfully. In short, navigational ability seems to be part and parcel of his reproductive success. Gaulin and Fitzgerald found that these polygynous male voles were better at learning lab mazes than would be a more monogamous species that does not rely on roaming as a means of reproductive success.

After X and Y chromosomes do their stuff in fetal development, sex hormones begin the sculpting of the sexual brain by influencing the way neural circuits are formed. But even sex hormones take a backseat to the brain's dominance of sex as time goes on, contradicting the classical dogma that having testes and testosterone is all one needs to get a male and that the absence of these is all that is needed to get a female. As David Crews says, "hormones very usefully organize our internal organs' readiness for all aspects of sex, but it's the brain that can exploit sexual opportunities and make sense out of random stimuli."

The pioneering mapper of the sexual brain, Arthur Arnold from UCLA, among others, has challenged the dogma of testosterone primacy. Arnold has highlighted the primacy of the brain as the major organ of sex with numerous studies of rodents and birds. Especially birds.

Contrary to popular wisdom that a bird brain is no brain at all, birds' neural wiring is both complex and informative. Differences between birds and mammals must be factored in, of course. In birds, for example, it's the female that has two different sex chromosomes, WZ, and the male the same ones, ZZ, the reverse of humans and other mammals. Thus, if the dogma about testosterone were right, then testes would secrete a hormone that masculinizes the brain of either sex. So Arnold and his colleagues gave baby female songbirds estrogen, a metabolite of testosterone, and the females did indeed sing because they had become masculinized. (Only male songbirds sing, normally. Bird-store owners have long known this and have often injected females with testosterone long enough to get them to sing and get sold to a hapless customer who, when the shot wears off, never gets the bird to sing again.) But the masculinized females *never* sang as well as any male. And when Arnold reversed the experiment and withheld the hormone or blocked it in the males, he was unable to prevent the masculinization of this bird. That boy sang. "The message is pretty powerful," Arnold says. "Testicular secretions are not the whole story in making males, in masculinization."

That's nature. What about "nurture"? The preponderance of research evidence shows that environmental experiences and pressures account for most sex differences, but they do so principally because of the impact of sex hormones on the shape, size, and organization of the brain. In short, the environment is working on differently wired brains from the moment of birth. And such differences, says neuropsychologist Doreen Kimura, "make it almost impossible to evaluate the effects of experience independent of physiological predisposition."

These differences are hardly subtle. UCLA's Arnold, for instance, has shown that certain brain cell clusters in songbirds displayed male-female differences so vast that they could be seen by the naked eye without a microscope. In canaries and finches in which the males do all the singing, he found song-control nuclei that were five to six times bigger in males than in females. Arnold demonstrated the point by handing me a class exercise he uses to teach students the differences in shape and size between male and female brains. The lesson plan contained magnetic resonance images of twenty-two healthy adult humans ages nineteen to thirty-five. Half were male and half female, but the images contained no indication which were which. Only Arnold had the "key."

The images I was to examine were "mid-sagittal," the kind of view you get if you draw a line between the eyes, down the middle of the brain from the top of the head to the bottom. The instructions were to measure the size and shape of a few brain structures, including the corpus callosum, the ropy tangle of fibers that connect the two hemispheres of the brain and allow communication across the spheres. My tools were low-tech: a millimeter ruler and a see-through "grid" made up of tiny squares. I also was to count the number of squares on the grid to get an estimate of the area of the corpus callosum and other structures, to "eyeball" the images and to rank the rounded back end of the corpus callosum, called the splenium, from least bulbous (1) to most (22) in the images.

Even to my naked eye, confirmed later at home using the millimeter ruler and grid, and the "key" Arnold provided, the differences in male and female were stunning.

Most scientists and certainly most nonscientists are reasonably comfortable with the idea that the brain, like some supercomputer, takes in signals from our five sense organs, organizes the information into a plan of action or reaction, then fires off a bunch of chemicals and electrical messages that tell various parts of the body to think or behave in a certain way and to interact with the environment in a certain way. Under this model of the sexual brain in action, finding some part of it—the hypothalamus, for instance—to be different in males versus females, homosexuals versus heterosexuals, suggests that the structural differences alone may be *causing* differences in masculinity or femininity, in sexual attitude and orientation.

But in fact, the brain is not only the "computer" or the body's operating system. It also is made of tissues, cells, and fluids; it too is an organ that can and does change its own makeup, its shape, its network of cellular connections. In this view of things, the sexual brain is an even more powerful force in our sexual lives long after its initial structures are formed than it was before—because it can *learn.*

"The human brain and the human brain alone," says neurologist-author Richard Restak, "has the capacity to step back and survey its own operation. . . . [O]ur capacity for rewriting our own script and redefining ourselves . . . is what distinguishes us from all other creatures in the world." In other words, not only does the pre-wired brain influence our sex culture, sex acts, and sex feelings, but the latter in turn can permanently change the brain and how it works. The brain can not only bring experience to bear on sex, but also undergo changes that account for the enormous variety of sexual experiences people have. Just as the sex differences in the structure of the brain are hard for some to swallow, the notion that what we *do* can influence what we *think,* and not just the other way around, is even tougher to grasp. But the importance of this endless loop with regard to our sex lives is enormous.

"Laymen may assume that a structural difference in the brain is the immutable signature of purely biological forces, but three decades of neuroscience research have made it clear that experience [and environment] can dramatically alter the structure and function of the brain," says Marc Breedlove at Berkeley, one of a group of brain experts tracking the neural changes triggered by sex activity. Working with rats, Breedlove developed an imaginative experiment to illustrate his point by comparing two groups of males he dubbed "copulators" and "noncopulators." With the help of testosterone-filled, timed-release implants, Breedlove made both male groups hormonally identical. And with the help of microsurgical techniques, he rendered the noncopulators unable to perform the sex act. The copulators were put in a cage for a month in the presence of fertile and sex-seeking females. The noncopulators were put together with females who had the rat equivalent of a monthlong "no sex" headache. Breedlove also devised a way to measure and evaluate the activity of nerve cells in the rats' spinal cords that control certain muscles that are needed to copulate, cells that are directly controlled by a circuit linked to the brain. After a month of constant sex (he kept replacing the receptive females in the copulators' cage), the copulators' spinal cord neurons were measured and found to be much *smaller* than those in the noncopulators. As one science wit described the findings, "Honey, I shrunk my brain!" Since sex hormones could not account for the difference (testosterone levels were identical, remember) the conclusion was that brain differences, at least among these males, may have been the result, not a cause, of sex behavior differences. It's likely, says Breedlove, that similar differences occur in females, and that many of the sex differences we see *between* the sexes develop the same way, in a complex loop of actions, reactions, signals, and sculpting of the brain.

In another widely publicized example, scientists studied oxytocin, a brain chemical secreted by specialized cells in the hypothalamus. Among the things oxytocin does is contract smooth muscle so that during nursing, it is produced in larger quantities to squeeze milk out of breast ducts controlled by muscle. In nursing rodents, oxytocin-secreting cells partner with other cells that have some of the same chemical-making machinery, to generate, temporarily, a tremendous surge of the hormone. Nonnursing females don't have such cooperative brain cell activity. Moreover, these very direct changes in how the brain performs biochemically is triggered not by some automatic switch, but by a very special environmental cue—the presence of a baby rodent. And when scientists went looking further into the brains of these females, they found differences in the circuitry not only of the hypothalamus but of the "thinking" brain, the cerebral cortex. Only in a nursing rodent, says Gillian Einstein of Duke University, who analyzed the research, do cells in the cortex expand in a particular area, suggesting that whatever is happening is under the control of experience, not hormones. "The brain is basically changing its circuitry so that it will devote more area to sensations of the rodent's breasts, and in fact some circuits have been identified that link this area of the cortex to skin sensations, perhaps those that are activated when a pup nuzzles the breast. When the rodent stops nursing, this area of cortex goes back down. There's a plasticity, an ability to change with stimulation and this stimulation is what we call parenting." It also is called evidence for the ability of experience to change our brains.

Recent discoveries of genes and chromosomes linked to human and other animal brain structures also have added strength to the conviction of most neuroscientists that the things we do sexually, regulated of course by hormones, have permanent effects on the brain throughout life. The classic example of this "plasticity" of mature brains is found in the seasonal variation in brain structure of songbirds, specifically in the birds' dendrites. Dendrites are simply the receiving ends of neurons—conduits for electrical impulses fired from an axon, a process that in turn forms a synapse. In the songbirds, dendrites preferentially elongate and form new synapses during breeding season, but create far fewer synapses at other times.

Similarly, in the gerbil, the volume of sex-specific clusters of cells in the brain can be altered by castration and hormone treatment. In humans, even in adulthood, the ebb and flow of hormones—driven by growth, strength, stress, joy, aging, and other outside forces—induce marked changes not only in the sexual parts of the body, but also, as any parent of a teenager knows, in behavior. And scientists have measured blood-level changes in androgen in adult rats and found them to cause changes in the actual size of neurons, the number and size of chemical and electrical signals between cells at the synapses, and even the lengths of dendrites that nerve cells send out to transmit electrical and chemical messages to other nerve cells.

Perhaps the most dramatic evidence for the impact of hormone-related behavior on the brain throughout life has come from a life span study of more than a hundred humans. The study focused on a region of the brain called the interstitial nuclei of the anterior hypothalamus, or INAH, and of examinations of a similar area of the brain in rodents called POA, SDN, or the preoptic area of the sexually dimorphic nucleus.

In 1985, Dutch investigators claimed that a nucleus, or bundle, of neurons in the anterior hypothalamus was bigger in men than women, the so-called INAH. Subsequently, Roger Gorski and Laura Allen at UCLA reported that two other such nuclei in the human preoptic area were larger in men than women and they called these INAH 2 and INAH 3, while their INAH 1 observations seemed to correspond with the overall finding of the Dutch group. Furthermore, INAH 3 in particular was much bigger in men than in women. And while males had some INAH 4, females had far more of these.

The scientists had gone looking in the preoptic area of the brain primarily because they knew that this part of the organ is heavily influenced by androgens, which are more prevalent in males than females. If they were going to find sex differences in small clusters of brain cells, this would be the logical place to look. Importantly, however, they also knew that hormones alone could not be responsible for making men walk tall and women talk like Mae West because for one thing, animal and human studies clearly showed that adding androgens like testosterone can't make every man a fighting machine who falls in love with Maureen O'Hara, no matter how hard he struts. And they knew that giving women testosterone would increase the libido, but not the *way* the libido is exercised. Mae and Maureen were not going to seduce each other just because of higher testosterone. Thus the scientists speculated it was more likely that the POA and the INAH were linked to sex differences because of forces beyond hormones.

In his landmark book, *The Sexual Brain*, Simon LeVay of the Salk Institute reports that in addition to the POA, another part of the hypothalamus, the ventromedial nucleus, plays a role in female sex behavior. Damage to this area has been shown to inhibit or damage the ability of female rats and monkeys to execute lordosis and other exclusively female sex "cues," including ear wiggling, while electrical stimulation of the area enhances them. Like the POA, the ventromedial nucleus is influenced by sex hormones, but more variably so, in keeping with the more complex chemistry of female fertility cycles.

People who have had INAHs surgically damaged or accidentally destroyed (West Germany, according to LeVay, performed deliberate psychosurgery on men with pathological sexual behavior in the 1960s), also show dramatic reductions in some masculine or feminine behaviors, demonstrating that this part of the sexual brain is responsible not just for instinctive, mechanical sex drives, but for desire and overall interest in sex. "If," he says, "the hypothalamus were a low-level center . . . one might expect such men to say that they continued to desire and seek sexual partners, but that once in bed they couldn't perform. In fact . . . what many of these men reported was an overall reduction in loss of sexual feelings. . . . imagery and drive, as well as of actual sexual behavior. This implies that an intact hypothalamus is required for the generation of sexuality in the broadest sense. Similarly, androgen-blocking drugs which are sometimes given to sex offenders, tend to reduce their sexual thoughts and drive, not just their copulatory behavior." Finally, he says, "it is important to bear in mind that the connections between the hypothalamus and the rest of the brain, especially the cerebral cortex, are two-way; the cortex can influence the hypothalamus and the hypothalamus can influence the cortex. . . . One may assume that [regions of the cortex] send signals to the hypothalamus that, combined with olfactory inputs, sensory inputs from the genitalia, and sex hormones circulating in the bloodstream, set the level of activity of neurons in the medial preoptic area and the ventromedial nucleus. These neurons in turn send signals down to the brain stem and spinal cord to influence the mechanics of sex, but they also send signals back to wide areas of the . . . cortex that very likely influence sexual [imagery] and complex sexual behavior [overall]. Thus, one can think of a cortex-hypothalamus-cortex circuit [combined with other parts of the brain] whose activity as a whole is the key to sexual life." The nuclei found in the POA and ventromedial nucleus could be the main switches of this circuit, especially with respect to arousal in males and arousal in females, although more research is needed before scientists can be certain.

What is certain, however, is that at least some circuits continue to evolve and change throughout our lives, influencing our sexual attitudes, thoughts, and acts, and in turn being influenced by them. Since the mid-1990s, for instance, LeVay and a few other neuroscientists have learned that INAHs are different between hetero- and homosexual men, as well as between men and women. Specifically, LeVay discovered that INAH 3 was two to three timers bigger in heterosexual men than in women and that INAH 3 was about the same size in gay men and women (see also chapter 12). But LeVay, Dick Swaab of the Netherlands, Roger Gorski, and Laura Allen all acknowledge that the difference could easily be the *result* of behavior and experience as the cause of any sexual behavior.

Studies of the higher brain, the cerebral or neocortex, have offered additional support for the idea that brain and body influence each other's development and function. The cortex also appears to be flexible. Infants deprived of touch, sounds, and sight fail to develop normal *brains*, not just normal behavior. Put a newborn in an environment without light, and the brain will never learn to "see," even if the optical nerve and the eye are perfectly normal. In children with congenital deafness, simply hooking up the auditory nerve to the brain via electronic relay stations, as with a cochlear implant, won't result in anything approaching "hearing" without extensive training because their minds haven't learned to discriminate sounds or make sense out of the auditory noises they now sense.

Yet even in adulthood, studies show, enriched environments stimulate increases in brain weight, cerebral cortex size, neuronal size, the proliferation of the brain's cellular "glue"—or glial cells—and the density of dendrites.

In animals, scientists can do more precise studies of this phenomenon by prodding brains with a variety of stimuli to trigger "evoked potentials," a jargony phrase for the way particular neurons fire or light up when stimulated. They especially like to work with the visual cortex, because it's relatively easy to set up experiments that track what an animal or human sees and, using combinations of electrical monitors and imaging machines, *where* exactly the brain "sees" something. Mark F. Bear of the Howard Hughes Medical Institute and Brown University's Institute for Brain and Neural Systems was able to demonstrate actual changes in the structure of the cortex, and the biochemistry of the brain, merely by manipulating what animals are given to look at. Such brains are clearly shaped by experiences, building connections among neurons as they cope with the world.

In 1996, Bennett and Sally Shaywitz and their colleagues at Yale published an article in the prestigious journal *Nature* that turned out to be one of the most important demonstrations of the impact of experience on the brain, with implications for the sexual brain as well. They asked nineteen normal male and nineteen normal female English readers to submit to functional magnetic resonance imaging, or fMRIs, which take pictures of the brain as it uses energy—in action. Specifically, the subjects were asked to decide if a pair of nonsense words such as "lete" and "keat" rhymed. Doctors have long known that women have an easier time recovering from strokes in the left hemisphere of the cortex, where language processing occurs, and men are more prone to dyslexia than women. What this study showed was that men performed this rhyming language task using a single area on the left side of the brain, while women used areas on both sides of the brain. Somehow, the women's brains had "learned" to process this information differently, suggesting that women's and men's brains develop in different environments. That's just another way of saying they have different experiences. This experiment also demonstrated that men and women use their brains differently in everyday life.

At Johns Hopkins, learning and behavior specialist Randy Nelson complains that the use of fMRI to interpret differences in female and male language abilities is "like trying to understand how men and women run differently by examining how hard they breathe." Many of his fellow scientists share his skepticism. Nevertheless, there is some striking evidence that the differences seen in language processing between men and women might well come because they not only have different brains; they use them differently. When they did the task, the fMRI lit up a half-inch-long area of the brain near Broca's area (named for nineteenth-century neuroanatomist Paul Broca), at the front of the left side of the brain at the temple. Women used this region, too, but also a comparable area on the right side of the brain.

What all of this suggests so far with respect to the brain is that experience counts, not only in priming us for sexual behavior, but also in influencing all of the precursors to coital activity—courtship and the mating dance, arousal, orgasm, love, and commitment. Many who have studied the cultural-biological aspects of sex in children believe that sexual teasing, roughhousing, contact sports, masturbating, kissing, cuddling with parents, and imitating the sexual behavior of adults, even without penile or vaginal involvement of any kind, is a rehearsal for the body and mind, necessary lessons for our brains as they prepare our bodies for sex. Given that left to their own devices, these kinds of behavior in children are almost universal, it seems nature intended that what we learn early about genital contact, love, tenderness, odors, and pleasure should inform our adult sex experiences. Studies also show that the absence of early experiences with intimacy and sexual, even erotic touching may handicap adults in the bedroom.

SEXUAL BRAIN MAP

Say you remove the testes or ovaries from animals at various times in their lives and then gave them—in a controlled and measurable way—the hormones these sex glands produce naturally. What might you learn about the real power of hormones *alone* to influence the whole spectrum of sexual behavior, from courtship to intercourse? Scientists have in fact used such experiments to tease apart areas of the brain specifically involved in mating. They chose as their subjects various species whose sexual behavior was distinctly either "male" or "female," then castrated them, then examined how their brains differed under various sex hormone and environmental influences.

Obviously, such experiments cannot ethically or safely be conducted on people, but the insights from the animal experiments have focused attention on some very specific structural and functional differences between male and female human brains that are linked to sex, on the chemical messengers that integrate brain and body, and ultimately on the genes that carry the code for them. If indeed the human brain comes in two flavors, male and female, then tracking down the neural origins of these differences would likely shed light on how nature engineered us to be attracted to our mates, to perform the acts that keep our species alive, and to develop masculine or feminine personalities, to *think and feel* "male" or "female."

The modern search for those structural differences has been led by a handful of courageous neuroscientists whose quest for the brain/mind roots of gender and sex behavior continues to be at odds with contemporary social, religious, and political agendas. Nevertheless, they have moved their studies forward, artfully selecting as their "test" models animals that were not only flamboyantly different in their male-female behavior, but also least likely to inflame antisex research sensibilities. Rats and other rodents made good choices. Before 1970, scientists were adept at injecting female rats with testosterone to make them unable to ovulate, for example, then restoring that function simply by stimulating parts of the brain—testosterone notwithstanding. They gave young castrated male rats ovary implants under their skin or in their eyes—before their testosterone had a chance to completely masculinize the brain—and produced males that ovulated, teaching them that it was the male's essentially "female" brain that could ride roughshod over testosterone. These scientists quickly came to understand that with mammalian brains at least, most if not all are inherently female, and that to produce male behavior something must happen to both turn on the development of the male brain and turn off the development of the female brain.

Birds, as we've seen, have been another frequent choice of sex brain scientists. Nobel laureate Ferdinand Nottebohm, at Rockefeller University in the mid-1970s, was the first to report microscopic differences in male and female bird brains. By 1976, Nottebohm and his group had completed an atlas of the canary brain and described networks of nuclei that controlled birdsong, only produced by males in this feathered creature. Meanwhile Arthur Arnold at UCLA was studying zebra finches for the effect of hormones on the song-controlling regions that Nottebohm had identified in canaries. And each noticed gross differences between male and female birds in the part of the brain that controls song. The excitement felt in the brain-science community at the discovery was electric, Arnold recounts, in part because up until that point, the last scientists to report any major sexual differences in the brain was the famed French neuroanatomist Paul Broca, 120 years earlier, who was the first to insist that the brain of a man was larger than a woman's.

One reason that birds provide such a grand model for sorting out how male and female brains might develop so differently is that the song center in the brain is unique; it includes a very distinctive set of brain cells. These clusters make it relatively easy to learn just what shapes them—sex hormones from the gonads, for example, or chemicals manufactured and distributed directly in the brain.

Gregory Ball at Johns Hopkins chose Japanese quail for his studies, which, in contrast to mammals like rodents and humans, seem to develop male reproductive organs without hormones at all. In fact, the female hormone estradiol secreted in the ovary of the female embryos actually demasculinizes the female, so the basic body plan in these animals is, unlike us, male, not female.

In zebra finches, estrogens given early in life demasculinize copulatory behavior in males but masculinize the vocal-control regions in the brain and singing behavior of females. Thus it is hard to understand how these behaviors happen so differently, given that normal males sing and copulate like a male and females normally never show these activities. It is quite possible that for birds at least, and at some level perhaps mammals, steroid hormones might not be greatly involved in some aspects of sex differentiation. Perhaps, says Ball, because the songbird vocal control system represents a uniquely specialized group of brain cells, brain hormones, not sex hormones, may be more responsible for the differences between the male and female bird brains.

Arnold and his colleague at UCLA, Roger Gorski, still worry about the public furor such findings have historically generated. Brain scientists, they say, learned the hard way—by having grants rejected and labs picketed—that much of the public, and many scientists as well, were uneasy about research suggesting men and women had different brain "talents" or capacities. As the forces of political correctness grew, especially on American campuses, they became—and remain—wary of even discussing "sex differences in the brain" without quickly qualifying every remark with the notation that nothing in their research should suggest in any way that one sex is smarter or better than the other. Discoveries such as theirs, said Arnold, "changed our lives, in that we all became very interested, for a lifetime, in explaining how those sex differences arise, and I've spent my own career since then investigating how hormones affect the brain." But at the same time, he cautions against those who interpret such discoveries as supporting any particular view of human brainpower or relationships.

One such difference was reported by Dr. Ruben Gur, director of the brain behavior laboratory at the University of Pennsylvania School of Medicine, and his coworkers, who used MRIs to look at metabolic activity. Gur observed the brains of thirty-seven young men and twenty-four young women at rest and not consciously thinking of anything. He found only one clear difference: in the part of the brain that governs feelings, thought, motivation, and behavior, both sexual and nonsexual. Men on average had higher brain activity in the older ancient primate part of the limbic system that is more involved with action. Women, on average, had more going on in the newer and more integrated areas of the brain that process symbolic activity. The difference may be why she wants flowers and he wants to grope. Differences in the brains have also been linked to differences in how men and women read facial emotions, a critical skill to be sure when flirting and seducing, or declaring love, fidelity, and commitment. Both males and females could tell a "happy" facial expression, and women could tell sadness on anyone. But men were not quite able to read sadness on a woman's face as on a man's. "A woman's face had to be really sad for a man to notice," according to Gur, who conducted the studies. "Subtlety didn't do it," again suggesting that men and women live in different "inner" worlds.

As it turns out, in mammals and humans, as well as birds, several brain areas stand out in the sexual arena because of their location, shape, or role as recipients or regulators of hormones and brain chemicals. The tale of the Japanese quail, as noted earlier, has provided a window on several of these areas, notably the preoptic area (POA). Ball and others have performed experiments in which they damaged or otherwise manipulated a portion of the POA called POM (nucleus preopticus medialis) that in quail almost completely controls the male's ability to copulate. By adding testosterone, other androgens, or estrogen to this area they radically altered male copulatory activity, changed the size of the cells in POM, and sabotaged the activity of a chemical helper called aromatase that converts testosterone into other sex hormones. Later, in an even more stunning series of experiments, Ball and his colleagues were able to make quail fall in and out of "love" by manipulating hormones that have nothing much to do with POM.

The key to the study was construction of the bird equivalent of a bordello, first devised by a University of Texas psychologist who was studying how quail learn. In his version, Ball took a large rectangular box or arena that has a sidecar attached to it, with a window between the two. The window can be opened or shut to allow the a bird to see in and out. When a sexually experienced male is in the big box, and the female is in the sidecar, the male can look through the window and see her. "Typically, if a female is in there, the male will look at her, but then wander off and not care a damn," Ball explains. "This is the equivalent of two strangers in proximity. There's some interest, but not a whole lot." But when Ball opens the sidecar and lets the female out into the male arena, the male will immediately try to copulate. "He doesn't know her from any other bird, but he's somehow primed for sex by her presence," Ball says, "and if he *does* copulate with her (some males will achieve this immediately, some a bit later) there is a remarkable change in the male. Right after copulating, in every single case, this male will then spend *all*, every bit of his time staring at the female through the window. For hours, or days. This bird's brain has learned something very clearly that the female's brain has not. We know that because the female could care less either way if you reverse the test. The males will stare all day, not even stopping to eat. Once they copulate with this female, they are permanently enthralled. And keep in mind this occurs in the complete absence now of any more chances to actually have sex. All the female has to do is *be* there. Most compelling of all is that the male's testosterone levels stay sky high. You can even put a different female in and he will still stare. You can manipulate the plumage and he'll stare. The male falls madly in love, whether the lady quail is a 10 or a dog."

The lesson from this experiment, says Ball, is that male *anticipatory* sex behavior is absolutely testosterone-dependent (because females don't exhibit it). In fact, when he blocks testosterone, the response disappears. But unlike the copulatory behavior in his earlier experiments, there is no obvious involvement of POM. "What this looks like is a second whole system for sexual behavior determination. If we look where the testosterone is acting in these love-mad quail, we don't find hormone receptors activating in the POM during this test, but we do think we have seen activation in other parts of the quail brain, such as the nucleus of the striae terminalis or incumbens." Ball's work suggests that different parts of animal brains evolved to handle very different aspects of sex and "love," but also evolved to act in tandem when it suited the time and place.

A cascade of discoveries like these have pulled together a picture of the sexual brain—structures, clusters of individual neurons, receptors for hormones and neurotransmitters, and the genes that generate all of these. Most important is the group of structures referred to collectively as the limbic system, or limbic brain, briefly described earlier. But the limbic brain's key role in sex can only be understood in the context of the entire brain's overlapping interests and its basic division into a primitive brain and a more modern one that grew and developed with the older brain as we evolved.

Think for a moment of the sexual brain as a three-dimensional structure with a stratified but interactive architecture whose spaces house the three forms of human sex. One kind is the "instinctual sexual drive," or what scientists prefer to call reflexive or "stereotyped sex behavior." It is the kind of drive that operates when every male quail grabs a female's neck before mounting for sex in exactly the same way, and that gets children to rehearse pelvic thrusts long before they can spell penis or vagina. Another kind of sex is emotional or intimate, the desire to not only copulate but also make love and stay bonded in a long relationship. Finally, there is the kind of sex we associate with the precursors, mood, and experiences leading up to intercourse and mating, the sexuality that most separates us from other animals, that lets us think about ourselves as sexy or desirable, that brings a sense of morality and responsibility to sex, that is aware and deliberate, that wants to "play" and improve our sex lives, that wants to watch in fascination how and why we act the way we do with sexual partners, and that wants to read a book like the one you have in your hand. It's the part of our sexuality that translates sexual desire and action into music, painting, sculpture, dance, and song, that lets us think of sex in the abstract and causes us to seek understanding of sex.

The earliest part of our sexual brain, that old "reptilian" brain, regulates what are known as instincts and, in the case of sex, the basic survival mode that operates the same as our drives to satisfy hunger, find the waterhole in the desert, and flee the saber-toothed tiger. We don't "think" about these actions so much as sense that we must pursue them. Our body responds reflexively. If we're hungry, our stomachs growl, our heads ache, and we literally can kill for food or salivate at the sight of it. Similarly, this part of the sexual brain responds automatically to a sexual stimulus, a fantasy, a woman, a man, a picture, an arousal, and it wants to satisfy the urge. It's partly—but be careful now—why we are aroused by pornography, odors, pheromones, even an *imagined* flash of pubic hair when Sharon Stone crosses her legs. It's what anthropomorphizes genitals—as in "Mr. Meat"—as if our sexual bodies and chemicals had "minds" of their own.

This deep primitive brain, also known as the hindbrain, includes—at the base of the skull and the brain stem at the top of the spinal cord— the old cerebellum, or small brain, whose job it is to control our posture, our muscles, and our balance. About the size of a peach, the cerebellum has its own cortex, or covering, and its two halves wrap around the brain stem at the rear. When a baby learns to walk, when we learn not to fall if we graze a table, the memory of how to do that balancing act is stored in the cerebellum. It enables us to do with our hand what the brain wants us to do, to pick up an egg with tenderness, deliver drinks without spills, and bend an iron bar in half with strength. Although it takes up only a tenth of the brain's volume, the cerebellum contains half of all its neurons. In recent years a few neuroscientists have begun to suspect that the cerebellum is an important player in the integration of motor skills needed in sexual attraction and arousal.

If you plucked out the cerebellum like the pit of a peach, you'd see the brain stem, a thick stalk of white fibers that carry messages to and from the rest of the body's organs, bringing in nerves that govern sight, taste, and smell. Also known as the medulla oblongata, the inch-long structure controls breathing, heart rate, vomiting, song, and speech. This level is where the things we sense overlap with how we move and act. It's also where the signals begin to cross over, so that if we feel a kiss on our left cheek, it's the right part of the brain, via the brain stem, that lets us know that sensation. Severe injury to this part of the brain usually results in death because this area governs the unconscious fundamentals of life, such as breath, blood pressure, and basic hormone output.

This voyage through the brain brings us finally to the limbic system, the group of cellular structures between the old reptilian brain and the cortex that evolved between 300 and 200 million years ago. Known also as the "mammalian" brain, the limbic system forms the psychological and behavioral brain that is subject in part to our will and to learning, memory, and reason but also is influenced by the primitive brain. It commands feeding, fighting, and fleeing. It is here, in the limbic system, that tranquilizers, antidepressants, alcohol, amphetamines, mood elevators, and pain relievers exert their influence. It is here that the sexual brain is most fully represented and expressed.

Its principle component is the pea-sized hypothalamus, which regulates sexual action along with hunger, thirst, fighting, and fleeing, and exercises command and control of the pituitary, the body's master gland. The hypothalamus is the "brain of the brain." At just one-sixth of an ounce, this organ situated just behind the eyes is nevertheless the link between our endocrine glandular system, which produces hormones, and the brain's instructional system. It is the single most important part of the brain, organizing male and female sex and reproduction, responsible for feelings of well-being, pleasure, and all of the emotions and actions involved in courtship, seduction, love, and arousal. And the hypothalamus does even more: Along with the pituitary, it has control over the sympathetic nervous system, the part of the autonomic nervous system that lets us cope with the dragon in the doorway and with sexual excitement by producing the stimulants adrenaline and noradrenaline. (The counterpart of the sympathetic nervous system is the parasympathetic nervous system, which produces our own "calm-down" chemicals and operates in large part through the vagus, a pair of cranial nerves that serve muscles of the pharynx, larynx, heart, chest, and deep abdominal organs. With the help of oxytocin and other brain chemicals, these nerves are also a major force in sexual behavior. In one set of experiments at the University of Maryland Institute for Child Study, Stephen Porges, now at the University of Illinois at Chicago, and his colleagues, for example, found that infants who had difficulties in regulating the vagal brake during social and learning tasks—who were, in other words, fidgety and frantic—had difficulties later in life developing appropriate relationships and social behavior. Because the vagal nerve system also operates in the pelvic region and may be involved in orgasm, it may have similarly profound influences on the way our sexual relationships develop.

In any case, the hypothalamus is the chief motivator of sexual behavior. During the dark heyday of psychosurgery, in fact, surgeons would damage the hypothalamus and markedly reduce the sex drive of patients who were compulsive rapists.

The geography of the hypothalamus also explains its dominant role in the sexual brain. It dangles down just to the front of and below the thalamus, which, although it serves as a *conduit* for lots of sensory information traveling from our outside to our brain, appears to have very little to do directly with our sexual appetite and behavior. Nevertheless, the lemon-sized thalamus in the direct center of the brain is so tied into the cortex and consciousness that our sexual impulses probably can't get anyplace without this juxtaposition between thalamus and hypothalamus. From the hypothalamus dangles the pituitary, the body's master gland, used by the hypothalamus to exert forceful if indirect influence over other glands as distant as the adrenals (near the kidneys), the ovaries, and testes. Notes Simon LeVay, most brain scientists "including myself until recently, prefer the sunny expanses of the cerebral cortex to the dark regions at the base of the brain. They think of the hypothalamus—though they would never admit this to you—as haunted by animal spirits and the ghosts of primal urges. They suspect that it houses, not the usual shiny hardware of cognition, but some witches' brew of slimy, pulsating neurons adrift in a broth of mind-altering chemicals."

The chemistry involved in the hypothalamic pathway is as critical as geography. This is by dint of its profound influence on the release of every hormone we have in our bodies. Because the brain is the actual target organ for many hormones, as well as "juice up," "calm down," and "feel good" chemicals, there was a need for a control center and the hypothalamus got the job. Its specialized clusters of cells regulate biological rhythms by regulating these chemicals. One part of the hypothalamus, for example, called the paraventricular nucleus, carries particularly high densities of cells that act as receptors for nitric oxide (NO), the neuro-transmitter shown by Dr. Solomon Snyder at Johns Hopkins and others to be closely linked to erection and sexual aggression. Mice with disruptions of neuronal NO synthase offer an indication of the importance of this molecule: They display grossly abnormal appearance, abnormal sex habits, and motor activity, and willingness to rape females until they shriek.

Eventually, of course, evolution put the neocortex in charge of our more primitive, older brain structures. We needed a center of reason to guide our feelings of intuition, fight, flight, smell, and emotional reactions to fear and pain. At the end of the decade of the brain, scientists are paying a lot more attention to the parts of our brain, that can do advanced algebra and read Plato. But everything that happens in the limbic system to regulate mood, drive, and emotional reactions is what actually creates our reality, our conscious world that we must deal with and live in. The limbic system helps us sift, sort, and remember what brings us rewards and punishments. It is here, among the limbic system's connections with the upper, more "intellectual" part of the brain, that we feel pain when we smell the perfume worn by an unfaithful lover, that the mild anxiety we might feel when our boss grimaces at us becomes panic if the grimace comes from our mate. The limbic system is how a couple making love can blot out the noise of traffic and neighbors but will react immediately if an out-of-context sound emerges—like a child's cry. It's how and why we lose all track of time when we're in the throes of sexual arousal; why the dripping faucet that drives us nuts is not even on our radar screen if we're having sex. The limbic brain is how a man may consider a woman irresistible Saturday night, obsessed with single-minded focus on the object of his sexual desire, then retreat on Sunday from a permanent relationship with the femme fatale who suddenly emerged as a bimbo with bad hair.

Also part of the limbic system is the amygdala, named for its eponymous "almond" shape. Resting on the hippocampal tip, from whence it organizes and regulates experiences and emotional expression, the amygdala contains a large number of receptor proteins that form the sites of expression for the body's own opiates, called endorphins and enkephalins. Under the right conditions, including those of "good" and "bad" stressors—and sexual desire and release—we produce biochemicals that produce feelings of joy, well-being, and even ecstasy. (Scientists in the 1970s searched for and found these chemicals and their receptors after reasoning that our brains could not become addicted to narcotics without them. In fact, it is precisely because man-made opiates fit like little keys into these receptor "locks" that they work. In other words, it's because heroin and morphine are chemically similar to what our bodies make naturally that they make us feel the way they do and that we begin to crave them.

If one could design a nearly perfect experiment to expose the value of the amygdala in sex, it might be one led by Herbert J. McGregor from the University of Cambridge in England, who infused beta endorphins into the amygdalae of male rats. The rats' sexual interest in female and their copulation rates dropped like a stone. But the fact that these males were *still* able to sniff out sexually "ready" females from their unready sisters suggests that the male rats' brains were functioning just fine—they found appropriate sex partners—but the endorphins gave them the joyous payoff that previously they could get only with sex.

In another study, scientists used chemicals to kill certain cells in the amygdala and significantly impaired male rats' ability to press a lever they had been taught to press to get access to a sex-ready female. Put another way, the males' motivation for sex was damaged, although the damage to these cells didn't seem to interfere with the ability of the males to have intercourse if the female was already present. This part of the rats' brains apparently was responsible for helping them "imagine" the joys of sex; when damaged, they could imagine nothing unless the females were right in front of them, giving them the smells and the female rat equivalent of the tease. Taking the experiment further, investigators injected these males with amphetamines that induced the release of dopamine in the amygdala, and they discovered that these injections overcame the motivational deficits. They had (rightly) guessed that because the amygdala is important for motivational components of sex, adding chemicals that increase the desire to concentrate on the pursuit of pleasure would help. Dopamine is a neurotransmitter known to be at the root of many drug addictions and to yearn intensely for certain experiences. As dopamine response in our brains weakens with age, sexual vigor also decreases, while depression may increase; and drugs such as L-dopa that boost dopamine have antidepressant effects. The fact that amphetamines, via dopamine pathways, can overcome motivational deficits may explain why "speed" makes some males hypersexual. In addition to the impact of such neurotransmitters as NO and dopamine, ordinary sex hormones work through the amygdala as well. Scientists at McGill University in Canada, for instance, have discovered that the sex hormone dihydrotestosterone, or DHT, can increase the kind of roughhouse sex play and aggressions found in young male rats—and presumably young males in general.

Wherever in the limbic brain large collections of endorphin receptors reside, neuroscientists have generally found links to sexual behavior and interest. For example, in the basal ganglia are tissue clusters that contain many cells that make dopamine. In one set of experiments, scientists at Purdue University in Indiana found that vagino-cervical stimulation during female sexual activity in hamsters and rats *increased* dopamine release in the basal ganglia. They measured dopamine in this area during mating by putting tiny "masks" across the females' vaginas, preventing males from inserting their penises. Dopamine significantly increased during the first half hour in copulation tests for the hormonally primed females without masks, while no increases were seen in those with masks. However, the total time spent in lordosis was the same in the two groups, suggesting that stimulation is necessary to induce increases in dopamine during copulations, if not to make the hamsters ready or eager for sex.

Such experiments remind us that while it may be useful philosophically to talk about sex in the context of "primitive" or "instinctual"— areas of our brains duking it out with our rational, politically correct, and more advanced thinking brains—the evidence suggests that such distinct separations are no part of nature's design. Beyond the anatomical tiering of the brain is a unity and coordination. Our sexual brains, most notably our limbic systems, recruit instinct and thought, emotion and experience for use in every act of desire and sex. Moreover, they demonstrate that mammalian brains are highly specialized when it comes to different aspects of courtship and copulation. Even within the limbic structures, very specific clusters of cells are selectively responsible in very precise ways for particular aspects of sex.

Remember the sexually dimorphic nucleus of the preoptic area (POA SDN) of the hypothalamus and the anterior hypothalamus that Roger Gorski studied in rats? He clearly identified the POA as critical to motivation for sex, desire, or lust, rather than merely coitus itself. But it would take an extraordinary set of experiments to demonstrate how exquisitely specific the activity of this cluster of cells is, at least in rats. In the experiments, conducted in the 1980s, investigators first trained sexually experienced male rats to press a lever to get access to a female in heat with which they eagerly mated. Because it's hard to tell what motivates a rat's lust, the experimenters needed a system to sort out motivation from everything else. They devised this by attaching a flashing light above the lever. Initially, each time they dropped a sex-ready female into the male cage, the light was set to automatically flash on, training the males to associate a flashing red light with a lusty female ready for love. Then the scientists injected chemicals that selectively destroyed cell bodies of the POA. The males continued to press the lever and flash the red light to get the girls in the cage—the scientists' index of masculine sexual motivation—but the brain damage dramatically reduced the proportion of males that copulated with an estrous female once they got access to the females by pressing the lever. In other males, lever-pressing rates dropped when the scientists damaged small areas of the amygdala or castrated them. Another group of scientists put small lesions in the medial preoptic anterior hypothalamic area of five adult male pygmy goats and reproduced the same results found in earlier studies with rats, cats, dogs, and rhesus monkeys: a marked decrease in the goats' ability to ejaculate, but not in certain aspects of sexual buildup in these animals, such as flehmen (nostril flaring) and penis licking. In a similar experiment, scientists castrated sexually experienced male rats and implanted them with testosterone-filled plastic capsules. This gave the investigators a test group of animals with an absolutely identical sex-hormone playing field and hormone levels not subject to individual animal habits that could be precisely measured. Then, separating them into two groups, the scientists removed the olfactory bulbs in one group, leaving them unable to detect odors properly. After they recovered, the males were tested for their ability to copulate on alternate days. Fewer than 40 percent of the animals who had their smell organs removed could ejaculate, while 80 to 100 percent of the control animals did just fine. The investigators then went looking in the amygdala, hypothalamus, and preoptic area for cells that bound androgen and found that removing the bulbs significantly reduced androgen binding in the amygdala and hypothalamus, but not elsewhere. The conclusion is inescapable: The brain's smell sensors, the olfactory bulbs, play a dominant role in nerve cells that concentrate androgen in just those parts of the brain needed for effective copulation.

Then there is the spinal nucleus of the bulbocavernosus, or SNB, another part of the brain that appears to very selectively govern the muscles attached to the penis. Males thrust penises and females don't, so for some scientists it stood to reason that there was a brain control center responsible for this difference, and that they might find it by examining the areas of the spinal cord that control muscles that attach to the penis, as well as the motor neurons that activate and control those muscles via the brain. The muscles in question are known as the bulbocavernosi, levator ani, and ischiocavernosus, all of which attach to the base of the penis. Both the bulbocavernosi and the ischiocavernosus are used in men to eject the last drops of urine, and the bulbocavernosi is attached to the base of the clitoris in women. The special "motor neurons" or nerve cells that control these muscles make up the SNB and are found, predictably, in greater quantities in male rats than in females. And they are larger, too. Motor neurons that control the external anal sphincter are also located in the SNB region. Female rats begin life with all three muscles attached to the base of the clitoris and the SNB operating them just as in males; but without male hormone expression, many of these motor neurons degenerate. Tests show that a single injection of androgen early in life can permanently salvage the SNB system in females, "masculinizing" the action of all three muscles. In humans, the SNB motor neurons are found in both men and women and look very much like those of rats, especially those supplying the perineal muscles in the pelvic floor. And as in rats, women have smaller bulbocavernosi and ischiocavernosus muscles than men.

Finally, there are the particular sex roles played by the cluster of neurons in the ventromedial hypothalamus (VMH) and the bed nucleus of the stria terminalis. The former has a lot of receptors for a female sex hormone called estradiol and for oxytocin, and experimental lesions made in this region of a female rat's brain disrupt lordosis. In the latter, a hypothalamic hormone called galanin influences a drugstoreful of sex hormones and neurotransmitters. Scientists from the University of Washington in Seattle have shown that the differences in galanin levels between male and female rats parallel differences in the number of cells that come under the influence of vasopressin, a small protein that, like oxytocin, is secreted from the posterior pituitary and is linked to orgasm and surges of emotion.

TWO MORE "SEXUAL BRAINS"

Along with the brain we all know and admire in our skulls, scientists also have identified two other brain systems that profoundly influence sex and are in turn influenced by it. One is the so-called gut brain, and the other is the vomeronasal system, an ancient cluster of sensors that "smell" chemicals the nose can't.

The gut brain, or enteric nervous system, and its role in sex came to light forcefully in studies by Barry Komisaruk of Rutgers University in New Jersey, which showed that sexual arousal and even orgasm can be achieved when the nerves in the spinal cord and brain stem are cut off permanently from the main brain itself (see chapter 9). It was long believed that without an intact spinal cord, messages from, to, and between the genitals, the muscles, and the area of the brain required for arousal, intercourse, and orgasm could not be transmitted. That remains true, but the gut brain appears to have a way around the "breaks." The gut brain may also be responsible for the "butterflies" we feel in our stomachs when we are in love, for the intense sick feelings we get when love is lost, and for the fact that Prozac and other antidepressant drugs make us feel good but also cause nausea, cramps, and diarrhea.

Central to the gut brain are the rich sources of nerve cells that line the fibers and tissues of the digestive system. Along with the neurons are receptors and cells that bind and make neurotransmitters and other proteins that communicate with each other, glue-like cells such as those in the brain, and nerve-cell connections that, although they bypass the skull brain, conduct a form of learning and give us what we call our "gut feelings" about Mr. Right or Mr. Wrong. Dr. Michael Gerson, professor of anatomy and cell biology at Columbia Presbyterian Medical Center in New York and an authority on the gut brain, says many symptoms and GI disorders and diseases, from colitis and colic in babies to irritable bowel syndrome, emerge in the gut brain. "Just as the brain can upset the gut, the gut can also upset the brain," he says. While most of what we "feel" from our gut is negative—pain, cramps, nausea—the fact remains that our bodies can and do recruit the gut brain to give us sensations, including sexual ones. And when the gut brain is upset, so are our sexual responses.

To understand how this occurs requires a more detailed look at the makeup and operation of the gut brain.

First, nearly every chemical that operates and influences what our brains and minds do—including what happens during sexual arousal, desire, and performance—shows up in the gut: serotonin, dopamine, glutamate, norepinephrine, and nitric oxide, along with tinier brain chemicals involved in pain, enkephalins (related to endorphins), and benzodiaze-pines, the major ingredients in mood- and mind-altering drugs such as Valium and Xanax.

That makes sense in the context of evolution. The reptilian brain or old hindbrain that would eventually evolve into the more mammalian limbic system had no thinking neocortex, so perception of such things as hunger, thirst, food, and water were distributed more widely in animals like snakes and lizards, all along their tubular nervous systems and their digestive systems. Over eons, the central nervous system we mammals developed gave us the means to search for sex as well as food more rationally, but for infants, whose higher brains are undeveloped at birth and who for years cannot forage and survive on their own, nature retained some kind of brain in the gut to guide their "instinct" for demanding and getting both food and the "sexual" feelings of pleasure, perhaps, that come with cuddling, stroking, and accidental rubbing of the genitals. When infants are hungry, they literally cry in "pain," and when they are fed, or stroked, they are comforted and feel wonderful. Such powerful reinforcers of the will to survive were preserved by nature in the enteric nervous system as an independent neural pathway, a "brain" that can operate on its own without direct involvement of our neocortex or limbic brain per se.

In a remarkable display of practical redundancy, evolution also gave the gut brain some of the same capacities for reacting to and processing chemical information as the head brain. Dr. David Wingate of the University of London reports that during sleep, the big brain in our skulls yields cycles of slow-wave sleep and brief periods of rapid eye movement (REM) sleep that produces dreams. The gut, at night, without food, does the same thing with longer cycles of contractions and short bursts of quick muscle spasm. If Grandma said indigestion and eating too much at bed-time would give you nightmares, she was probably right. And if attracting or losing a lover gives us cramps, the gut brain may have a role as well.

Gut brain reactions, sexual or otherwise, depend on the connections between our two brains. When our minds are stressed out, excited, fear-ful, anxious, or happy, we release stress hormones and other chemicals that produce reactions in the stomach and gut. We feel "nervous" and get that feeling in the pit of our stomachs. Fear can trigger the vagus nerve to release more serotonin, putting the gut into overdrive and diar-rhea. When our love is lost or a child dies, we "choke" in part because esophageal muscles are driven to shut down in a spasm by the over-production of adrenaline and other stress chemicals.

This communication is why we have trouble eating when we are in the throes of new love or love lost. Extreme stress, for good or ill, may well trigger the release of our gut brain's store of histamine, prostaglandins, and so-called P, or pain-producing, substances that inflame the tender tissues of the digestive system. Evolution makes sense here as well. The old reptilian and gut brain alerted an animal to danger of injury in part by shutting down the digestive system and releasing immunological chemicals that mopped up toxins that might be released from the intestine into our inner organs. But the chemicals also cause pain and diarrhea because of their potency and sudden release.

The gut brain is of course, as noted earlier, connected to the main brain in humans and other higher mammals. The connections are a set of fibers we've already met, the vagus nerves. It once was thought that whatever sensory nerves and motor cells were in the gut were directly wired via the vagus nerve to the brain and that the gut was just a digestive tube with primitive reflexes that jumped to the brain's tune. But developmental biologists have recently tracked the gut brain's development to an early formation of nerve cells called the neural crest, which appear in embryos. One part of the crest develops into our brains and spinal cord and nerves and another into the enteric nervous system. Moreover, the gut has been found to have not just a few but 100 million neurons, more than the spinal cord, while the vagus nerve only projects a few thousand nerve fibers there. It turns out that the brain sends signals to the enteric nervous system by sending signals to a relatively few neurons that, like in a relay, turn these signals to nerve cells in the gut, which then carry critical messages around. These two sets of nerve cells, which Michael Gershon has termed "command" neurons and interneurons, are widely dispersed through gut tissue in an outer and inner layer called plexuses. The command neurons control the action in the gut while the vagus nerve plays its role by increasing or decreasing the rate at which it fires its electrical impulses.

Finally, there is compelling evidence for the gut brain's capacity to learn and to become "addicted" to our own opiates, as well as heroin, just as our main brains do. As anyone who has used codeine, morphine, or other narcotics after surgery can attest, these molecules in the brain give us euphoria; in the gut, constipation. This means the gut brain must at least have receptors for these chemicals, and possibly producers of and receptors for benzodiazepine or Valium-like chemicals, which relieve anxiety. Just as nature didn't give our brains opiate receptors to addict us to heroin but to respond to the body's own sources of opiates, so it did with receptors for benzodiazepine. The calming influence of our gut brain's own Valium-like chemicals also affect sexual mood and performance.

THE SMELL BRAIN

The other unusual "brain" with a prominent place in sex is actually a knob of sensory cells that can detect odors regular nerve cells in the nose and their connections to the olfactory system in the main brain do not. Called the vomeronasal organ (VNO), or erotic nose, this structure primarily is thought to detect pheromones, known throughout the animal world to trigger sexual arousal and behavior.

Rather than routing odors along the usual nerve fiber avenues to the brain centers responsible for higher awareness in the cortex, the erotic nose appears to have a connection to primitive areas of the limbic system, especially the amygdala, part of the old, nonconscious brain, where perception of odors and "instinctive" behaviors are based.

Anatomically, the VNO consists of a primitive clump of tissues located behind the nostrils in the bony wall that divides the nose and meets the hard palate. Inside are sensory receptor cells directly linked to the brain that get social and sexual information from odors in fellow members of the species at hand. All of this has been known for some time to exist in nonhuman mammals. There is increasing evidence, however, that humans have this capacity as well. While these are not odors that we consciously smell, they clearly produce reactions in animals—and possibly in humans—that are real, if subconsciously triggered.

Most of the VNO-pheromone system appears to be at play in sexual attraction and sex appeal (see chapter 5). But it clearly has other roles as well. In rodents, signals from the VNO can bypass the main olfactory bulb of the nose and, according to Sally Winans of the University Michigan, go directly to the part of the brain that controls maternal love of pups and reproduction. Dr. Michael Meredith of Florida State University and Charles Wysocki of the Monell Chemical Senses Center in Philadelphia have found that in virgin animals, VNOs are important in triggering first sexual encounters. If a virgin male hamster's VNO is removed, he won't mate with a receptive female even if his other nose is OK; but if he's already sexually experienced, the effect is less drastic.

That the VNO would have sexual implications is no surprise to scientists, given that sex hormone receptors abound in the structure, that
nerve cells project between the VNO and such "sex brain" structures as
the POA, and that the male animals, which depend heavily on sniffs and
smells to select females and initiate sex, have far more neurons in the
VNO "pits" deep in their nostrils. When a typical male hamster investigates a female in heat, for instance, he licks her anogenital region and her
vaginal secretions before mating, then with a single mount inserts his
penis, thrusts, and ejaculates in several minutes. When scientists have
interfered with the VNO and the smell nose by plugging them, the hamsters wouldn't mount at all.

Additional insight in the value of the VNO comes from studies by
Hopkins's Greg Ball and his studies of the brains of female prairie voles.
Ball investigated them when they came into heat right after giving birth
for clues to what triggered their sudden renewed interest in sex. He knew
that the presence of males could induce estrus, but learning *how* was the
challenge. Using means of detecting the expression of genes called Fos,
which are dispersed throughout the brain and profusely during reproductive activities, Ball found a significant increase in the number of Fos
cells in the smell brain and the limbic system of the females within twelve
hours of giving birth. What's more, the pattern of Fos expression he
found looked very much like that found in females after exposure to male
urine. He concluded that the biological "cue" triggering heat in these
animals may well be contact from a pheromone from either a male or
newborn pups.

To fully appreciate the function of the VNO, it may help to understand the role of odors and smell in general in the story of sex. Odors,
consciously or unconsciously, repel or lure us to each other because our
brains evolved to make us sensitive to them. The olfactory system at the
base of our brain contains 5 million specialized nerve cells dangling from
the nasal cavity to transmit smell messages to the brain, but they also are
connected to the limbic system, those primitive structures in the middle
of the brain that regulate sexual pleasures and arousal. And because the
limbic system also houses our long-term memory, we rarely forget an odor,
at least not as fast as we forget a face or a sound.

Nature in fact invested heavily in our noses. At the conscious level
alone, we can detect more than ten thousand different odorants. (We
should not feel too smug about this, however. Hamsters, like many mammals, are capable of telling a fellow hamster's sex, species, and individual

identity on the basis of its odors alone, something we think infants can do with mothers, but that not many older children or adults can do.)

The links forged by odors between hormones, brain chemicals, and experiences explain how musky smells move from offensive when we're children to the basis of expensive perfumes (and beer) at just about the same time teens and young adults begin to think in "sexy" terms. Those same links tell us a great deal about how guys move from hating girls to "love at first sight," and from labeling body fluids "disgusting" to discovering the pleasure of "French" kissing and sucking and eating raw oysters.

Consider the sexual power of scent and season, informed by studies of everything from food aroma to blood flow, and the way the brain's memory centers are wired. Research has demonstrated that the scent of cinnamon buns, licorice, and fresh doughnuts all increase blood flow to the penis in men young and old, and that the fastest aromatic way to an erection seems to be a combination of pumpkin pie and lavender, a consequence of the combination of sensing a smell at the same time a strong memory is hardwired in the brain. Neuroscientists have shown, too, that sensory experiences, including smells, need never go to the "thinking" part of the brain but straight to the amygdala, which controls bodily responses to fear or excitement. Thus, a person who experienced sexual arousal at or near a time when they smelled lavender or lilac can experience the increased heart rate and dilated blood vessels that engorge the vagina or penis decades later when a date wears lavender perfume. Even a poem that evokes *thoughts* of scents and romantic experience can produce a predictable physical and physiological response in those who have become sensitive to the poetic stimulus.

The story of how the deer and the antelope play the odor game may also shed light on the role of pheromones and our two smell brains. The male of the species display an instinctive behavior called flehman, in which the males curl their lips and raise their heads—but only after sniffing a female's urine. Similarly, females of a mouse-like mammal, the prairie vole, in spring produce more of a protein that signals the opposite sex that nerve cells essential for mating have been activated. Scientists have found that the female voles' manufacture of this protein is triggered by exposure to the urine of the male. The male chemical serves as a chemical spark converted in the brain into a signal that helps the release of sex hormones in female voles, and a cascade of other mating activities.

For more evidence of the relative importance nature gave to smell over, say, our sense of sight, and to its heavy influence in sex, consider that neuroscientists have identified one thousand genes devoted to smell and just three to sight. From the time naturalist Jean Henri Fabre captured the cocoon of an emperor moth, left it overnight inside, then found forty males flapping at his window, we've known that creatures put out attractants or secretions that are sexual lures.

Men and women have sweat-producing glands in their armpits, groin, and around the breasts that only become active at puberty; at some level, our brains must find them attractive. Napoleon is reputed to have written to his lover Josephine, "I will be arriving in Paris tomorrow evening. Don't wash." Mideastern men, Greeks particularly, wipe their armpits with colorful scarves and offer them as sexy tokens to women who grab them to dance. In 1986, Winifred Berg Cutler and others at the Monell Chemical Senses Center asked male volunteers to wear gauze pads in their armpits for several days a week to capture "male essence." When this "essence" was mixed with alcohol, frozen, and dabbed on the upper lip of women three times a week, the women—all of whom had similar fertility problems—said they smelled nothing but alcohol, but after several months of this treatment, their monthly menstrual cycles became normal.

Richard Axel, a molecular biologist and neuroscientist at Columbia University and the Howard Hughes Medical Institute, an expert on the smell brain, says the neurons that process odors may seem identical in size and shape, but all detect separate odorants. Endlessly responding to odorants and forming circuits, cells responsive to one odorant, say jasmine, project to certain cells in certain places in the brain, while those responsive to say, roses, project to another. Axel believes that these projections come together to form a total perception of an odor, so if you smell your lover's cologne or the roses he brings to you, the odor activates a combination of locations simultaneously in the brain's olfactory bulb that converge in the thinking brain, where memories are stored. In this way, the possible combinations of what could be smelled are almost endless, but evolution conserved only those that mean survival, such as sex-related odors.

Very clearly, people are sexually influenced by what they smell. But do humans also have a VNO that responds to pheromones? If such a nose exists in men and women, says Axel, it's by definition a perceptual, nonconscious nose. Proving its function will therefore be difficult. Dr. Bruce Jafek, an ear, nose, and throat specialist at the University of Colorado in Denver, and others have found tiny openings in the nasal septum of patients, however, and in 1994, Thomas and Marilyn Getchell of the University of Kentucky College of Medicine found that cells in these areas have molecular makeups in common with smell neurons in the nose.

One way to determine if we have a VNO is to find the chemicals in humans that may be sensed by such an organ. Benjamin Sachs, the Connecticut specialist, says evidence for "aphrodisiac odors" in humans has been notoriously difficult to collect, and what there is remains preliminary. Moreover, it's unclear even if these odors are "volatile"—and thus detectable in some way by the main olfactory system, or molecules that only can be detected by a human version of the VNO. "And even if we should find these chemicals," he adds, "there's no reason to believe that they play the same role in humans they do in animals, that the absence of them can account for sexual dysfunction, or that adding them to perfume or hair spray will increase or enhance sexual performance."

Nevertheless, the evolutionary and circumstantial evidence for the existence of a human VNO and pheromones is mounting. For one thing, it's unlikely that humans would have had a VNO and its biochemical receptors, then totally lost them. Evolution is a conserver, not a spendthrift, particularly with systems that work powerfully.

Recently, scientists have confirmed that in rats and mice and other animals, pheromones are sensed by cells in the VNO while other smells are perceived in cells that line the main olfactory bulb in the nose that leads to the brain. Receptor molecules trap both pheromones and odors, and these receptors, as soon as they are activated, trigger the opening of little channels in nearby cells that let in positively charged atoms, called ions. This in turn sets in motion nerve impulses that travel to the brain. In a string of experiments, Catherine Dulack of Harvard and Drs. David Corey and Emily Liman of Massachusetts General Hospital believe they have found the ion channel that rats use to smell pheromones in the animals' VNO. The gene that creates this channel exists intact in fruit flies, but in the human genome it doesn't work normally, suggesting that if humans have the ability to detect pheromones at all, the capacity resides in some other ion channel.

THE SEXUAL BRAIN'S GENES

Beyond brain structures, hormones, and clusters of neurons, scientists have looked for and found some of the genes that are the blueprints for our brain's sexual encounters and operations. Dean Hamer of the

National Institutes of Health, a behavioral geneticist, has linked "novelty seeking" genes with certain sexual behaviors and brain structures implicated in some sexually dimorphic behaviors.

Michael Baum, with his colleagues at Boston University Department of Biology, did imaging studies using immunological tags on the protein products of certain Fos genes in the brains of male rats, both after the animals copulated with females in heat and when they didn't. These genes made Fos proteins in greater numbers during certain reproductive activities.

In related studies, another variety of the Fos gene called FosB appeared to control whether a mother mouse nurtured her offspring. In one of the first pieces of direct evidence that nurturing in animals is genetically based, Jennifer Brown and her team at Harvard Medical School created a knockout mouse that inactivated the Fos gene and found that the mothers stopped loving and caring for their pups. They tried everything to get the mothers to nurture, without success. The only defects Brown could find between these mice and normal mice was in the POA of the hypothalamus. Brown found that when they exposed normal female mice to pups, production of the FosB protein was turned on in the POA, which couldn't happen in the knockouts because they lacked the FosB gene. It's highly likely that FosB is what is inducing nurturing behavior by acting specifically within POA neurons.

At the Jackson Laboratory in Bar Harbor, Maine, where inbred strains of mice offer models of human disease and behavior and whose genomes and environments can be manipulated at will, Elizabeth Simpson and others tied a gene found on the sex and nonsex chromosomes of certain mice to some extremely aggressive sex behavior. An equivalent gene is known to exist, moreover, in humans.

Specifically, they produced a knockout mouse, a strain without this particular gene, but otherwise with all of their genes intact. The male knockouts immediately began to bite their brothers and sisters and were so agitated and frantic that they would squeal, startle, jump, and bite more often than litter mates. They literally could not do enough to get away from themselves and others. When these males are put in cages with females, they bite the females, often to death. "My technicians," said Simpson, "kept finding dead females in the cage and I'm sad to say it took a bit of time to figure out what was going on. Every single one of the male knockouts maimed or killed the females they were to breed with. With careful watching, we got a few pregnancies, but the females were

always injured." After the animals matured, Simpson and her team examined the brains of these animals for clues to their sexual aggression and
violence and found that the knockout brains were atypically small, especially in the cerebrum and olfactory lobes, where learning, emotion,
memory, and smell are facilitated. In female knockouts, moreover, these
same brain changes were linked to attacks on males. "She fights enough
to cow him, steps on his back and may be attempting to mount the way
a male does in sex. These females," Simpson said, "also got pregnant and
gave birth but abandoned their babies and refused to nest or feed them
very well."

THE MEANING OF IT ALL

At the start of the twenty-first century, the brain is clearly understood to
be in charge of sex. It is equally clear that scientists have only begun to
understand all the ways this is so. But there is a sort of unified theory of
it all, in which personality and mood are believed to be as much rooted
in biology and physiology as a female rat's automatic back-arching when
she's ready for sex, or as sperm manufacture and erection. The brain and
mind are no longer viewed as human compartments separated from the
rest of the body, with the former based on some mysterious psychological
construction and the latter on genes, molecules, and cells. Whatever else
the brain is, it is an organ, and a gland, that puts out and uses biochemicals. There is a biology of the sexual mind as well as the sexual
brain and body, the theory holds. If the scientists are right, it would
appear conclusive that people, perhaps uniquely, are able to *control* their
sexual behavior, despite the instinctive drive for sex and the enormous
power of sex to drive our behavior. For without this capacity for control,
we would be so overwhelmed by sexual desires that the higher brain
functions we alone in the animal word developed would have no time to
work, if in fact they would have developed at all. Our brains, operating
as our minds, are what make it possible to put sex in perspective, to let
us love and care for many people, and to bring the power of sexual
attention and sexual love to bear only when it makes sense to do so. Thus,
while we appropriately pay homage to the hypothalamus, the VNO, the
pituitary gland, and INAH neurons—along with neurotransmitters rush-
ing through the brain or to neuronal circuits carrying electrochemical
messages to our hearts, lungs, and genitals—to consider the physical and
chemical brain so far studied by scientists to be the be-all and end-all of

brain sex is to trivialize the wonder of sex. Infatuation and love are not just a case of our limbic system kidnapping our cortex and taking our oxytocin out for a spin. When we seek dates and courtship, we are not merely attraction junkies, responding with reddened skin, heavy breathing, and sweaty palms to some automated switches fired by our brains. Emotions and experiences are mediated by genes, chemicals, and brain cells, but we still have tremendous *conscious* command over our choices. We still have choices in where we go, with whom we choose to go, and what we do with each other. The time may come when there will be a unifying biology of consciousness as well. But that time is not yet.

5

SOME ENCHANTED EVENING

"The ruling passion, be what it will,
The ruling passion conquers reason still."
—ALEXANDER POPE

She was so extraordinarily beautiful that I nearly laughed out loud. She . . . [was] famine, fire, destruction and plague . . . the only true begetter. Her breasts were apocalyptic, they would topple empires before they withered. . . . her body was a miracle of construction. . . . She was unquestionably gorgeous. She was lavish. She was a dark, unyielding largess. She was, in short, too bloody much. . . . Those huge violet blue eyes . . . had an odd glint. . . . Aeons passed, civilizations came and went while these cosmic headlights examined my flawed personality. Every pockmark on my face became a crater of the moon.

So Richard Burton famously described his very first glimpse, in a restaurant, of a nineteen-year-old Elizabeth Taylor. The Welsh actor never reported the details of what happened next, but most scientists who study sexual behavior would bet their research budgets that sometime shortly after that breath-catching, gut-gripping moment of instant, mutual awareness, Liz tossed her hair, swayed her hips, arched her feet, giggled, gazed wide-eyed, and perhaps flicked her tongue over her lips while extending that apocalyptic chest. Dick, for his part, no doubt arched his back, stretched his pecs, imperceptibly swayed his pelvis, swaggered, laughed loudly, and clasped the back of his neck, which had the engaging effect of stiffening his stance and puffing his chest.

What eventually pulled these two strangers together from across the fabled crowded room was what does it for all of us—attraction enhanced by instant flirtation. All of us have the capacity to rapidly and automatically turn our actions into sexual semaphores, instinctively and predictably signaling interest in the opposite sex. And we all have too a finely toned ability to read these signals, however subtle they may be.

What is it that so attracts us and keeps us attracted to someone, ringing internal bells and sending total strangers vibrating with "lust in their hearts" to each other's rhythms? Long before a couple gets to sex or married with children, even before they get to serious, prolonged flirting or overt seduction, something moves us physically and emotionally to the point of infatuation—and sometimes to the prospect of joining two gametes. Is it something she says or the way he looks at her? Something she splashes on at fifty dollars an ounce, or his Patek Philippe watch, power suit, and *Wall Street Journal*? Is it champagne and music?

We might be tempted to think it's some modern scheme inflicted by Madison Avenue and romance novels if we were the only animals to succumb to sudden attraction. But birds, snakes, rodents, and even fruit flies parlay chance meetings into parental destiny, and engage a lot of energy-intensive plots and plans to do so. It would seem, then, that living things, including people, must have a lot of biological and chemical machinery devoted to instant enchantment with a member of the opposite sex. In fact, the silent language of attraction and its accompanying gestures are so universal, and so successful it's now axiomatic, that the plan has become embedded in our genes and brain's operating system along with every other sexual trait. Darwin knew this as long ago as 1871. Describing the case of a female duck who seemed especially attracted to a particular male duck, he noted, "It was evidently a case of love at first sight, for she swam around the newcomer caressingly. From that hour she forgot her old partner." In modern parlance, Darwin was saying that some kind of emotional system was activated when a bird or a mammal became attracted to a specific other. "Who you choose is governed by a host of cultural, biological and environmental factors," says Rutgers University anthropologist Helen E. Fisher. "And in humans your childhood experiences . . . play a role in who you find attractive. But once that preference has been made it is my supposition that it activates a specific emotion system we call attraction."

These days, then, sex appeal, "love at first sight," and all manner of coquettish behavior in men and women alike—long trivialized and even demonized as shallow, callow things—are gaining scientific respectability as studies increasingly reveal these behaviors to have biological as well as psychological and social roots. Scientists now recognize that whether erotic, platonic, gay, or straight, sex first *requires* two individuals to recognize and submit to some irresistible force that pulls them down the road to what Virgil called the "awful madness." Indeed, getting sex is now thought to be wholly dependent on attracting attention and being attracted, of using flirtation to focus the attention of a specific member of the opposite sex. If our ancestors hadn't done it well enough, we wouldn't be around to discuss it.

Fortunately for the human species, it turns out that every come-hither look sent and every sidelong glance received are not only mutually understood signals of transcendent history and beguiling sophistication but also emblems of a built-in biological and psychological wisdom. We don't have to reinvent this with each generation in order to survive. As with every other aspect of sex, our brains and bodies have within them capacity to trigger the desire to not only partner but get us out looking for Mr. and Ms. Right, to hunt for the mate of our dreams and be receptive to that eureka! moment. In short, attraction and flirtation are nature's solution to the problem every creature faces in a world full of potential mates—how to choose the right one. We all need a partner who is not merely fertile but genetically different, as well as healthy enough to promise viable offspring, provide help in child rearing, and offer social compatibility and emotional bonds. Our animal and human ancestors had to have a means of quickly, safely, and reliably judging the value of potential mates without "going all the way" and risking pregnancy with every possible candidate they encountered. Later in a relationship, our rational higher brains may take center stage and inform the decision to love, honor, and obey. And undoubtedly the limbic system's connections to the thinking brain keep primal urges from going on permanent tilt (or in the case of Dick and Liz, prevented an X-rated scene on the restaurant floor). But for the first moments of attraction, bodies and senses are *meant* to fall hostage to the more ancient parts of the instinctual and impulsive brain. And what attraction and flirtation accomplish is a hardwired, relatively risk-free set of signals with which to sample the field, try out sexual wares, and exchange vital information about candidates' general health and reproductive fitness.

Like any other language, attraction and flirtation may be deployed in ways subtle or coarse, adolescent or suave. Nevertheless, they evolved just like pheasant spurs and lion manes: to advertise ourselves to the opposite sex. The next chapter describes in detail the biology of extended flirtation—what we often associate with dating. Here we'll deal with the nature of those instant "on" buttons experienced by Taylor and Burton, as well as mere mortals.

Psychologist William James was the first to propose that when the brain commands us to feel emotions about something, it acts out those emotions in the body, and that the changes in the body that result loop back to be represented in the brain and perceived by the conscious mind in some fashion. Not until two generations later was neuroscience and physiology able to fill in the "how." In his book *The Emotional Brain: The Mysterious Underpinnings of Emotional Life*, New York University brain biologist Joseph LeDoux makes the point with a description of a person walking across a field who panics when he gets a sudden sense of a snake or some wild animal nearby. Instantly, his body prepares to rush away, his breathing becomes heavy, his stomach is in a knot, his fists clenched, his throat locked. Even after he sees that what's in the grass is a stick or a boulder, it takes some time for his heart rate to drop, the sweating to stop, his voice to return. For LeDoux, this is compelling evidence that the neocortex and the limbic system operate, at least for brief periods, independent of each other. When our survival may depend on it, our bodies and brains react without even a neural nod to logic: stopping to "think" could mean extinction. If it turns out that the snake in the grass was a twig, no matter. Whether it's the threat of a predator or the survival of our DNA, we can't afford *not* to react. We have evolved the means of letting emotions kidnap our more rational minds before we even know what is happening. Similarly, when the matter is sex—surely another situation on which our survival depends—we also react without even a neural nod to thought. Instead, the sex appeal operating system appears to kick in without conscious consent.

LeDoux and Antonio Damasio, a University of Iowa neurologist (and author of two acclaimed books, *Descartes' Error: Emotion, Reason, and the Human Brain*, and *The Feeling of What Happens*), believe that our emotional systems evolved this way in part to keep us from endlessly rationalizing everything. If couples stopped at the moment they met to consider all of the possible outcomes of a relationship, they would both be old before they got to the parking lot. Damasio studied people with damage to the connection between their limbic brain and higher brain and found that they were smart and rational but unable to make certain emotion-laden decisions. They brought commitment phobia to a whole new level. So in a way, the moment of attraction mimics a kind of brain damage in that chemistry alters limbic brain–higher brain communications. What happens to a couple isn't an integrated rational event but a "gut" feeling of butterflies and giddiness and sweaty palms and flushed faces brought on by their emotional brains. We suspend intellect at least long enough to get ourselves to the next step in the mating game. Moreover, our arousal mechanisms also come to full alert. We are on the lookout. Like a preschooler suddenly "ready" to read through exposure to alphabet letters, language, pictures, and parental coaching, our minds and bodies are ready not only for the prospect of sex and love but also for the brand of sex and love that we instantly determine *only one* other is capable of giving. As Tom Stoppard wrote in *The Real Thing,* "I remember the first time I succumbed to the sensation that the universe was dispensable minus one lady."

The power of a Dick and Liz moment is as recognizable to a psychoneurobiologist as to a lovesick teenager in its ambiguous, irrational effects created by instant attraction and infatuation, morphing sights into sounds, feelings into colors, passion into pain, and love into violence. And the body and brain of any one of us even minimally lucky in love will follow the same biological rules of sex appeal and attraction.

First, our minds and bodies assure that the sex appeal cascade turns on very fast. The language of love and romance is spoken over weeks, months, and years, but the vocabulary of sex appeal must be instantly understood and answered. Proof is to be found among the distinctive— and potent—physical and biochemical markers of sex appeal, from color, odor, size, and strength to shape and shimmer. Male birds attain and use bright plumage that female birds watch intensively; female baboons develop bright red behinds that male baboons respond to hormonally. Frogs, insects, and birds sing, whirr, buzz, whine, and croak, and mammals use scent markers, in urine, feces, and skin. Whatever navigates a turtle three thousand miles through the sea from South America to the Ascension Islands to another turtle is no casual piece of flirtation but a highly conserved book of communication. For silkworm moths the mechanism is well known: they carry bombykol, a chemical attractant so powerful that 1.5 micrograms of it can attract a billion male moths. But a closely related canon of our biology is that signals develop reciprocally. That is, if one sex develops a trait, it's highly likely the other sex has coevolved some very special capacities to recognize and respond. This is particularly so if the trait in question has to do with attracting a mate or keeping one. Hence, the male moths are able to detect the odors with olfactory hairs whose proteins pull bombykol into the brain and use the signals to navigate long distances to get to the female.

Second, animals from simple to complex have evolved biological systems for attraction that are *separate* from the systems they use to actually mate. For people, in other words, attraction and sexual intercourse operate at very different levels of body and brain. What happens at the moment of enchantment isn't about courtship, copulation, or orgasm; that's later, in what the neuroscientists call "consummatory" behavior. Attraction, our instant "on" switch, is different. It puts us on alert. It brings us the desire for desire. The hormones, brain chemicals, and body parts we engage and recruit for copulation are often the same ones we recruit and engage for attraction, but they are used at different times, in different proportions, and for different purposes. Studies reveal that in infatuation, for instance, we make greater use of the body's natural amphetamines, its "uppers," while for committed love and parenting, when the excitement dampens, we rely more on oxytocin and endorphins to calm us down and give us contentment and joy in attachment. Over time, we can become "tolerant" of all that feel-good, calming chemistry, just as opiate addicts become inured to the effects of heroin. And it's then that we become primed for the so-called seven-year itch and go looking for thrills again. Our sex hormones, too, play into this scenario of attraction. Starting with the notion that one of the most important reasons mammals have estrogen receptors in their brains is to use estrogen from the ovaries to facilitate sexual behavior, Emily Rissman and her colleagues at the University of Virginia studied the phenomenon in genetically engineered mice. Rissman used mice whose estrogen receptor genes were bred out of them to see if being interested in copulation was the real turn-on for a male mouse, or if the capacity to attract and be attractive were a separate biological operation. First, she took out the ovaries—the main source of estrogen—in the "knockout" mice and some normal female mice. Then she gave carefully controlled doses of an estrogen compound to each and tested the females for their willingness to accept male mounting and copulation. Compared to normal mice, with their estrogen receptors intact, knockouts were in no mood to fool around, whereas the normal rodents were more than willing to have sex. But the telling result is that males' behavior toward the knockouts and toward the normal mice were the same. The guys found the ladies equally attractive and were interested in sex with both. The females without receptors had differences in their brains in areas that are linked to sexual activity, but, according to Rissman, "while estrogen receptors were required for sexual receptivity, they had little to do for female attractivity." There are two distinct systems at work, one for sex and one for sex appeal.

Third, what men and women look for, what attracts them, is *specific* to their sex and gender. If Darwin was right, then, females will be more interested than males in recruiting a mate's resources to care for offspring that will burden her for a long time with demands for nutrition and nurturing. Men, on the other hand, overall will be more interested in recruiting as many chances as possible to fertilize eggs. If men are programmed to get eggs and women are programmed to choose the best man to get at her eggs, then all the things we do and feel to get us attracted to another human being should somehow be connected to these basic strategies. Moreover, it follows that if men are biologically programmed to seek women who are fertile, men must "know" somehow that a woman is good mother material even if he can't tell whether she's menstruating or ovulating. Similarly, if women are programmed almost universally to have clear preferences for powerful, wealthy men, certain looks and actions must be just as readily figured out, and their underlying meanings understood, when a woman first sees a man. For us, as among Stone Age men and women, whose biological legacies we express and most resemble, beauty in a woman, and therefore attraction, tended to relate to her fertility; beauty in a man, and therefore attraction, tended to relate to his status in the community, his wealth, his power, and his ability to provide resources to her fertilized eggs.

The study by New York psychologist Glenn Hass that rated forty photos of opposite sex strangers (see chapter 3) offers dramatic evidence that the internal "programs" people have to play out "Some Enchanted Evening" are indeed matched up very closely to the differences between the two sexes in their reproductive strategies. Recall that the women in his study were more likely than men to rapidly equate physical attractiveness with "successfulness" while men were more likely to spontaneously process physical attractiveness of women in terms of their fertility and sexuality. Further, his test results showed, a woman's immediate sexual availability isn't part of the mix of what the men deemed "attractive."

What got their attention is what evolution wanted men to attend to: women's potential fertility. In those first moments, men are sizing up not sexual availability but the *likelihood* that women can come through childbirth and pass on good genes to mix with his. Being sexually receptive in those first moments of attraction is neither prologue nor proof of attraction.

And finally, if attraction and flirtation are a form of self-promotion, nature demands a certain amount of truth in advertising. "For a signaling system to convey something meaningful about a desirable attribute, there has to be some honesty," says Steven W. Gangestad, an evolutionary psychologist from the University of New Mexico in Albuquerque, "so that if you don't have the attribute you can't fake it." Just as the baboon can't fake its rosy rear, humans have sex appeal signals that can't be faked, either, such as the waist-hip ratios described by Singh in chapter 3. In simplest terms, says Gangestad, this ratio is an honest indicator of health and fertility. Studies have indeed shown that hourglass-shaped women are less likely than pear-shaped ones to get heart disease or diabetes. They are also most likely to be able to bear children well, as hips take their shape at puberty from the feminizing hormone estrogen. "The literature," he says, "shows that women with a 0.7 waist-hip ratio have a sex-typical hormone profile in the relationship of estrogen to testosterone and that women with a straighter torso have lower fertility. It appears that males have evolved to pay attention to this cue that ancestrally was related to fertility."

The virtually visceral responsiveness to physical features in flirtation may also be as good a guarantee as one can get that a potential partner shapes up on a hidden but crucial aspect of health: immunity to disease. Scientists know that the testosterone that gives men jutting jaws, prominent noses, and big brows, and, to a lesser extent, the estrogen that gives women soft features and curving hips, also suppress the ability to fight disease. But looks have their own logic, and bodies and faces that are exemplars of their gender signal that their bearer has biological power to spare; after all, he or she has survived despite the biochemical "handicap."

Return for a moment to the case of such elaborate male ornamentation as peacock tails and stag antlers. In the 1980s, as we've seen, evolutionary biologists William Hamilton and Marlena Zuk linked such features to inborn resistance to parasites. Antlers and tail feathers are known to be attractive to females of their species and are major machinery in the sex-appeal works. But developing and maintaining such extravagant equipment is costly, taking huge nutritional resources and even slowing the animals down, making them more vulnerable to predators. The only animals that can "afford" such ornamentation, it follows, are those with the best constitutions. Facial traits that we have come to prefer have evolved, according to this idea, just like pheasant spurs, to advertise fertility and genetic quality. In this way, sex hormones, in balance or not, honestly advertise genetic health; they mean these individuals have enough resources, genetic and otherwise, to sustain themselves even if they are "handicapped" by the immune suppression. It's worth noting that the value placed on physical attractiveness in the choice of a long-term mate in each sex correlates well with the prevalence of parasites across human societies. So, like big bones, big horns, big tails, and big spurs, jutting jaws are honest markers for a healthy immune system. So much so that they are respected by other men as well as attractive to women.

Studies of ancient and modern societies have verified the claims. They show that tall, square-jawed men, for instance, tend on average to achieve higher ranks in the military than do those with weak chins, and that taller men are overrepresented in boardrooms as well as bedrooms. On the female side of that equation, it may be that a woman's jealousy over other women is not likely to occur unless the other women are attractive.

What else makes up the science of attraction would have delighted Darwin. Psychophysiologist Nicholas Pound of McMaster University in Canada has evidence that our attraction reactions to couples and singles are part of our biological makeup, related to strategies that enable us to quickly read signs that say "I'm available" or "I'm not available." Does a woman hanging on a man's arm make him more or less appealing? Does a woman's venture into a party alone make her appealing to a man, or mark her in a man's mind as a "loser"?

Using Adobe Photoshop computer graphics, Pound developed an interactive photographic story, then constructed a five-minute point-and-click test about the story that he administered to one hundred undergraduate males in a darkened room. The story had twelve images of a fairly typical heterosexual couple, beginning when the boy, "Steve," meets the girl, "Lucy." Lucy and Steve have jobs, they travel, they fall in love, and they marry—a quick virtual relationship. Lucy liked holidays on the beach, however, and Steve wanted to fish and liked to drink with his buddies now and then. After establishing the scenario, the test then presented two scenes. In the first, Lucy is in a bar with Steve and another man. In the second they are in their bedroom with another man, "Dave," and suddenly Dave is holding Lucy.

Pound then asked the test takers some questions, not—as they probably expected—directly about fidelity or infidelity, but about how *attractive* Lucy was to these test takers. How sexy, how desirable. (Researchers can be tricky.) Said Pound, "I asked them, 'If the opportunity arose for *you* to be sexually involved with Lucy, how angry would you be if she refused sex with you now that you've seen these scenes. How desirable would she be as a wife or for a one-night stand?' "

It turns out that Lucy's instant desirability to these computer voyeurs went into the toilet among the test takers who saw her in the bedroom and bar, being coddled by Dave, but her appeal as a sex partner actually zoomed up for a short-term affair. The men were more likely to want sex with her than they imagined Steve would want sex with her after she sat on Dave's lap in the bar.

The single-guy test takers saw Lucy as much more attractive and sexually desirable in these cases, but they were less likely to want sex if they saw her in the bedroom with Dave. Those test takers who were dating were *more* likely to want Lucy, evidence that single guys are more ambivalent about a woman's infidelity than men in a relationship. On the one hand this might sound quite normal, but looked at another way it might seem odd. Why would a man care if a woman is attached or married if he can get her in bed, anyway? But while men might indeed like the *thought* that they might be the beneficiaries of a woman's infidelity, once *interested* in a woman, they don't much like the idea that she would sleep around—even with themselves! They want her exclusively to themselves. The classic double standard, yes, and its origin is classic biological doublethink.

For single test takers asked to give their personal perspective about Lucy, more were likely to want sex with Lucy, but would applaud Steve for being less likely to want sex with her.

There are other lessons from such experiments. One is that people's capacity for attraction, feeling it and responding to it, seems to be a reflection of far deeper reproductive strategies; at some level, equally mature people appear able to read these strategies and adjust their own to match the need.

Pound, for one, believes that this failure to exhibit a drop in sexual desire in the face of a woman's proven willingness to sleep around is a clear case of sperm competition and the mating race that occurs even at the first glance. He also concludes that the emotional responses linked to sexual interest are likely to be linked to hormonal changes that regulate the male's reproductive system and even alter the amount of his ejaculate compensation, and how soon after orgasm he can have another one. Such hypotheses are a stretch, but there is nothing illogical about them and studies are under way to try to support them. If indeed the test takers who were in dating relationships were consistently more likely to want sex with Lucy than the singles, it would appear that men who have had relationships in the past have a better sense of the fragility of such relationships and may indeed be psychologically and biologically primed to ignore the lessons of infidelity, their own or others.

Let's imagine a chance meeting between a man and woman in a café. The fellow arrives with another date, not the love of his life, but a good sex partner whose company he genuinely enjoys and whose great looks, sexy clothes, and obvious infatuation with him is the envy of his friends. Suddenly, he looks around and makes instant eye contact with a woman absolutely different than the blonde model on his arm. They smile at each other and he notices her empty glass. He finds himself compelled to leave his date, who is chatting with friends, and get to the bar, thinking only of getting the *stranger* a drink. He guesses red wine, gets his date a white, sort of pretending the red is for himself, and walks to the stranger. She declines the drink, but smiles at him again. Without thinking, and without a word, he makes an exaggerated bow and kisses her hand, then returns to his date, to whom he remains attentive enough. But he finds himself fantasizing about having sex with the stranger, and his eyes keep finding hers. When she leaves, he has the urge to follow her to the parking lot.

Meanwhile, for her part, the woman he can't seem to take his eyes from has come to the chic café alone, feeling rejected and lonely in her quest for Mr. Right. Her mood was so bad that she was about to leave when she caught *his* eye and suddenly felt gorgeous and desirable. His voice, heard only at a distance, sounded to her like Tom Brokaw and she kept looking for him in the crowd, finding his eyes so often that she shyly turned away after the third or fourth contact. His bearing reminded her of her dad, who had been an air force officer. She saw his beautiful date sending "daggers" to her periodically and found herself irrationally angry that he didn't immediately abandon his date and pursue her further after she rejected his drink.

According to Pound's research, the man in the scene was playing out a universal response to sex appeal. The date on his arm made him even more likely to want sex with a sexy stranger. Moreover, in the first moments of mutual attraction, the face and figure, the overall gestalt, the impact of the whole picture are what get the juices going. Who in particular we choose to actually mate with ultimately involves much more than just attraction (remember the neurochemistry of attraction is separate from that of choosing a mate). There are biological and emotional battles and conflicts, duels and contracts to be worked out. But that first attraction between two individuals depends on "superficialities," and our capacity to respond to them is as hardwired as the circuitry that carries our hormones.

When our imaginary couple first met, he could have been an ax murderer and she a thief. He could have been already married and she engaged. He might've been a con artist and she a loser. But for that moment, that instant, their psychologies and physiologies were in instant "synch" with and for each other.

It follows that sex appeal must be a sort of shorthand for what will become more apparent, appealing, and important as a relationship moves to more durable or more permanent levels. Whatever it is that attracts us to someone, and makes us attractive to someone, is the "halo" we apply to whatever else comes later.

Nowhere, of course, do nature's rules say both members of a pair must be equally or simultaneously attracted to trigger desire and mating. Nor is the halo always destined to stay shiny. If it's too tarnished by realities that aren't apparent in the first moments of attraction, we move on, albeit often with hurt feelings, depression, or sorrow. But whether it's a teenager swooning at Tom Cruise, a nine-year-old "in love" with his teacher, or your best friend falling for someone without morals, our sex appeal gear makes us *willing* to feel the pull. Attraction happens, and we keep letting it happen, even when our experiences demonstrate it can't always lead to "happily ever after." As Helen Fisher reminds us, "attraction has a dark side." A fourth of American homicides are perpetrated by the victim's spouse, sexual partner, or sexual rival; more than half of all college women report harassment from a rejected lover; and suicidal depressions are common among those whose attractions are not met with equal passion. Yet sex appeal, and opportunities to attract the opposite sex, are as sought after as they ever were.

Of course people can and do fall in love, mate, produce children, and stay married without attraction. For centuries, arranged marriages were all the rage. The species survived, presumably without many enchanted evenings. But even in societies where marriages were (and still are) arranged, a bit of careful observation (and lately the advent of DNA paternity testing) attests to the fact that teenagers and adults become attracted to each other, sneak off together, and somehow succeed in influencing their parents' choices of mates for them. The ubiquitous popularity of paperback romances built around the theme of the mean patriarch's rebellious daughter escaping the arranged marriage to a vicious, rich scoundrel and finding love in the arms of a poor but gorgeous knight who never knew what hit him is no surprise to sex-savvy publishers.

If it's true that nature and nurture have firmly conserved our ability and our instinctive *need* to light up someone's eyes, then each of us must come equipped with the means to compose an advertisement we hope will draw the attention of a potential buyer of our genes. And the "copy" must be written in ways that we can read it fast, in our emotional brains and in our sensory systems, where we perceive feelings, smell, taste, touch, sight, and sound. If the point of sex ultimately is to reproduce our DNA, and the joy of sex is in the arousal and the act itself, then the role of attraction is to get us going in the right direction and keeps us always on the lookout for the right sexual partner. The most reproductively successful of us (evolution didn't count on artificial contraception) would be attraction "junkies," always yearning for knights in shining armor and sex kittens, for the perfect 10 in our mind's eye.

Humans, and other animals, for instance, have a propensity for being attracted to people who are not relatives and not exactly strangers, either, individuals who resemble or remind us in some fashion of those we already know and trust. It's not an accident that he wants a girl just like the girl who married dear old Dad. (Or, if Dad was a painter, she finds the bohemian lifestyle very appealing in other men.) The resemblance needn't be physical, but reminiscent of whatever it is about other people that makes us comfortable: a certain posture, look, or even scent. Parents, the initial and most important love objects of our lives, are good models. And modern psychobiologists, contrary to psychoanalytic theory, say attraction to parent-like traits is not evidence of emotional immaturity or pathology but evidence that who we're attracted to is a complicated mix of what we are born with and what we learn. It's both logical and practical that women are often attracted to traits that evoke Dad—personality traits especially such as warmth or authoritarian bearing. And men may subconsciously be moved to be attracted to women who laugh or walk like Mom. In these affinities, we are genetically and biologically programmed, in the same way that females are attracted sexually to males and males to females.

Nature plays its role in all this by giving us brains and bodies that learn and retain memories of particular traits, actions (such as smoking or eating chocolate cream puffs), even music that made us feel good or sexy or wanted or loved. Parents play a role, too, with their ability to train children, impose a desire for certain traits on their offspring, and literally teach them whom—and what—to find attractive. Such desires for particular traits, moreover, may be passed on in families for generations almost as if they were inherited genetically. "There is more to heredity than genes," says Dr. Eva Jablonka of Tel Aviv University. She and her colleagues point for evidence to the fact that human (and animal) mothers will care for nonbiological offspring, in which there is no incentive with respect to passing on one's own DNA. They do this because it gives them a way to stamp their own preferences and personality on the foster kids and impress on them the desire to mate with women like them (for their foster sons) or men they like (for their foster daughters.) They can also become matchmakers for safe biological matches with their natural children (the foster kids aren't biologically hers, so there's no inbreeding problem), giving the woman a means to influence her biological children's destinies by influencing their choice of mates. A foster mom gets to promote the perpetuation of traits she likes the best—her own—and to promote her choices of partners for her offspring as well. Animals learn a lot from their parents, from whom they should or should not mate with to which hunting grounds are likely to feed them best.

Anthropologist Helen Fisher's research has affirmed that attraction is a universal mating emotion and uncovered some of the physiological and psychological roots of lovemaps. To an astonishing degree, Fisher reports, people in the throes of attraction and infatuation focus on a love object's good points, even though when pressed almost every one of them can cite chapter and verse about what they don't like about the person that they're infatuated with. In addition, they display a widespread pattern of physiological symptoms including exhilaration, euphoria, buoyancy, sleeplessness, loss of appetite, shyness, awkwardness, trembling, flushing, stammering, butterflies in the stomach, sweating palms, weak knees, dilated pupils, dizziness, a pounding heart, and accelerated breathing. Overall, she believes, the principal reason for the evolution of the attraction system was probably to focus attention on a specific individual to the exclusion of others.

Using a questionnaire as a screening instrument to find out which of her subjects were in a state of attraction at the time of the study, Fisher and her colleagues have now begun to put subjects into a functional MRI machine to see which areas of the brain and which brain chemicals are activated in such a state. She has preliminary evidence that some of the brain circuits associated with attraction may be in those areas rich in receptors for serotonin and dopamine, the brain's punishment and reward molecules.

Considerably more controversial, some studies by psychologist and molecular biologist Dean Hamer of the National Cancer Institute have tracked these neurotransmitters, and perhaps some of the genes that make them, to certain complex behaviors, some of which are crucial to attraction and sexual interest. Among only a handful of scientists attempting to figure out what biologically makes us do the things we do, Hamer has conducted studies on twins, families, and the human genome. All have led him to conclude that between a third and three-fourths of all the differences in personality traits and manifest behavior are inherited. Most of the scientific community remains unconvinced of so broad a thesis without considerably more evidence. Nevertheless, his peers have found some of his work credible as well as provocative.

With Dennis Murphy and Klaus-Peter Lesch of Germany, for example, Hamer investigated the regulation of the serotonin transporter gene and found that it comes in two major forms, long and short. In a landmark paper in the journal *Science* in 1996, Hamer and others reported that the short variety reduces the production efficiency of a promoter of the transporter gene, resulting in decreased transporter gene expression and serotonin uptake in cells. As Hamer described it, this gene is a "key player" in *how* serotonin is delivered to brain cells. The protein it makes hooks onto free serotonin between neurons, literally binding it and locking it out of the action in the cell.

Hamer believes that his work is making it clearer and clearer that some people really are, from infancy, more—or less—adventurous than others; that thrill seekers' basal thrill rate is set higher than usual. Results of several different personality tests, combined with DNA testing of the transporter genes, have compellingly demonstrated that people with the long form are different from those with the short form on every one of the mood traits measured. If Hamer could investigate the genomes of our imaginary couple, or anyone looking for sexual or other thrills, it's a good bet he'd find the form of the dopamine transporter gene associated with novelty-seeking behavior. "We just know of people," says Hamer, "who love avant garde art, paradoxical jokes, hippies, TM, drug use, anything for a thrill. These people are far more likely, as well, to be sexually pro-miscuous and engage in risky sex, kinky sex." Why would nature have favored and conserved such personalities? From an evolutionary stand-point, Hamer says, it isn't too hard to imagine why there might be genetic conserving of a gene that promotes risk taking and novelty seeking. Such a trait would have helped animals, including early hominids, take chances in hunting dangerous game, or venturing into unknown territory to search for food or shelter.

"The obvious evolutionary benefit is that those with the inclination for adventure would seek more sexual partners over their lifetimes and such behavior, from a reproductive fitness point of view, would have given advantage overall for the production of children," says Hamer. Moreover, such a trait may have given an especially good edge to females. While it seems clear that men's testosterone and brain chemistry make sexual promiscuity a more dominant trait in them than in women, women must have the temperaments to turn on to such male promiscuity or else these men with all of their sexual thrill seeking would have nowhere, literally, to put it.

In women, the novelty seeking may be just as strong, but less apparent in the beginning stages of a sexual relationship. One reason may be that for women, due to her Stone Age biological need to assure resources for her offspring, a man's sex appeal must eventually involve his willingness to both go beyond attraction through a courtship and stick around after sex. These needs give her a powerful incentive to be a little less willing to seek sexual thrills right away.

"It may seem at first blush contradictory to say that if you're more happy and less neurotic, you have less sex, but that's the way it is," says Hamer. "You can thank your serotonin. Perhaps when you're anxious and worried about getting partners, you work harder and look for it more. When you are happier and content and calm, you aren't as interested. So, relatively speaking, you can be neurotic and sexy or happy and less inter-ested in a new exposure sex appeal."

A further paradox of sorts, says Hamer, is that a person with higher brain serotonin levels will have lower anxiety and a lot of self-confidence and self-esteem, seem more vivacious and outgoing, more carefree and alive and downright happier, while also having relatively more inhibited sexual appetites.

As an evolutionary biologist might see it, those males and females who wandered out of the home cave a few million years ago to search for new infatuations spread around more of their genes than those who stuck around forever, with the consequences that we favor brain chemistries that let us ride the roller coaster of behaviors we experience when we glance across a crowded room and fall madly in love in that instant. Even if we later realize we fell for a toad, we become willing to try it again. And we seem to have brain and body chemistries that give us the huge amounts of energy we need to keep at it. Energy and stress, after all, are markers for attraction in every species. We literally get frantic when sexual mood is upon us. Adrenaline and other stress hormones like cortisol, DHEA, LHRH (which triggers testosterone production), and phenylethylamine, or PEA, a potent stimulant chemical, are all activated during attraction and infatuation.

Although there is no proof that PEA directly plays a role in human attraction (it may be that it is dopamine receptors that are affected by PEA), what evidence there is comes from scientists who have studied the pathologically infatuated who flit from one hot romance to the next, in a sort of attraction sickness. In a famous experiment, New York State Psychiatric Institute doctors Michael Liebowitz and Donald Klein worked with a group of people whom they called "attraction junkies." These individuals are exhilarated with each new encounter and thrown into despair as soon as they are rejected, only to try again. Suspecting that their behavior was driven by their brains' abnormal desire for a PEA "high," they gave their subjects antidepressants that block monamine oxidase, an enzyme that breaks down PEA, norepinephrine, dopamine, and serotonin. The net effect was to *increase* the amount of PEA the limbic brain had around, reducing the subjects' need to find a way to turn up the PEA stream indirectly with endless romantic encounters. The subjects calmed down and began to choose partners more rationally. Studies by other psychiatrists have found that PEA levels are high in the urine of people who report they are in love and low in those going through separation and divorce. MAO (monoamine oxidase), sometimes dubbed the "risk taking" chemical, is also reported to be at lower basal rates in women than men, suggesting one chemical reason why more men than women are always on the sexual prowl, looking for a new adventure. In another study, psychiatrists found that levels of PEA metabolites in urine were higher in couples in love, and lower in a couple going through a divorce. Mice injected with PEA jump and flip out, and monkeys who get PEA in their food have been known to go into overdrive, pressing levers in labs to get more—a sign of addiction to the brain's own infatuation chemical.

If the emotional and chemically primed brain is the ammunition of sexual attraction, the trigger is what the mind and brain can quickly "take in"—physical shape, size, color, odor, and overall appearance. Somehow, our minds and bodies "read" the obvious to decode the subtle. We may *think* attraction semaphores are all about "sex now!" but subconsciously and evolutionarily, they are signals of long-term reproductive interests and status.

Predictably, the signals are as varied as nature: Plumage color, the size of some physical features, vocalizations, and smells dominate the animal world. Scientists at Lund University in Sweden studying pheasants mounted radio transmitters on adult birds and tracked matings, broods, and which cocks the hens went for. They found that hens preferred males with long spurs, and they seemed to care little for body mass, wing length, "social position," wealth, or territory, all traits that the investigators were privy to. But to further test what the real attraction was, the investigators randomly selected males and artificially changed the spur length with prostheses—artificial spurs. Males with these enhancements did indeed get the girls more often; but the ladies were more savvy than superficial, it turned out. Males with longer spurs lived longer and had more offspring than short-spurred ones. Spur length, like all good attraction signals, reflects genetic quality, and females were unwittingly using this trait to choose fathers with good genetic traits. In line with work from Hamilton and Zuk, long-spurred males clearly carried genes that made them more fit than males with shorter spurs.

In humans, the face, the physique, the beauty, the buffed body, the "look" that signals youth, fertility, and health all take the place of spurs. Beauty may be in the eye of the beholder, but the beholder's vision is sharply and quickly focused by millions of years of "preferences" that signal reproductive fitness. Just as every human being has automatic systems for identifying a threat or a potential sex partner, and for doing so emotionally and instantly, so we have them for identifying fertility potential. If we see that snake in the grass, we don't "think" about the threat, we react, instantly, by leaping out of harm's way, even if that "snake" turns out to be a stick. Whenever we see a face, or a body shape, the amygdala, in concert with other parts of the brain, sizes it up and prompts us to respond. If the face we see is smiling and lit up, we respond one way; if it's scrunched and scared, another. Neuroscientists learned this in part from observations of those with damaged amygdalas, who are as likely to laugh as to cry over a tragedy. Psychosurgeries of decades ago deliberately damaged amygdalas in order to attenuate a lot of emotions, and in the process flattened responses to danger signals, threats, and erotic stimulation.

Still other evidence has come from those who have tracked facial recognition systems to certain parts of the brain and demonstrated how carefully nature built us to be attracted to some faces and not others. In a study aptly titled "Here's Not Looking at You, Kid," neuropsychologists at the Veterans Administration Medical Center in West Haven, Connecticut, found that when you are looking directly into the face of another person, an area in the visual part of the brain is activated more when the person is actually looking away from you rather than toward you. They conducted the study by placing thirty-two electrodes on the scalps of fourteen college students to record the electrical activity of the brain, while the subject sat in a quiet recording chamber and looked at images of faces presented by a video monitor controlled by a computer. Sometimes they were facing the viewer and had eyes looking directly at the viewer, and sometimes they were facing elsewhere. The scientists concluded that there is a very specialized part of our brain's visual system that deals mainly with detecting eyes and that activates even more when the eyes are turned *away* from our direct view. Moreover, the two sides of the brain react differently, with the left side responding more when the whole head of the person you are looking at is turned away, whereas the right side responds more when only the eyes are averted.

Why would our brains be interested so specifically in the position of another person's head or eyes? Monkey studies offer a clue. Again using tiny electrodes, this time planted directly into the monkey's brain to record activity of individual brain cells, investigators found that some cells in the visual portion of the brain respond best when the monkey looks at a front view of human or monkey faces, whereas other cells respond best when the eyes or head are looking away. Clearly something about eye and head position is important from an evolutionary standpoint, since that something is retained in humans and our closest biological relatives and therefore isn't simply a random or "modern" phenomenon. Integrating these findings with what is known about monkey behavior, researchers have concluded that both monkey and human visual systems are monitoring eyes and faces closely because the ability to "read" eyes and faces bears directly on their survival. Monkeys, for instance, regard a direct stare as threatening and monkey leaders—the alpha males and females—will attack or punish subordinates who try it. Similarly, they regard lowered or averted head and eyes as evidence of submission.

Because survival is so closely linked to sex, and sex to attraction, it's likely that at least part of the reason faces play so big a role in sexual attraction is because those animals best able to read faces get a reproductive advantage. A woman who "shyly" averts her gaze when she flirts probably regards—at some deep level of her brain chemistry—a stare as unpleasant or threatening, especially if the other person is a stranger. Humans, like monkeys, will avert or tilt their heads or eyes as a sign of friendliness or submission. (A bow, say anthropologists, is a ritualized form of the same social signal, and making a deep bow over a woman's hand and kissing the palm is considered erotic.)

The part of the brain that pays attention to eyes, faces, tilts, and direct or indirect gazes may not be so much involved in determining who a person is (for example, John) as where that person is directing his attention (John is *not* looking at me). There's little mystery here about why such information may be important in attraction and in determining the intentions of a member of the opposite sex.

In a related study, Valerie E. Stone of the University of California at Davis noted that an animal able to anticipate the behavior of another member of its species by reading facial expressions and translating them into markers for certain emotions has a significant adaptive advantage. Having a poker face or an emotionless, flat affect is not the stuff of successful lovers; on the other hand, people who learn to "lie with their eyes," to smile when they are not pleased, can be Lotharios you would rather not deal with. Since there is no evidence that the most recently evolved part of the human brain, the neocortex, has any specialized areas for processing perceptions of facial expressions, this ability to make instant judgments about people's most innermost emotional and physiological foundations seems rooted in the older, more primitive and instinctive parts of our brain.

If indeed attraction is chemical, if we are in fact "set up" by evolution to react with thunderous emotional and physical chemistry to particular kinds of faces and bodies, it also appears to be true that our brains selectively remember looks to reinforce our desire for both *particular* faces and bodies and *ideal* faces and bodies. Twenty-five men in the village all fixated on just one female might be the stuff of female fantasy, but it's no way to perpetuate a species. Thus, we need an attraction system that embraces variation, that increases the probability that there's more or less "somebody for everybody" but still has enough universality to keep us interested generally in the opposite sex.

As it turns out, nature has figured this one out, too. For whatever the particulars of the face or body that turn on your attraction apparatus or mine, a common denominator appears in all of them: symmetry, the quality of evenness and balance. Gangestad and Randy Thornhill, a biologist at the University of New Mexico who has pioneered the study of symmetry, believe it is a marker of "developmental precision," the extent to which a genetic blueprint is realized in the flesh despite all the environmental or cultural perturbations that can throw development off course.

Symmetry is why Elizabeth Taylor, Denzel Washington, and Nefertiti are universally recognized as full of sex appeal, even though individually they look nothing alike, and why (to the sexually ordinary) cripples, giants, dwarves, and crones are not. Although men may crudely joke that with a bag over her head, *any* woman is sexy, and although women may indeed fall in love with ugly men, both sexes value facial symmetry and, if only subconsciously, equate an "even" face with sexual health and desirability. Why else would children as young as two months show distinct preference for an even face over an uneven one, demonstrating a biologically built-in preference for symmetry before they can tell Mom from a monkey?

"Bilateral symmetry is a hot topic these days," says Thornhill. It's hot because increasingly, geneticists and neuroscientists have been able to track *asymmetry* to developmental instability, to deviations in the biological blueprint animals are born with, and ultimately to their reproductive fitness and sexual prowess.

Developmental instability is measured in units of "fluctuating asymmetry," or FA, small, random deviations from a balanced face and body. Because genes (DNA) translate into "phenes" (expression or external shapes), biologists believe that fluctuating asymmetry may be the most useful "quick and dirty" measure people have of a person's underlying health, strength, and sexual fitness.

Recent studies conducted by Gangestad and Thornhill demonstrate not only that women prefer symmetrical men, they prefer them at a very specific time—when they are most fertile. "We found that female preferences change across the menstrual cycle," Gangestad reports. "We think the finding says something about the way female mate preferences are designed. Because the preference for male symmetry is specific to the time of ovulation, when women are most likely to conceive, we think women are choosing a mate who is going to provide better genes for healthy babies. It's an indirect benefit, rather than a direct or material benefit to the female herself." Symmetry also appears to be a marker for biochemical sensors made manifest in our ability to detect biological cues that may not be consciously detectable.

In their study, fifty-two women rated the attractiveness of forty-one men—by their smell. Each of the men slept in one T-shirt for two nights, after which the women were given a whiff of it. Prior to the smell test, however, all the men had undergone careful calipered measurement of ten features, from ear width to finger length. Those whose body features were the most symmetrical were the ones whose smells were most preferred, but *only* among women who were in the ovulatory phase of their menstrual cycles. At other times in their cycles, women had no preference for either symmetrical or asymmetrical males. What's linking symmetry and smell is probably not the *attractiveness* of the smell per se (his symmetrical "handsomeness" has taken care of the initial attraction already) but the women's subconscious use of the smell as a further marker, or reinforcer, of health and fertility.

If a preference for symmetry only occurred in humans, that would be interesting but not especially important in the story of sex. Of course, that is not the case. Thornhill first stumbled upon symmetry two decades ago, during experiments with scorpion flies in Australia, Japan, and Europe. He noted that females chose particular male flies on the basis of the level and quality of so-called nuptial gifts, nutrients passed to the female during courtship and mating. (In some insect species, in fact, the male itself is the edible gift, submitting to death by female consumption.) "That was the first inkling I had that insects were very sophisticated about their mating strategies," Thornhill recalls. But the more time he spent recording the sexual lives of scorpion flies, the more he realized that the females were selecting partners long before they sampled any gifts, and they were reckoning by the symmetry of the males' wings. "I discovered that males and females with the most symmetrical wings had the most mating success and that by using wing symmetry, I and presumably the fly could predict reproductive fitness better than scent or any other factor." With another colleague, Anders Pape Moller of the Zoological Institute in Denmark and the Marie and Pierre Curie University in Paris, Thornhill analyzed fifty-two studies among thirty-six species, including insects, fish, birds, and mammals, and found that male mating success and female mate choice is highly dependent on how symmetrical is the face and body of a potential mate. Thornhill and Gangestad have now conducted dozens of separate tests of human symmetry, in everything from eyes, ears, and nostrils to limbs, wrists, and fingers. Even if they never speak a word or get closer than a photograph, women view symmetrical men as more dominant, more powerful, richer, and better sex and marriage material. Moreover, symmetrical men view themselves the same way! Men, for their part, rate symmetrical women as more fertile, more attractive, healthier, and better sex and marriage material, too, just as such women see themselves as having a competitive edge in the mating sweepstakes.

Although reading these attributes seems an enormously easy task for people, it is a complex one for scientists. Thornhill and yet another colleague, Karl Grammer, developed a technique to measure precisely fluctuating asymmetry that calculates the distance of lines connecting corresponding points on the two sides of the face and body; they then calculate the midpoint of each line. "If you look at the midpoints of the lines between the outer edges of eyes, between the ears, the sides of the jaw and nose, and so on, you can see how symmetrical, or not, a face is. In a perfectly symmetrical face, all of the midpoints are on the same vertical line," Grammer says. People whose eyes or nose edges are on different planes demonstrate higher levels of FA. Similar measurements of body symmetry assessed seven to eleven traits, including ear breadth, ear length, elbow breadth, wrist breadth, ankle breadth and foot breadth, and distance between fingers. (The investigators needed computers for all of this work; all we humans seem to need is a quick look.)

It's a common observation and obvious assumption that good-looking people get more dates, more mates, and more attention from the opposite sex. It doesn't take a scientist to tell us this. But why should this be true? What is it, exactly, that is considered "handsome" or "beautiful"? It turns out that not only are attraction and flirtation most successful among the most symmetrical, but so also is actual mating success. Further studies on four hundred men by Thornhill and Gangestad have found that men's bodily symmetry matches up with the number of lifetime sex partners they report having. Symmetrical men also engage in more infidelity in their romantic relations—"extra pair copulations" in the language of the lab. "In men, symmetry significantly predicts the men's number of 'extra pair copulation' partners who *themselves* are stepping out on their mates. Our results show that it's the symmetrical guys that are pulling women from committed relationships like marriage and thus symmetric men may be cuckolding the rest of us out there," says Thornhill. "Oh well," he adds.

It gets worse. These same studies show that symmetrical men also get to sex more quickly after meeting a romantic partner, and they tend to lose their virginity earlier in life, too. And for the coup de grace, they also satisfy women more than asymmetrical men. Gangestad and Thornhill surveyed eighty-six couples in 1995, for example, and found that symmetrical men "fire off more female copulatory orgasms than asymmetrical men." In fact, women with symmetrical partners were more than twice as likely to climax during intercourse.

Thrills are only a short-term payoff, however, since female orgasm also is potentially a shill, more or less for female fertilization, pulling sperm from the vagina into the cervix. (See chapter 9). Robyn Baker and Thornhill feel this finding is a significant part of our biological capacity for making "cryptic" choices of sexually fit partners. Why? Because orgasm in women also is associated with the release of oxytocin, the social bonding hormone, so it appears that female orgasm is a cryptic or hidden "sire choice" mechanism. A man who is symmetrical (and presumably more sexually appealing to a woman) is likely to be more successful at bringing a woman to orgasm, thus increasing the chances that his sperm will be retained and will fertilize her eggs. Such a man is also more likely, thanks in part to the release of oxytocin, to be "loved" by a woman, long after the first attraction and fulfillment of sexual desire fades. That's good for his offspring—and hers. So while it's true that symmetrical men play around more and are more promiscuous, this doesn't seem to bother women from an evolutionary perspective; they still want the symmetrical male because it signals honest health and an ability to fertilize eggs.

In biology, as we've seen, nothing is ever simple. As successful as symmetrical men are at attraction and flirtation, the same studies that persuade us this is so show that women definitely do *not* prefer symmetrical men for long-term relationships even though they may continue to find them sexually desirable. That's because, once again from an evolutionary perspective, there's a definite downside to getting someone with really good DNA into your bed. Why?

Because symmetry, Gangestad explains, affords those men who possess it to take a dastardly mating strategy. His studies have shown that symmetrical men invest less in any one romantic relationship—less time, less attention, and less money. And less fidelity. They're too busy spreading around their symmetry. They also tend to treat women more as mere sex objects than as individuals, he reports. "It may be that males who can have the most access without giving a lot of investment take advantage of that." The men who will stick around and help out with parenting is on most women's wish list of qualities in a mate right at or near the top. Concedes Gangestad, "I wouldn't exclude the possibility that men have been doing some direct parental care for some time, and so a preference for that might also have an evolutionary basis." But also on a woman's wish list from an evolutionary standpoint would be someone who is going to provide good genes for healthy babies. For good or ill, the Albuquerque researcher says, "what can and does happen in a mating market is that those things don't all come in the same package."

All told, Gangestad and Thornhill have now collected FA data on more than a thousand young adults, including more than two hundred who were in a romantic relationship. And they've even conducted studies on how much "lying" goes on in these surveys to correct for the possibility (or the likelihood) that men and women would exaggerate or minimize the number of sex partners they've had. "We discovered that for women symmetry does not predict lying at all, so asymmetric women lie about the same amount as symmetrical women, which means whatever results we get on the self-reports, lies cancel themselves out."

However, it turns out that symmetrical men lie *more* about their intended involvement in the relationship and they lie more about their faithfulness in relationships. But men's lying does not correlate with self-reported sex partner numbers. "So it doesn't look like men are lying in a way that would influence any links between symmetry and sex partner number. In short, we don't think that lying is a factor in either sex in terms of symmetry."

In their studies of women, Thornhill and Gangestad have also learned that unlike symmetry in men, symmetry in women is not significantly related to sex partner number, infidelity, how fast they get to sex after meeting a romantic partner, their investment in a relationship, or their orgasm frequency. In yet another affirmation that evolutionary biology is behind our attractions, it appears that women's face and body symmetry are closely related to how attractive they and men think they are.

These results tie nicely back to that initial moment of attraction, when a man and a woman size each other up. Thornhill, Gangestad, and other experts on symmetry also have evidence that men's physical attractiveness may signal ability in physical protection. Symmetrical men think of themselves as more protective of their mates, and more able to protect them. Their romantic partners claim that, too. So being handsome, and physically attractive overall, may signal physical protection potential in a man.

If physical protection is what women—or our female ancestors, at least—want in a man, such a connection would add more support for the notion that symmetry is an instant signal not only for "handsome" but for "great mate material." And so it seems to be. In one experiment, Thornhill and Gangestad explored this theme in 203 romantically involved couples. They obtained self-reports and partner reports of men's physicality—their muscularity, robustness, and vigor—their social dominance and resource potential, along with reports of the men's body mass, narcissism, sexual assertiveness, ratings of facial attractiveness, number of sex partners, and fluctuating asymmetry.

FA predicted not only how physically handsome and fit the men were but also their ratings of social dominance, vigor, resource potential, and capacity to protect their women. In fact, these variables accounted for 70 percent of the relationship between FA and the number of sex partners. And, as we've seen, although more symmetrical men give less time and attention to a partner—they're too busy exploiting and spreading around all that symmetry—"they are judged by women to provide better physical protection."

Putting all of this together, what is it that women really want in a mate? As anthropologist Donald Symons of the University of California at Santa Barbara has noted, beards can disguise a weak chin and moustaches a thin lip, just as bleaching creams can hide a woman's overproduction of testosterone. Hundred-dollar haircuts and lipstick can promote the small jaws and lush lips that evolutionary pressures have made men favor in women. But in our evolutionary past, what passed for "natural beauty," the "real thing," was whatever sent a message that the bearer of a trait was young, strong, healthy, fertile, and available. And those best able to assess those traits were better able to get their DNA to survive than were their contemporaries who were more cerebral. We have a mutual passion for symmetry—and for looking for it—because it's a logical tautology. We enjoy reading a lot into each other's appearance because one glance tells us instantly rewarding things about the health, wealth, fertility, and orgasmic skill of the one we're ogling.

Thus, men who prized beauty (read, "fertile") over brawn in a woman won out over men who prized women's brains in the absence of beauty. Men who prized women who preferred "sensitivity" and protectiveness to nonstop aggressiveness won out in the long term over men who prized women who preferred aggression. What women want are symmetrical men and men who exploit that advantage, but also men with no "extremes" of height or weight, or overt behavior—men with prominent cheeks, angular jaws, broad foreheads, prominent brows, and strong chins, but only slightly above-average upper-body muscles, a waist-to-hip ratio of 0.9, and bony, symmetrical wrists and ankles.

An amusing contemporary affirmation of the answers comes from even a cursory analysis of romance novels, which are written by women for women. When describing men, they use "muscles" or "muscled" 39 percent of the time, and "hard muscled" 21 percent. They use various permutations of "broad-chested," "hairy," and "broad-shouldered." They describe "bunched" muscles, flat abdomens, fine musculature, and hard nipples. They draw word pictures of "hair roughened skin," virile chests, hard contours, hard rippled muscles, tight buttocks, lean muscular legs, rippling flesh, rippling pectorals, and on and on. Rarely do the writers dwell on the men's faces, clothes, voices, emotions.

Turning the same evolutionary camera on the opposite sex, what does a man really want in a mate? Answer: symmetrical women of average height, with large eyes, small nose, full lips, delicate jaw, small chin, firm symmetrical breasts, smooth, pale hairless skin, and a waist-to-hip ratio of 0.6 to 0.7. In sum, Elizabeth Taylor.

As Devendra Singh's studies showed us in chapter 4, "the silhouette of the female body is even more of a giveaway of fertility and sexual health than the face. A wider pelvis is an honest signal of ease of childbirth." Moreover, he adds, while you have to get very close to see the details of a woman's face, you can see the shape of her figure from quite a distance, in dim light and even on the run, signals that would have been of distinct advantage at times in evolutionary history when stability and time to linger in the singles cave were rare commodities.

In subsequent studies of his classic work with Barbie dolls, Singh took his set of twenty-six fat and thin plastic bimbos, varying them in every possible way—with and without Barbie-cute faces—to men of all ages, races, social and economic ranks, cultures and creeds. Just looking at the dolls, the male subjects assumed that the Barbies with bigger, thicker waists were *older*, which pretty much matches the fact that most women, by menopause, store fat more readily in their middles. A pretty face was enough to overcome the thickened waist–older age label as long as the waist was not bulging. Once again, what men see when they look at an hourglass shape is a clear glass view of her sexual biology and brain at work.

Singh's devotion to Barbie doll research has also lately affirmed the New Mexico research team's conclusions about symmetry and infidelity. Among American men, at least, those who liked the typical Barbie waist-hip ratio also "decided" that a real-life doll with such proportions was likely to be sexually unfaithful. "They voluntarily attributed this infidelity to Barbie dolls," says Singh with head-shaking disbelief. In fact, he adds, men ascribe all kinds of emotional and psychosexual personalities to Barbie-like women if given half a chance. He and his colleagues made up scenarios about their Barbies, including marriages to Ken dolls and affairs with other male dolls. The male subjects who were told these scenarios said that if Ken were having an affair, the most curvy Barbies wouldn't forgive him. The male perception is that if Barbie is attractive, he wants to be her mate, but he understands that if he fools around, she won't forgive him. However, men did not perceive the Barbies with more masculine figures to have these same attributes at all.

Women in developed countries these days are far more likely than women in less developed countries, or in past generations, to favor Barbie's proportions; the skinnier the better. From the ultra-thin "social X rays" popularized in the fiction of Tom Wolfe, to pronouncements about the glories of being thin and rich, thin is in. And Singh's Barbie doll studies also shed some light on a possible reason.

Singh proposes that men's preferences for women's bodies, particularly body weights, may be one mechanism women can manipulate and exploit to space children and avoid pregnancy. For instance, where women lack economic power, marriage is heavily favored by women as a means of getting resources for herself and her future DNA. Being curvaceous helps in this quest. In cases where women achieve economic independence, or have opportunities to do so, women favor slenderer appearances as an advertisement of their own wealth and power, and their power to decide for themselves when they want to bear children. They can bring home the bacon, fry it up, and feed their own offspring if they have to.

Historical changes in "fat" and "thin" as aspects of sex appeal may be commonplace for another reason. Arthur Campfield, an expert on the genetics of obesity, starvation, and reproduction, says that "throughout evolutionary time, the real threat was never overweight but starvation, so animals, including humans, developed a genetic makeup that was conserved not for weight loss but for the ability to accurately assess low energy. For example, a gene for a protein called leptin, when activated, destroys the animals' ability to "know" that it has to get food immediately or it will starve. That news potentially makes the weight-loss industry dance for joy—a product that could turn off the desire for food, although early studies based on leptin have so far proven only that overcoming obesity will be even more complicated.

In any case, the diet industry is not why we have leptin genes. Food, energy, and nutritional status, says Campfield, are absolute requirements of the reproductive system; humans would have evolved the means of knowing when they were at risk of losing reproductive fitness because of diminished nutritional status, as well as noticing very clearly who is reproductively healthy and who is not, just by looking at the body of a man or woman. If leptin works in people the way it does in mice, it could be the key to energy balance, an important signal and a brain mechanism involved in maintaining weight and sexual fitness. Of course, ten thousand years ago, as Campfield notes, periodic starvation was a major fact of life. Getting fat on a kill and then (relatively) starving for a while would have been routine, and an activated leptin gene could have been a problem. Most likely, what leptin biology is telling us is that nature's plan is for us to be flexible enough to eat when we have food, to not die of starvation (immediately) if we didn't, and to continue having sex most of the time in between.

Alongside the waist-to-hip ratio and fat-thin considerations, women's breasts, apocalyptic or otherwise, are enduring icons of sex appeal in a way that perhaps no other human organ is. Women throughout history have sought to enlarge, paint, and otherwise decorate their breasts; and men, for their part, have similarly used codpieces, penis extenders, and all manner of penile prostheses and prominence builders. But unadorned, the breast appeals to our Stone Age brain as an honest advertisement of reproductive and sexual fitness in ways that a penis never can. (Despite its importance to men, penis size is of little consequence to women, in general.)

Why breasts, especially large ones, take on such importance in the annals of sex appeal is less clear. As with the penis, size has little to do with function. As previously noted, the breasts of our closest primate relatives, chimps, gorillas, and baboons, are relatively tiny, but monkeys and apes nurse their young just as well as humans.

Thornhill and his colleagues have found that overly large breasts have higher levels of fluctuating asymmetry—unmatched breast sizes—than small ones and that asymmetries are higher in women without children. Thus, breast symmetry, if not size, is a reliable predictor of fertility, and choosy males who prefer females with symmetrical breasts get not only a direct fitness benefit in terms of increased fertility, but also an indirect one in terms of attractive or fertile daughters.

In addition to what we can see, nonvisual signals form a substantial part of sex appeal's machinery as well. Consider the sexual power of scent and season, informed by studies of everything from food aroma to blood flow and the way the brain's memory centers are wired.

As suggested earlier, it's not good to be too choosy from a practical standpoint, or to let endless evaluation and indecision, driven by our rational brains, run out our body clocks. But even with our more primitive brains in gear, we are "reading" certain signals that, while not cognitively processed, are being processed carefully nevertheless. And one of the most intriguing of these signals is odor.

Odor is so strong and important a signal that an infant finds its mother's breast by smell alone, and a mother can pick out her newborn based only on the baby's odor. More relevant to sex, psychologists at Duke University have found that couples who dislike each other's smells don't stay together. And, of course, the perfume industry's $6 billion a year intake is a financial monument to the fact that we humans consider smell a powerful attractant.

The role of odors in mate selection, which has been found in insects, birds, and mammals from mice to humans, serves as a potent reminder of how subtle biological capacities can heavily influence the most important decisions we make. Writing in the *Proceedings of the National Academy of Sciences*, Kunio Yamazaki of the Monell Chemical Senses Center and his coworkers reported that they were able to block pregnancy in mice by exposing the females to two males that differ genetically only in a portion of their Y chromosome. This small genetic difference cannot be detected in the shape or behavior of the animal, however; rather it confers a very slightly different odor on the male.

Other researchers have shown that a female mouse, given a choice of two mates, will tend to pick the one that is not as closely related to her biologically. The way she knows this is not by doing DNA analysis but by sniffing. If a recently mated female is caged near a male not closely related to her, and therefore smelling not quite like her mate, she often will spontaneously miscarry, probably using her sense of smell and the instincts it triggers to anticipate danger to her unborn pups from a strange male who is neither brother, father, nor grandfather.

So closely tied are odor and sexual *attraction* that when scientists knock out odor-perceiving genes in the brains of fruit flies, parasitic worms, and other animals, they can completely destroy their ability to find and select mates. Researchers have altered the smell center of a male fruit fly's brain, causing the fly to become bisexual, approaching male and female flies alike for sex. At Johns Hopkins, investigators found that even sperm rely on a kind of smell and sniff means for finding an egg to fertilize. On their surfaces, sperm cells have the same odor receptors as those found on the nerves of skin lining the olfactory bulbs.

How smell works its sexual wonders in the attraction and flirtation system is only now revealing itself to anatomists and biochemists and neuroscientists. The process involves both the traditional sense of smell based in our noses and brains as well as the unconventional ways that animals detect volatile molecules—those that travel by air.

The human nose, as neuroscientist Solomon Snyder likes to demonstrate with a forefinger imbedded in a nostril, is the only part of the sensory system whose direct connection to the brain can be easily touched. Nature clearly intended the sense of smell to be a powerful one, to help us in the game of survival. Animals can, literally, for instance, smell fear. People don't have to "learn" to associate bad odors with danger—from rotten food to poisons. And given the vital role of sex and sex appeal in our survival, the nose plays an important part in these aspects of our lives as well.

In people, the nasal cavity's tissue has about a thousand or so receptor proteins that, much like a lock, bind with the keys of different odor molecules in the air or propelled from the throat during chewing. Once a molecule connects, the receptor amplifies it and changes its shape, altering the property of the nerve cell, which in turn makes the nerve fire off a signal to the brain that says "chocolate" or "cinnamon." Odor signals traveling from the nose to the brain become more organized and refined, beginning with alerts to small receptors around the whole nose cell landscape, then focusing into more defined areas of smell memory or smell maps. In this way, although each nerve cell in the olfactory landscape only has one type of receptor, there are connections that can be made anew each time we smell a different smell. These connections can become permanent maps that let us, once we have smelled it, a remember a previously unfamiliar odor forever. Overall, our brains can detect and tell the difference among ten thousand different odorants, even though we don't have ten thousand separate receptors. We can tell lavender from cherry lollipops and artificial from real lemons because in our brains, an odor activates a small series of overlapping cellular events with just enough differences in pattern to tell them apart. The number of possible connections, according to neuroscientist Richard Axel, could be in the billions of billions, but because the human brain is finite in size and in the number of neurons it can spare for smells, natural selection has come up with the ten thousand. (In the smell department, rats and dogs do much better, possibly because their brains have more resources to bring to odors, given they don't need to have much for such things as algebra and poetry.) But even a thousand receptors and ten thousand connections account for a substantial part of our brain's activities. Nature gave our smell brains high priority precisely because we probably need it to accomplish sex and survival.

But as chapter 4 hinted, the newest part of the story of sexual attraction and smell is to be found outside of our classic smell brain and in the vomeronasal organ, or VNO. Musky, sexy perfumes, the yeasty scent of wine, the pungent seduction of garlic—these still are a large part of our sexual attraction system. But pheromones—protein chemicals made, released, and detected by the VNOs of thousands of animal species through specialized glands and organs—are increasingly thought to play a role in humans as well as animals.

In their studies of pheromones in prairie voles, Greg Ball and his colleagues at Johns Hopkins discovered that with the coming of spring sunlight, a female vole shows interest in mating only after she is exposed to a pheromone, which Ball has tracked to the place in the animal's brain where chemical conversion of a protein she makes triggers the release of female sex hormones that say "Mate now!" Following a string of clues, Ball's team further found that the pheromone that triggers this cascade is excreted in the urine of male voles. In adult humans, as already noted, the VNO is suspected to be a layer of tissue just above the palate in the nasal cavity that is connected to various parts of the brain. This connection is somewhat parallel to the connections between nerves in the nose that process smells we actually smell and the "thinking," "instinct," and "memory" regions of the brain. If, in fact, a human VNO plays a significant part in the sex-appeal operations, it is its links to the emotional, or limbic, brain that would make it possible for the emotional and instinctive brain to overwhelm that smart neocortex and, almost in spite of our rational brain, let us become infatuated so quickly—and often with the "wrong" person.

Benjamin Sachs, a behavioral endocrinologist at the University of Connecticut, has loosened if not untied the Gordian knot of sex appeal, smell, and pheromones. Working with hamsters, gerbils, and rats, and studying their sex habits for years, he reports that the range of odors and pheromones he has found makes it quite possible that the odors animals (including humans) can smell and those they cannot are part of modern sex appeal.

For instance, female hamsters in estrus have a "very strong and clearly detectable rotten odor to their vaginal secretions," which he can use to instantly assess which hamsters are ready for mating, but in rats, there is no such detectable difference. (Sachs declines to elaborate about how he inhales.)

At his University of Connecticut laboratory, Sachs has searched experimentally in all of these animals for what he calls "aphrodisiacal odors" and found molecules in estrous female rats that are ubiquitously made and sufficient to induce sex. "There's hardly a place on the female rat body that doesn't make pheromones," he says, "and there's a huge range and complexity about them. Males make them, too, and we see that their habits of sniffing flanks, backs, the anogenital region, and vaginal and clitoral areas are all involved. Even eye glands, in gerbils, make them and grooming spreads these volatile compounds all around. It's likely that even skin pigment has something to do with it, since in hooded Norway rats, albinos don't show the same erotic responses at all."

To test the attraction power of the pheromones he has identified, Sachs designed an animal wind tunnel with three chambers. With a male in the middle and a female either upwind or downwind, and air blown each way by a computer fan attached to either end, he is able to project presumably odorous pheromones upwind or downwind of a male at the flick of a switch. When the female was upwind, half the males responded, and when she was downwind, none did. Five to ten minutes later, without ever seeing the females or getting close to them, males who were downwind had erections. With just a whiff of whatever pheromone "she" was emitting, response rates in the males ranged from 60 to 90 percent even when the males couldn't see the females or get near them.

Sachs thinks that the rat pheromone response system is closely tied to sex hormones and a variety of brain chemicals that influence libido and attachment, including oxytocin, serotonin, and dopamine. In fact, a recent study of his shows that lowering serotonin in some animals can increase response to pheromones.

Moreover, when by damaging the preoptic area of the hypothalamus in these animals he eliminates their ability to actually copulate, he leaves their motivation for sex, their erections, and their response to pheromones intact. "They're motivated, but not aroused anymore," he says, suggesting that whatever the pheromones do, they seem to inspire desire quite separate from any ability to perform.

It's unlikely that so sophisticated a chemical attraction system in mammals would have completely passed humans by on the way to the Stone Age. And John Dwight Harder, a neuroendocrinologist and expert on mammalian sex strategies, is among the handful of scientists who have gone hunting for the VNO in higher mammals and humans. Harder has studied the VNO and pheromones in opossums and found some suggestions that whatever it is that happens in animals happens in people, too.

Opossums are marsupials and from an evolutionary standpoint something of an unmissing link, a creature with traits that fall in a muddled middle between animals that lay eggs and mammals that have live births. This odd placement on the evolutionary ladder gives marsupials hormonal and brain chemistries shared with both humans and much less evolved animals. If marsupials possess some iteration of a VNO and pheromone sex system of attraction, it is harder to argue that such a system is peculiar only to less evolved animals and possibly bred out of humans. But more importantly, an opossum VNO-pheromone system would give clues to where and under what circumstances the system arose.

The pheromone he and his team are working with induces females into being receptive to sex; in other words, it makes males attractive to them sexually. The male deposits the pheromone on the female and identifies her scent mark. She then nuzzles him, picks up the pheromone, and that turns her on to sex with the male.

Some animal pheromones isolated and tested turn out to be metabolites of DHEA, considered the most abundant hormone in the human and the precursor of many others, and it may be that personal scents mark us the way they do animals via this chemical. DHEA can act as an antidepressant, fueling energy and perhaps influencing the choice of who we find attractive.

The sexy cigarette exchange between Paul Henreid and Bette Davis in *Now Voyager* is a possible case in point, since cigarette smoking increases DHEA. More significantly, so do puberty, vigorous exercise, pregnancy, and hormones required for breast feeding, such as prolactin.

Karl Grammer, whose life's work assumes a human VNO and argues forcefully in favor of his observation that our limbic brains and our VNO have evolved ways to read almost subconscious signals, mourns the fact that attraction scientists have generally neglected the VNO and its operation in human studies. For him, the phenomenon we know as "instant attraction," the enchanted moment, is chained to our ability to detect certain odorant chemicals, consciously or unconsciously.

Grammer says that some of the main components of male pheromones are unpleasant odorants that cause women to reject men. In studies in which he passed around trays with vials of androgen-influencing or androgen-making chemicals, he found that women changed their attitudes toward these smells depending on the phase of their menstrual cycles. It's possible, he claims, that men go around "stinking" and only when at the right part of their menstrual cycles—their fertile periods—women aren't put off by it. "Male pheromones thus could function as a chemical device for the detection of ovulation and the induction of ovulation itself." Grammer speculates that females may also have developed chemical "weapons" that influence men's decision making. These Grammer calls copulines, which are deodorants that equalize male ratings of female attractiveness. They make ovulating females less attractive, but balance the loss by increasing male testosterone levels at the time of her cycle— midcycle—when she is most fertile.

Men exposed to the ovulatory odors or copulines of women show increases in testosterone, and Grammer postulates that this may be one way relatively unattractive women induce a male sexual response similar to that induced by attractive women. If copulines, combinations of acetic, butyric, and other acids, are indeed human female sex pheromones, they would ebb and flow during the menstrual cycle. And indeed, Grammer has found that oral contraceptive users produce no copulines. In other studies, Grammer had males inhale copulines for five minutes, then collected and tested their saliva for testosterone. He found that copulines raised levels of male-female attraction, as well as male readiness for sex.

Whatever is in copulines has physiological effects on male competition and sexual arousal, and it follows that women could "use" copulines to exercise choice of men by exploiting men's smell brains and influencing their thinking about attractiveness on a level that men can neither assess nor see. In such a scenario, pheromones are likely a part of any woman's feminine wiles, even if they are odors men can detect only in the primitive parts of their brains.

At the end of the day, what people consider attractive or sexy is almost endlessly diverse and the result of odd and very particular combinations of biology and culture. The basic framework is there—we want to attract quickly and to be quickly attractive. But how that plays out often leaves us scratching our heads.

Steatopygia, the protruding large buttocks found in some African tribes, is an inherited fat deposit considered the alpha and omega of sex appeal among the tribesmen. Beau Brummel's satins and laces would not have done a thing for them. Lip plugs, nose rings, tattoos, beards, shaved armpits, penile extenders, long nails, and full lips are fairly modern inventions and all part of someone's sex appeal armamentarium. Color preferences, especially for reds, also are a factor of sex appeal. Red is the standout color of the sexual "skin" in baboons and other apes, as is the red-pink color of human sex organs (rosy nipples, vagina, vulva, ear lobes, blushes, and flushes).

During menstruation females often have higher coloration and more reddened areas of the face and sexual skin. Cosmetic pigments, in use by early humans and today, may have evolved as a means by which women could mimic these features and lure potential mates. Conservation of such a trait can be traced to the fact that in humans alone among mammals, the periodic bleeding of menstruation (or its colorful effects on the visible skin) remains the only reliable signal that a woman has the capacity to ovulate and the only true signal of imminent fertility (once menstruation ceases) because ovulation itself is hidden. Men have an interest in detecting menstruation in females, but in modern times that isn't easy. Females of many species have been able to competitively manipulate menstrual signals to attract high-quality males, but at the same time, no female would want her man to run off and leave when some other woman was menstruating. Thus, it becomes plausible that women would find ways to copycat or mimic menstruation so that men would remain confused during that first attraction and dating ritual about just who was fertile. As one feminist biologist has put it, "concealed ovulation allows females to be selectively promiscuous, not obligingly monogamous."

*   *   *

As much as we may paint, powder, and perfume to promote our sex appeal, our built-in biology long antedates our ability to modify our appearance, or "improve" on nature's means of signaling health, youthfulness, fertility, absence of disease, stamina, good genes, and intelligence. If you are feeling romantic and flirtatious, it's possible that both the "novelty seeking" serotonin system and a natural amphetamine of sorts, perhaps PEA, are at work to trigger exhilaration and even euphoria. With your PEA rising, a favorite romantic piece of music you happen to be hearing at the moment you see a member of the opposite sex may drop your levels of cortisol, the stress hormone, because you are feeling mellow. Your DHEA, made by ovaries, testicles, and adrenal glands, may be primed, and dopamine, the pleasure chemical, may be flowing. Oxytocin, the "cuddle chemical," may move in, too, encouraging intimacy. You dance. You start to sweat, perhaps as a result of all this hormonal activity and the press of bodies; and pretty soon, vasopressin, a diuretic, is telling your brain to cool it. Sweating and possibly exhaling pheromones from your every pore, you get more intimate. Perhaps endorphins begin appearing in the brain, or oxytocin levels rise. And suddenly, you're "in love." Estrogen and testosterone will kick in when cued and full chemical lust is on the horizon.

Yet words of caution are warranted here, because none of this will happen in any way you can *predict.* Returning once more to our imaginary couple, we can only sum up their "chemistry" with a nod to both what is in their genes *and* what is in their minds. The recipe for their mutual attraction was there, but without a lot of other factors—including a chance meeting and the experiences they brought to the café—nothing might have happened at all. Notes Matt Ridley, author of *The Red Queen* and most recently of a book about the blueprint of the human genome, our genomes are "a sort of autobiographical record, written in 'genetish,' " that unlike other autobiographies "contain stories about the future as well as the past." That is because while our genes may determine what we are, they cannot predict what we will *become.* "The weather," Ridley has written, "is deterministic, but it is not predictable. Like the weather, we are a complex amalgam of feedback systems; genes in our brains switch on in response to our behavior as well as vice versa."

So, if life were an electrical hum, and sex amplified hard rock, then attraction would be the beat that gets everyone to pay attention. But the beat isn't pounding *all the time*, or in any way we can count on. The nature that gave us the capacity to respond to the subtle, subconscious allure of a breast under a sweater, a certain smile, the glint of sunlight on blonde hair is not at all implacable or free of free will. Though it may be hard to acknowledge that the survival of the human race could depend on something as unpredictable as a chance encounter of two determined genomes, so it is for us and for every animal that sexually reproduces.

> *And I have known the arms already, known them all—*
> *Arms that are braceleted and white and bare*
> *[But in the lamplight, downed with light brown hair!]*
> *Is it perfume from a dress*
> *That makes me so digress?*
> *Arms that lie along a table or wrap about a shawl.*
>   *—T. S. Eliot*

6

THE DATING GAME

"Courtship to marriage, as a very witty prologue to a very dull play."
—WILLIAM CONGREVE, *THE OLD BACHELOR*

In the marketplace of mating, the cost of flirtation and seduction for casual sex is quickly negotiated—and generally small. Not much invested, not much lost if things don't work out. And even if they do work out, we expend relatively little energy on a passionate one-night stand and can easily move on to another enchanted evening. So he really wasn't the master of the universe he pretended to be? And she wasn't exactly the innocent, twenty-something schoolteacher she presented to him? Big deal.

For those whose mutual attraction moves them along to formal dating and courtship, however, the costs of commitment—and fakery—are as high as the potential rewards and risks. Consequently, the process takes on the complexity and subtlety of waging a war and winning the peace. Because after that first meeting of eyes across a room, courtship—whether it lasts hours, days, or years—is when earnest negotiations for long-term sex and commitment occur. The process becomes at times Runyonesque, with guys and dolls regressing to every trick to win the biggest hand of all.

Let's return, briefly, to the imaginary couple. After their chance encounter, and a few lengthy phone calls, their mutual attraction grows. Three months of nightly phone calls, e-mails, and dates follow. "I'm falling in love," she thinks. She vows to do nothing to "scare him off," and he's given no hint of lost interest. They've never had an argument, and she has been careful not to start one, despite her itch to counter some of his political cant. Instead, they have "debates," as he calls them, with lots of mutual head-nodding and polite agreement to disagree. They have been unswervingly nice to each other, complaining about little, willing to try the museums, restaurants, and movies the other selects with bedrock good humor and good will. But suddenly, he stops calling. For five days, the messages she leaves on his answering machine go unreturned. He explains his absence on day 6 as an unexpected business trip three time zones away, which made a call to her "difficult." He's apologetic enough when she tells him how "worried" she had been, but he also senses her anger and feelings of betrayal. She has cried herself to sleep, alternately jealous of some imagined competitor for his affections and furious at his lame excuse for not calling. Instead of telling him any of this, however, she accepts his invitation for a "special dinner" at his apartment billed as a way to make it up to her. She decides to stifle her natural inclination to punish him with a bout of coolness, but plots to let him know she had filled her time happily during his absence. In fact, she had accepted a date with a man she had seen a few times. She had also slept with him, regretting it the next day and avoiding the three messages from him on her answering machine. Tonight, she won't mention the sex. She doubts she ever will. Meanwhile, as he finishes setting the table with candles, wine and crystal, he feels annoyed, guilty, and excited. With respect to his failure to call her, he tells himself, "If she really cares for me, she'd understand." There had been no plot to upset her. The faint strain in her voice underneath their upbeat conversation worries him, though. Maybe he should have called her to let her know he was going out of town, but once he made the move without doing so, he let it go. His excuse to her was mostly true. It *was* a business trip, although he spent time with a woman business acquaintance he had once dated but now considered just a friend. He resents the idea that he owes her a call every night. She is starting to have expectations and complicating his life. At the same time, he thinks about her more and more, devising plans for trips and more dates. Tonight, he plans to seduce her all over again with coq au vin, champagne, Bach, and the lapis and gold beads he bought for her on his trip, which were boxed and waiting, along with the yellow rose, alongside her plate. A snarl of anxiety rose in his chest as he waits to see her again.

What's going on here is so familiar it's cliché. Two individuals making serious evaluations of a potential partner in life and parenting. Compared to the relatively simple song of attraction, active courtship among humans and our animal cousins is operatic. If love at first sight is an internal sign of interest, and flirtation our first deliberate signal to the one who triggered that interest, then courtship is a fiber-optic load of messages, trafficking back and forth in pulsing waves of subtle and not so subtle communications at every physiological and psychological turn of events.

Past the initial glow of new love and lust, courtship depends on extroversion, not fear or submission. Dominance may get you noticed, but it won't, in the usual course of things, get you the girl. Success in courtship ultimately depends on cooperation and a willingness to stick together, at least for a while. In extended courtships, therefore—those perceived by a couple as a possible prelude to lifetime or at least long-term mating and parenting—the signals and rituals become ever more complex and subtle. The role of anxiety in courtship is a prime example. Decades of research have proven that anxiety disorders are several times more prevalent in women than men, that shyness is more prevalent in girls than boys. As all the theories about the biology of sex predict, traits that differ between men and women tend to have some fixed biological foundation that has been conserved for a purpose. So what would be the genetic and biological benefit to women of being more prone to anxiety?

Elizabeth Young and Randolph Nesse propose that the real question is why men don't show enough anxiety, since they take more risks, are involved in more accidents and injuries, and have a shorter life span because of these. If the true function of anxiety and fear is in fact to signal an unsafe or risky situation, then why would men have a built-in deficit in this regard? They conclude that decreased anxiety in men is a trade-off that has been shaped to sacrifice defense benefits for the sake of mating advantage. If, in fact, nature evolved men and women the way they are (he less anxious, she more so), it would predict that females prefer males who show less anxiety than would be in their self-interest, so that the males can take risks that are needed without expending a lot of anxious energy over them. It's also possible that because anxiety is a control switch, and with a less sensitive switch a man is more likely to take risks to defend a mate and her offspring, less anxiety is a better protector.

Of course there is modern fallout. What woman hasn't wondered—usually out loud—why the man of her dreams drives too fast or too aggressively, or refuses to ask for directions? Klaus Atzwanger of the Ludwig-Boltzmann Institute for Urban Ethology in Vienna did a study in which drivers were videotaped when they tailgated other cars and sorted by their gender, race, age, and type of car. Biological theories predict more risk-taking behavior in younger men than in women, more aggression in anonymous situations and dominance displays of higher-ranking individuals. Their findings were indeed that men, especially young men, drove closer and faster than women. Individuals who drove alone were more aggressive than those who had others joining them. Drivers of more expensive cars drove more aggressively than others, and fast, up-close driving was used to force others out of the way. The next time your date or your mate drives aggressively, it's possible that it's his biology that's covering up feelings of panic and making you a white-knuckle passenger. He's sending you signals of manly willingness to take risks; you're sending him signals of favoring his manliness and—unless you jump out at the next light—a great tolerance for his lunacy behind the wheel, and for him.

Most powerfully of all, when people become romantically involved they change their behavior in an effort to continue the attraction that began when they first laid eyes on each other. The risk-taking, novelty-seeking parts of his character that attracted her may become threatening backdrops to her need to test his interest in settling down to tender parenthood and faithful husbandry. The sexy, flirtatious, flattering woman who first attracted him now becomes backdrop to his inherent concerns about his ability to acquire the resources needed to care for a family. As David P. Schmitt and David M. Buss, evolutionary psychologists at the University of Michigan, put it, extended courtships offer much needed incentive "to minimize the impression of their negative qualities while maximizing the perception of their positive attributes."

As modern sophisticates we may resent or laugh at such cultural oddities as chaperones or dads with a Dun and Bradstreet report on daughter's latest Romeo, but courtships also provide time for much needed reality checks on our passing passions. With time and caution, sooner or later she'll discover if he can really afford the symbols and signals of wealth, and he can figure out if she really has the attributes he wants in a mate. Ideally, and most of the time, inevitably, courtship gives us time to get past the Porsche, the Harley, the former lovers, and the instant hormonal "yes" to more thoughtful evaluation of ambition, IQ, humor, kindness, and commitment. The most important information they want and need to make a "rational" decision about keeping this relationship going is whether the object of their affection can solve problems, cope with stress, and empathize. These are as much a part of genetic "fitness" as overall fertility and health.

How we get this information can be as straightforward as asking and checking out claims. But nature gave us, and all animals, a much more powerful set of "fact checkers." These are our brains, genes, and bio-chemistries, which, coupled with experience, operate as mate "bullshit detectors." They read emotional reactions, facial gestures, deep-seated attitudes. They can sense depression and subtle traits that suggest genetic weakness. In short, we are armed and equipped to uncover basic instincts in one another, traits that can't be covered up or covered over, at least not for very long. The way a lie-detector test makes use of unbidden galvanic skin responses to uncover truth, so does our evolved ability to read faces, bodies, smells, and secrets uncover the truth of our inner lives and our ability to form bonds with others. After all, the world is full of hot-shot courtiers who go to any length to woo the opposite sex. They sing or play endless love songs, display exhausting feats of physical strength, collect and offer elaborate gifts, organize romantic getaways, and fight off competitors.

And that's just the animals. Female gray tree frogs only turn on to males that sing all night long, exhausting their cardiovascular systems to demonstrate healthy genes and power. Female birds resort to food beg-ging normally seen in the young to signal how friendly and non-demanding and "helpless" they are, and males often respond by actually regurgitating food to them. Male spiders give nuptial gifts and spider lovers show the same kind of physical moves that spider parents exhibit when they bond with their babies. The courting male firefly, *Photinus consimilis*, gets to a receptive female more often if his flashes consist of repeated bursts that take much more energy than single long flashes. Female bower birds of New Guinea go for the males who have the largest collections of bits of glass, plastic, sticks, feathers, seeds, and beetle shells—preferably blue—collected to signal to the ladies that he is such a strong, resourceful fellow that he has the time and energy to collect all this junk *and* protect it from piratical competitors. Among the Siamese fighting fish *Betta splendens* that live in the marshes and rice paddies of Thailand, courting males first confront each other, darkening their skin color, spreading their fins, facing off in a frontal display of gill-cover erection, in which the gill covers swing out until they are at right angles to the body axis. They flip around to lateral displays, fully spreading their fins, beating their tails at the opponents. They but and thrust and bite and tear off each other's fins.

Are humans really any different? Fireflies flash lights; men flash gold cards. For if we resort to bringing bouquets and chocolates, offering diamonds to the showgirls and pizza to the kid brother, renting the limousine for the big dance, or schussing her off to Switzerland to ski, it's fair to conclude that we, like all other animals, have a lot of built-in biology and psychology devoted to the rituals of courtship. Scientists have identified many of the sequences that produce prolonged flirtation and courtship rites, leaving little doubt that what we think of as post-attraction "romantic" rituals evolved in our brains and bodies as elaborate responses to signals given and received by the opposite sex when choosing a mate is the goal of an encounter between the sexes.

One long line of evidence comes from the fact that serious scrutiny has shown these signals to occur almost universally. Some thirty years ago, Irenaus Eibl-Eibesfeld, now honorary director of the Ludwig Boltzmann Institute for Urban Ethology in Vienna, had become familiar with the widespread courtship dances and prances of mate-seeking animals. Then he discovered that people in dozens of cultures, from the South Sea islands to the Far East, Western Europe, Africa, and South America, similarly engage in a fairly fixed repertoire of gestures to more seriously test sexual availability and interest.

Having devised a special camera that allowed him to point the lens in one direction while actually photographing in another, he "caught" couples on film during their extended flirtations or dates, and discovered, for one thing, that women, from primitives who have no written language to those who read *Cosmo* and *Marie Claire*, use nonverbal signals that are startlingly alike. Pigeons swell their chests, codfish thrust their pelvic fins, and women, courtship scientists say, twirl their hair, toss their heads, sway their hips, arch their feet. On Eibl-Eibesfeld's screen flickered identical messages: a female smiling at a male, then arching her brows to make her eyes wide, quickly lowering her lids and, tucking her chin slightly down and coyly to the side, averting her gaze, followed within seconds, almost on cue, by putting her hand on or near her mouth and giggling.

Regardless of language, socioeconomic status, or religious upbringing, couples who continued this behavior eventually and repeatedly placed a palm up on the table or knees, reassuring the prospective partner of harmlessness. They shrugged their shoulders, signifying helplessness. Women exaggeratedly extended their necks, a sign of vulnerability and submissiveness.

For Eibl-Eibesfeld, these gestures represented primal behaviors driven by old parts of our brain's evolutionary chemistry and memory. A woman presenting her extended neck to a man she wants is not much different, his work suggested, than a gray female wolf's submissiveness to a dominant male she's after.

Researchers building on these ethnographic studies turned up the intensity, looking, for example, at compressed bouts of flirting and courtship in their natural habitat—hotel bars, lobbies, and cocktail lounges. From one set of observations at a Hyatt hotel cocktail lounge, scientists have documented an eerily predictable set of rapid attraction-through-courtship signals among business travelers seeking one-night or same-time-next-year affairs. For those who want to get through dating to mating fast, the giggling and soft laughs are followed by such semaphores as hair twirling and head tossing (hers) and body arching—leaning back in the chair, arms behind the head, like a pigeon bloating its chest (his). If all goes well to this point, the couple progresses, like actors in a script, to "lint picking"—the first tentative touch. She whisks an imaginary mote from his lapel; he brushes a real or imaginary crumb from her lips. Their heads come closer, their hands are pressed out in front of them on the table, their fingers inches from each other playing with salt shakers or utensils. Whoops! An "accidental" finger touch, then perhaps "dirty" dancing, more touching, cheek to cheek. By watching the body language alone, the investigators could pretty well predict which pairs would ride up the very public clerestory Hyatt house elevators together and which would not.

What's pertinent for human courtship's biological roots in all this is that very few words are actually spoken; the context and the signals are the universals and thus top candidates for behaviors grounded in our genes and hormones. Social psychologist Timothy Perper, an independent scholar based in Philadelphia, and anthropologist David Givens spent months camped out in bars and cocktail lounges in Seattle and New York watching men and women engage in courtship rituals. The women smiled, gazed, swayed, giggled, flicked their tongues over their lips, and wore high heels (invented by Renaissance princesses in the sixteenth century) to arch their backs, tilt their behinds, and extend their chests forward. No less than ear wiggles, nose flicks, and lordosis in rodents, these signals said "come hither." The men arched, stretched their pecs, and almost imperceptibly swayed pelvises in a tame form of an Elvis imitation, by alternating foot dominance. Instead of just lighting their cigarettes, they made a grand gesture of whipping out lighters, pointing their chins in the air with the butt in their mouths after the light, exhaling a huge puff of smoke, and making a wide arcing motion to put the lighter away. They swagger, laugh very loudly, tug their ties, clasp the backs of their necks (which has the effect of both stiffening the stance and puffing out the chest) in an urbane pantomime of male baboons and gorillas who prance and preen in a similar fashion. Man or monkey, the signals all say "look at me, trust me, I'm powerful but I won't hurt you."

In compressed courtship, intense conversation is followed by small talk, an ebb and flow of intimacy that again signals, "I won't hurt you. I don't want anything much . . . yet." As courtship encounters proceed, periods of complex talk increase. Finally, all the swaying, leaning, smiling, bobbing, looking, gazing, leg crossing, and vocalizing leads to fully aligned faces and shoulders, full frontal alignment, and simultaneous touching of everything from eyeglasses to fingertips. Says Timothy Perper, if you see this sequence of signaling, you can just about make book that the next stop is the bedroom. "This kind of sequence—attention, recognition, dancing, synchronization—is fundamental to courtship and the physiological and psychological outcomes are linked to what is hardwired in all of us. From the *Song of Songs* until today, the sequence is the same: look, talk, touch, kiss, do the deed."

Another line of evidence in support of courtship's biological imperatives is that we, like other animals, have in us the means to easily handle the expenditures of energy, tolerance of sleeplessness, anxiety, desire, and decision making that come with romantic courtship and mate choice. And to do it again and again, sometimes under great duress. Any trait or behavior that proves advantageous enough to be conserved, generation after generation, as Kent State University anatomist C. Owen Lovejoy points out, is guaranteed to have something to do with sex, including all of those aligned with dating and wooing.

Consider, for example, studies by Yonie Harris from the University of California at Santa Barbara Department of Anthropology of the wining, pining, and dining scenarios that mark most romantic courtships. In an analysis of a sample of forty-two primitive hunter-gatherer societies— those most like our Stone Age ancestors still living today—she found that in all instances, men and women had at their disposal ways to make sure their personal ("romantic") mate choices were known, regardless of the rigidity of "arranged" marriages in the villages. And if their romantic choices were ignored by family and society, their social system allowed enough wiggle room to leave a couple the means to enjoy extramarital affairs with their chosen loves.

Romance novels for women, most of them set in the distant past, play to this theme so predictably and with such universal appeal among women of all social ranks that it's irrational to consider these emotional tales as mere modern "fiction." Much more likely is that women recognize the evolutionary "truth" in stories about women pushed into loveless marriages by mean old guardians, but then "saved" at the last minute from this "fate worse than death" by their ability to seduce some hapless but alpha male into an adventurous—and usually somewhat scandalous—rescue. In some of these romances, the courtship lasts hundreds of pages before consummation, reflecting the power and persistence of courtship even when sex *never* occurs and is never likely to. Lancelot and Guinevere are the poster couple for this genre, products of the time when knights were expected to fall in love with and romance ladies who were already spoken for or in some way unattainable for sex. The reward was not in winning the lady but in courting her and winning her emotional favors. Its allure has informed our romantic dreams for centuries, with myths and legends that testify to courtship's power to arouse and satisfy whatever part of our minds, brains, and bodies are programmed for romantic love and courtship.

A third line of evidence for a biology of courtship comes from displays of "dirty dancing" and "dirty talk," almost universally considered "romantic" among young couples in love. Allowed only when intimacy and trust have been at least minimally established, both are potent signals between the sexes that courtship is welcome. The brain centers that control language and movement almost certainly developed both the sensors for these signals and the capacity to use them at very specific times in the mating game.

Similarly, food preparation, gift giving, and even walking upright, according to one theory, likely have their roots in the needs of mate choice and getting to "yes." Our female hominid ancestors, burdened by the lengthy restrictions forced by years of infant care, spent most of their time gathering fruit and bugs, nursing and carrying the babies, and being unavailable for intercourse in order to haul around the kids and put them down every so often to forage and feed themselves. That, says Lovejoy, may have been the incentive for males to stand upright to free *their* hands for finding and carrying food to the females in exchange for sex. (Pygmy chimpanzees do that today to assure sex and secure long-term bonding as well.)

Courtship is an endless pursuit of sex with one individual perceived as the potential best person to have sex with. It is prologue to the grandest shopping trip of all, for healthy genes and the wealth to sustain their trajectory into the next generations.

Perhaps the most significant evidence for a biology of courtship was the discovery at the end of 1996 of a master gene in fruit flies that controls the preliminaries, as well as the follow-through, of male sexual performance. It was the first time researchers pinpointed any gene in any species that is responsible even in part for the complex precopulatory or "dating" game. Many consider the finding the tip of the iceberg in behavioral genetics (the branch of science that seeks to link up universal acts and emotions to the biochemical activities of particular genes, especially the almost 3,200 human genes already known to be involved in the brain's development). As the work of Dean Hamer and his colleagues at the National Institutes of Health has suggested (see chapter 5), "behavioral" genes, unlike those that direct physical traits like eye color or foot size, don't directly cause such acts as courtship, homosexuality, anxiety attacks, impulsivity, shyness, gregariousness, or hostility. Indeed, no one yet knows how the combinations of adenine, guanine, cytosine, and thymine that make up the DNA sequences of genes control behavior, create perception and consciousness, or coevolve with culture and experience. But certain genes *have* been naturally selected by evolution to bias people in favor of specific personal and social temperaments and behaviors, to exert influence on the mind and body, massage the psyche in subtle ways so that individuals react to others and the world around them in different ways.

In fruit flies, as in people, courtship involves a series of actions each paralleled by visual, auditory, and chemosensory signals between males and females. The more visible and active parts are carried out by the males. First the male aligns himself with the female about 0.2 millimeters away, taps her on the abdomen with a foreleg, and follows her if she moves. He displays his wing and flutters it to make a sound she interprets as a love song. Then, if she's still willing, he unfurls his proboscis and licks her genitals, then mounts and copulates. "Rape," or forced copulation, is unknown among fruit flies, and the insects only mate when males go through this elaborate courtship routine and females agree to stand still for it.

Seymour Benzer of the California Institute of Technology, Jeffrey Hall of Brandeis University, and Ralph Greenspan, one of Hall's former grad students, reasoned that any group of behaviors so ritualized must be driven by genes, and so they went looking for these genes in fruit fly brains. In fly embryos, it turns out, sexual development is not a matter of XX or XY as in humans, but controlled by the dosage of X chromosome in each cell: cells that have one X give rise to male anatomy and behavior, and those that have two X's lead to female anatomy and behavior. These differences occur because single X (male) and double X (female) cells activate separate sets of sex-determining genes. Benzer, Hall, and Greenspan reasoned that if it were possible to construct colonies of mixed gene flies—flies that carried different dosages of X—then any flies they found that carried *mainly* female cells but had male cells in a particular site of the brain and showed male behaviors would clearly implicate the male pattern of gene expression in that part of the brain. This would greatly narrow the search for such genes.

The mixed-gene fly colonies, or "mosaics," were made and Hall set out to watch the courtship routine, making detailed notes of what happened and in what time frame. Then, painstakingly, he and his colleagues froze the flies, tediously sliced them into twenty sections tip to tail, and looked at the distribution of male and female cells. From this kind of yeoman science, Hall concluded that initiation of courtship (those first male steps including tapping her belly and extending his wing) required male cells on one side or the other of a small part of the fly brain that integrates signals from the fly's sensory organs. What he found, in short, were brain areas that trigger courtship in males but not females.

Further, he found that later steps in mating require male tissue in more parts of the brain, and that courtship "songs" and sounds are produced in cells in the body where the fruit flies' version of a spinal cord exists. Hall immersed himself in studying the "clicks" and "whirrs" that woo females. The male fly sings by vibrating a wing at a female. Using an array of mutated flies and something called an "Insectavox," a tiny microphone placed under a mesh-enclosed cylinder containing the flies, he has found that in certain mutants there is cacophony and dissonance, such as abnormal whirrs and disrupted rhythms of clicks that completely upset courtship behavior. "Without normal songs from the male, it's as if the female doesn't quite recognize the male's intentions. A tone-deaf male with no rhythm can still mate, but it takes him longer."

Greenspan and his colleagues also discovered regions of the fruit flies' brains involved in sexual *preference,* the ability to tell males from females.

The discovery that so many specific and different areas of the brain are involved in male courtship not only narrowed the search for genes that drive this process but also led to finding more than a dozen of them. The so-called *fruitless* gene, for instance, is involved in sexual preference, and damaged copies of it make males seek homosexual relationships. Another gene called *period* is the one involved in the male's love song.

Scientists also have demonstrated the biological or genetic control of courtship by examining the mate choices made in the presence or absence of healthy genes in the opposite sex. For example, if females were shown to choose mates on the basis of selecting good genes over bad ones, or on the basis of which choices give their sons and daughters a survival edge, that is powerful evidence indeed that the most elaborate courtship rituals are as natural and inevitable as breathing and sexual arousal.

Evolutionary biologists David Haig and Robert Trivers, among others, have looked at what behaviors have been conserved in humans and other animals and theorized that it would make sense for nature—bent on finding the means of perpetuating DNA—to have given females the means of choosing mates that give their *daughters* a slight survival edge over their sons. In their view, the battle of the sexes is played out in such "fetal attraction," in which mothers want to give extra nutrients to their daughters even at the expense of their sons. So, if during evolution a female developed a gene complex that helps her *choose* males whose genes improve her daughters' odds of surviving by, say, 10 percent, while decreasing her sons' odds by the same amount, then over time her daughters, who later mated and had other daughters, would conserve this gene. This would ultimately help out her sons, too, because it would lead to more females than males, giving males more opportunities to mate.

Conversely, their theory said, a complementary gene in males that gave *sons* an edge would actually not make life good for his sons, because there would be too few females to go around. Here is a case where the existence of genes that help females choose male genes that help her sons at the expense of her daughters is mutually disagreeable.

Scientists have tested the "fetal attraction" hypothesis in some female wild mice who seem to be able to literally smell males carrying a mutant gene that can lead to premature death of female fetuses. The males of this species get more copulations when female mice are given a choice of wood chips doused with the odors of males who lack this mutant.

Geneticist Chung I. Wu has also found that sexually experienced fruit flies of a particular species choose males that are not likely to be carrying an occasionally appearing "selfish gene" (found on an X chromosome) that thwarts nature's random throw of the dice with regard to sex ratio, or the more or less equal distribution of males and females in any given generation of flies. In these flies, males who accidentally get this X only make daughters, all of which choose males also carrying the sex ratio gene. Of course, that would eventually lead to the complete eradication of males, and the reason this hasn't happened in the flies is that only about 20 percent carry this "mischievous X." But what this also means is that the lady flies must have some means of choosing males with or without the X, because it certainly would not be in her interest to only choose males *with* this X.

What Wu found was the fruit fly equivalent of females who do better if they've been around the courtship block a time or two. He found that virgin females could not tell which males had the X but sexually experienced ones could. Wu's explanation of this is that virgins are programmed to get sex fast, since their breeding lives are short, while those who already have some sperm inside can be more picky, and that it is the second matings for these females that are most prevalent and the most "selfish," keeping the sex ratio healthy and giving her daughters at least an even break in the DNA games. When the experienced lady finds a willing male, his drive is to displace the sperm of the guy who took her virginity, and Wu's experiments show that the sperm displacement power of males with the X is weaker, while that of the non-X bearers is stronger. He believes that X-carrying males are weak and may give off a smell or pheromone that females can detect.

Of course, females of any species who are *too* choosy may get left behind. Like the woman in our imaginary courting couple who accepts her courtier's unwelcome political views, females try to keep their biological eyes, at least, on the real prize. Ken Kanishiro of the University of Hawaii has studied vinegar flies and verified what many humans have learned. Having too high a requirement hurdle for a mate may result in a situation where no individual can manage to clear it. Less choosy females probably have a somewhat different set of genes than their choosier sisters, genes that when passed on to offspring in a stressful environment may give the next generation a survival edge. In this scenario, the "best" males don't win, but the females and the whole species do.

The nature of this competition is often, however, misunderstood. Indeed, among the most forceful recent discoveries about courtship is that it is not merely about male competition for the best female, but about competition among females, among generations; and most of all it's about male-female competition for very different goals. As Shakespeare correctly noted, the course of true love never did run smooth, and ultimately, the reason is that "Momma's baby and Papa's maybe" remains a driving force in mating strategies. *Her* Stone Age biology and psychology still negotiate for long-term commitment of his resources to raise any offspring. Even in the age of women's liberation and paternity tests, *his* Stone Age biology and psychology still negotiate for some guarantee that any offspring she has after mating with him are *indeed his*. She needs his resources, he needs her fidelity. Thus, their strategies for finding a mate—what we call courtship—are different and often in conflict, but the *point* of court-ship is to get to "yes." It's wise, therefore, to read into courtship's moves and countermoves a script designed to reduce hostility between potential sex partners with appearance, persistence, appeasements, persuasion, and deception. And the more extended the courtship pas de deux, the more intricate and complicated it gets, as the signals the couple transmit and receive grow more powerful with repetition and time.

David Buss has even developed a "taxonomy" of courtship and mate-keeping tactics that even casual observers of human nature will be able to identify: vigilance, jealousy, concealment, emotional manipulation, der-ogation of competition, money, seduction, submission, apologies, decep-tion, violence, threats, and stares. But how each uses them, in what measure, when, and in what sequence as a relationship builds depend on the person's sex and circumstances. "Displays of love, commitment and devotion are powerful attractions to a woman," says Buss. "They signal that a man is willing to invest his time, energy and effort to her over the long run. . . . and one strong signal of commitment is a man's persistence in courtship." Whether the persistence takes the form of daily calls, nightly dates for months on end, or extended time together, interspersed with gifts, it works in winning women as more or less permanent mates. Using a seven-point scale, Buss found in his studies that persistence was more effective in getting her to accept a marriage proposal than getting casual sex from her, because persistence is a strong signal for more than a casual interest in sex. In some ways, *all* displays of devotion, deception, persistence, inconstancy, interest, kindness, respect, tenderness, dominance, fear, trust, arrogance, wealth, promiscuity, and fidelity are like proffers and promises in all realms of life: They carry the capacity to fulfill or destroy mutual dreams and goals. And whether courtship lasts minutes or months, natural selection favors males and females of every animal that court best. Those who get the best mates and keep them happiest are more likely to win the DNA lottery, and the result is an evolutionary wallet full of physiological and psychological credit cards we use to buy our way through courtship. Ever clever, social scientists have even analyzed mate preferences by analyzing personal ads in magazines and found consistent differences that mirror evolutionary strategies for mate choice. "SWM, 35, seeks SWF, 22 to 25, w/athletic interests and blonde hair" is a far more likely ad language than "SWF, 32, seeks SWM, 22, with athletic interests and blond hair." Except in the late twentieth century, men in all cultures have preferred submissive women and women in all cultures have preferred dominant, powerful, and wealthy men. It may not be pretty, it may not be politically correct, but it is our history.

The tactics employed in our courtship playbooks are destined to fill relationships between women and men with fear and tension. Says Mary Batten in her book *Sexual Strategies*, "Virtually every woman at some point in a relationship fears that her mate may be using her for sexual gratification, that he may desert her for another woman. Men, on the other hand, fear that women are out to manipulate and control them. . . . If we were biologically literate, if we accepted ourselves as members of the animal kingdom, we could see that sexual differences lead to patterns of behavior that we share with other animals. . . . that sexual differences are merely biological facts. Males make sperm; females make eggs. The trouble is that humans have imposed value judgments on these . . . facts."

Not only have we imposed value judgments, but also—until fairly recently—the values we brought to the enterprise were those of the mostly male scientific establishment. Predictably, most of the members of this establishment saw courtship strategy differences almost entirely in terms of male dominance. Few went beyond the idea that vicious fighting between male animals for access to a female was all that was needed for successful liaison. Today, evidence is plentiful that female biology and courtship behavior are just as strategic and competitive and that men and women coevolved the biology and psychology of their mating rituals. Her decision-making apparatus is as hardwired as his competitiveness. In his experiments with fruit flies in the late 1940s, British geneticist A. J. Bateman showed that female fruit flies allowed to freely choose mates consistently chose males that produced more offspring. Bee drones literally tear each other apart to compete for the virgin queen, who may wait days to choose her mates. Female elephant seals play hard-to-get for days as bulls bite and wound each other for control of beach nests and wait for the females en masse to choose only one of hundreds to give them his favors. The females are indeed a harem, but *they* choose which bull gets that harem.

In another search for clues to how mate selection preferences may be hardwired into our biology, Joanna Scheib studied how women chose sperm donors, analyzing information demanded by single women seeking the best sperm they could imagine. "It occurred to me that mate preferences might influence the choice of sperm donors and if so, the choice would say something about the value of particular attributes. If women didn't think certain attributes were important, they wouldn't show up in such a study and that would tell us a lot, too."

To set up her study, Scheib worked with undergraduate women and compared their preferences for husbands and for sperm donors (a metaphor, in some ways, for having offspring without the burden—or benefit—of sex and a husband). Each coed was given two hypothetical scenarios involving such attributes as health, physical looks, IQ, and character. The women were asked to rate how heritable they thought each trait was so that Scheib could also correct for those who might only care about heritability in a sperm-donor scenario.

Results showed that for these women, health, looks, and IQ were considered heritable and desirable in a donor. "Character" was more important in a mate than in a donor, but still high in a donor, too, even though the women understood the extremely low likelihood of it being passed on to the child through a donor. "Mate preferences clearly influence donor preferences," Scheib concluded.

In a further study in a Canadian sperm bank that offered women lots of information on donors, Scheib studied catalogs of donors and profiles of donors, which give occupations, race, height, weight, education, and interests. From this she developed a "character score," with independent judges rating each profiled donor. With married women in the market for sperm donors, Scheib found that no specific set of traits predicted their choices (no surprise; they matched to their partners). But among the single women seeking donor sperm, after race, donor's body mass affected choices most clearly; the desirable donors were mesomorphs, neither under nor overweight, and that cues for "health." Women also chose tall donors, around six feet but not much taller; character did not seem to affect choice of donor. To test this further, she gave these women at the sperm bank the same catalog of descriptions of potential sperm donors and husbands as she had given to those in her undergrad study. Results: There was complete overlap in preferences after race. As with the under-grads in the first study, these women too preferred taller people, of normal weight. Conclusion: Women choose sperm donors the same way they choose someone they'd like to date, *not* necessarily the way they would choose someone to marry.

Think back for a moment to the courting couple we've been following and to the ease with which each is willing to tell at least white lies about their feelings or behavior. Their motives are varied and complex, and they change the courtship stakes in enormously subtle ways that each seems to understand. The beat goes on. Now no one has yet overheard an animal say, "Of course, I'll respect you in the morning," but deception is not only built in to most courtships, it appears to be necessary and therefore built in to our sexual biology.

One line of evidence for this emerges from studies of male birds who build elaborate bowers or develop enormous, colorful plumage. Why do they do it? One school of thought says that if females are to have some means of sorting out "good" males from "bad ones," and "deceptive" signals from "honest" ones, then such traits are a biologically driven "handicap" that nature developed as "honest indicators" of male quality. Thus, "cost" is high but incidental because the stakes are higher.

But Gerald Borgia of the University of Maryland, an authority on bird courtship, takes a different view. Such courtship traits *are* costly, but not in the way we might think, and not equally so, and it's the *differential* in costliness that is both rooted in the animals' genes and hormones and the major factor determining mating success or failure.

"*Honest* indicators," he says, "emerge in many ways, not all of them costly, and not all of them equally costly to all males. Moreover, females know this and have coevolved a complex accounting system that "costs out" a male's quality based not solely on the gaudiness of his traits, but also on where females perceive him to be in the overall hierarchy of his male society. The ladies are not gold diggers, but they are meticulous bookkeepers with built-in bullshit detectors.

Why all this folderol is so important emerged from Borgia's long study of many species of these bower birds—so called because they build elaborately decorated structures that are *not* used as nests for offspring.

First, Borgia indeed found that when females assess males, they seem to pay attention to the bowers. The birds are extraordinarily resourceful in building their structures, with each species preferring a particular color scheme (blue and red are popular). Some will or will not use man-made or heavy objects. But because the number of structures—along with colorful plumage—are fairly evenly distributed among males, the presence of a bower alone could not explain the large "skews" in male mating success; some males do incredibly well in attracting females, but the majority do poorly. Something else, then—or several other things—must account for the differences in courtship success.

What Borgia learned in a seven-year study of thirty males was that females were seeing something in some bowers, but not others, that they could use to assess a male's propensity to give help in rearing young or nest building. The females, in other words, were reading non-resource-based mating signals into the bower structure. To conduct the study, Borgia put cameras in bowers and monitored them continuously to find out just how often the most frequently mating males were mating. (The pairs almost always mate in bowers and females mate only about once.)

Borgia found that four males accounted for half of all matings. Using DNA markers, Borgia was also able to affirm which birds actually got to *fertilize* a female, as opposed to just sexually mating with her (by examining the markers in the offspring). "It's the equivalent of fingerprinting, only better, because these markers establish paternity without doubt." Recording how often each male mated, Borgia then related the number of matings to the number of decorations in each of the thirty bowers. Then he removed decorations from some bowers and measured the number of matings again. These calculations established beyond a doubt the relationship between fewer decorations and lower male mating success.

An easy conclusion would have been that the females were able to measure the degree of "handicap" that a male was willing to take on to attract a mate; that they figured the more decorative the bowers, the more willing a male was to seek and secure resources over time. Presumably the more elaborate bowers, for example, took more time to build.

But that turned out not to be the case at all, Borgia believes. Because what *also* became clear to Borgia as an evolutionary biologist was that all of this bower building was not necessarily *costly* to a male. The activity was not an "honest indicator" of genetic fitness. Some of the males in his skewed population didn't even build their own bowers; they were clever "liars" that borrowed or stole them, and some declined to build any until they were eight or nine years old, suggesting they weren't ready to automatically "spend" resources they didn't have just because it's mating season. (These birds can live for seventeen years.)

Like other biologists, Borgia also knew that the best predictor of future mating success among most animals is current mating success, suggesting that displays of wealth, even if necessary to prove one's worth *once*, should not be necessary in subsequent years. Yet the bowers were built year after year by top males. "You just wouldn't expect costly displays every year," Borgia says, so he had to conclude that these displays couldn't possibly be very costly. "They may be relatively costly to less successful males, but not to everyone."

If the bowers built by Borgia's birds aren't honest advertisements of a male's ability to "handicap" himself and thus indicate his genetic superiority, what was the role of these structures in courtship?

When he went looking for answers, Borgia found a whole other story, in which females are able to figure out, from experience, which males are likely to "cheat" on them and which are not. In short, which will give a female an "I'll respect you in the morning" line, and which will mean it.

Borgia had already observed that male bower birds were able to draw lots of females to their bowers. He also knew that males were incredibly aggressive and that females responded positively to the aggression. In bower bird sex, the male typically jumps on a female and copulates while holding her by the neck, although most species are a bit more gentlemanly. But the bowers are built in such a way, all of them, that the males have to do a lot of steeplechasing, running around barriers, and while this is going on, females can escape. In bower courtship visits, only one in ten ends in a successful mating—and it can be as low as one in fifty—so females are running away from a lot of courtships. But if a female is to truly "choose" her mate, she must be able to get away, otherwise choice goes out the window. So if you are male, and you want success, you first need to give a female a safe place to let her see you, and you have to do this so much that it makes sense for nature to make it low cost for males to get females to come and see them. The males sometimes let females view themselves from a fern or tree, but they are better off if they build something females perceive as "safe" themselves. It may seem illogical that males build a structure to reduce their own opportunity to copulate, but if the bowers give them increased female *traffic*, that compensates very well and the males are ultimately better off.

This fits with what Borgia observed among these animals. The top bower builders are the most successful; they get one in four matings. So what they gave up in forced copulation they made up in volume of traffic, especially since females—even if they mate with one male—can go off and mate with others. In evolutionary terms, once one male had a bower, then no male could afford not to have one.

Bowers exist, therefore to prevent *forced* copulation, the bird version of rape. With enclosed bowers, the males have to court on the floor of the bower and chase females around nearly on their bellies. They can't fly or clip above her to copulate. That usually requires her willingness to move into more open space. It's a classic case of boy chasing girl until she lets him catch her.

Borgia concludes that bowers are probably a way for males to display vigor and aggression, but *not so much so* that they scare off the females. "Females won't stay around the hyperaggressive males," Borgia points out. So one strategy would be to put some females in a situation for a while where they can "watch" aggression but not experience it. "Males are manipulating the courting circumstance," he says.

Borgia's studies have not settled the disagreements about what "cost" really is in this or any mating system. But he believes he has substantial evidence that what female bird biology is assessing during the rituals of courtship is how vigorous her *female* offspring are likely to be. If you want to look at what benefits females, you need to look at female off-spring. If you look at what benefits males, you must subtract costs. The ideal is to have sons who survive without paying excessive costs, so that both male and female offspring get benefits.

If a lot of human courtship behavior seems just as convoluted and manipulative, well, in Borgia's opinion, that adds to evidence of its evolutionary origins. Sooner or later, he believes, genes linked to bower building and similarly programmed behavior will be found. In the meantime, for our imaginary couple, Borgia would probably say, courtship involves an internal conflict in females as to whether to choose material or genetic benefits, a conflict that must be balanced. David Westneat and Patty Gowaty, for instance, have done studies suggesting that females of many species bond and mate with one male but copulate with many (see chapter 10). This is because females who decide in favor of the genetically better fit males may lose out if they are too choosy, so they hedge their bets.

Urban anthropologists see the same strategy among poor women, who choose sex with men who offer flashy displays of wealth even though they are more than likely to abandon her and any offspring they father.

Other evidence for a built-in proclivity for deception has emerged, some of it in humans. Consider this vignette, used in a clever experiment conducted by Canadian researchers Jeffrey McNally and Michele Surbey: You and a classmate, of the same sex and in many of your classes, both plan on attending the same law school and both have the chance to meet with an attractive person of the opposite sex. This person wants to meet both of you at the library to get some study help on Tuesday night. Both you and the classmate would each like to date this person and you agree to all meet at the library. Tuesday is the best time for you both, since you both work at the cafeteria on Monday until late in the evening. On Monday afternoon, you hear that this attractive person will also be at the library Monday night. You and your classmate have supper breaks, each an hour long, at different times and you begin to consider the idea of meeting this attractive person alone on your break. It is also possible that your classmate, who also knows that this attractive person will be at the library on Monday, may go there on his or her break, too, an hour after yours. If you go to the library on Monday, you *could* make an edge for yourself by getting to him or her first and perhaps disparaging your classmate. You also realize, however, that if both you and your friend do this Monday night, the person might get pissed off at both of you and date neither of you. It's also possible that if you both go Tuesday night and make each other look good, the person might date both of you. But since you know that your classmate would never know that you went Monday night to meet him or her, you are considering it.

When male and female undergrads were asked what they would do in this situation, results showed not only that overall, men were indeed much less cooperative than women, but also that high self-deceivers cooperated more than low self-deceivers. More important to the story of courtship and sex, men and women alike were equally willing to deceive or not deceive. When it comes to courtship, both sexes seem hardwired to let truth be the first victim in the battle for a mate.

By now, scientific studies have demonstrated that courtship is partly programmed in our genes and memes, and part of it is not. They've also shown that every one of courtship's rich array of signals—visual, tactile, auditory, olfactory, and intellectual—coevolved so that for (almost) every reasonably fit sender there is a reasonably fit receiver. And, that the rules—and especially the *signals*—of courtship evolved to level the playing field for the courtship game, assuring some balance between the heavy investment of females in pregnancy, birth, and mothering and the risks to males who (until DNA testing, at any rate) had no way of making sure it was *his* DNA surviving in her offspring unless he made some heavy commitments and investments of his own. "Throughout much of evolutionary history," notes primate expert Sarah Blaffer Hrdy, author of *The Woman That Never Evolved* and *Mother Nature: A History of Mothers, Infants, and Natural Selection,* "the uncertainty of paternity has been one of several advantages females retained in a game otherwise heavily weighted toward male muscle mass." Female primates, and presumably humans as well, evolved a variety of strategies to pursue this advantage, including concealed ovulation, so that they could manipulate males, or at least keep them guessing long enough to get resources and care while infants were carried, born, and reared. The same natural forces of evolution worked on males to support this female behavior, making him a sucker for her signals but giving him much in return—the capacity and willingness to support infants that were possibly his own. Signals given and received, between deceiver and deceived, make it happen.

As couples invest more of their personalities and resources in courtship, the stakes rise and the consequences of both success and failure become more exhilarating and more fearsome. Evidence for this bedrock pattern comes from studies by Douglas Kenrick of Arizona State University, who asked undergrads to indicate the minimum requirements for someone they'd like to date in a variety of scenarios: a onetime date only, a sexual relationship, steady dating, and marriage. As predicted from evolutionary theory, women were far more choosy than men at every level, especially with regard to a sex partner. Men were far less choosy than women on variables such as IQ, wealth, ambition, friendliness, kindness, and emotional stability. Such preferences have been uncovered by so many social scientists that it's hard to argue they are not part of nature's design to get us through courtship and into the arms of suitable long-term mates.

From an evolutionary perspective, Kenrick notes, one might think there ought to be benefits to a dominant man who chooses a dominant female to pass his genes through, on the notion that his sons would then have a double dose of dominance and a better edge in the mating game, but there are equally compelling biological reasons why that's not a good idea. Unlike a man's resources, which must continue to appeal to a woman throughout courtship because of the costs of a pregnancy to her, a woman's appeal as courtship moves on must be less threatening, and her personality more compliant and tolerant. The signals of courtship, in other words, take on different meanings as courtship progresses to the "serious" or "marriage possibility" stages. Thus, the cosmetically enhanced, expensively dressed, bejeweled woman may get the initial attention from the man with the Rolex, the alligator briefcase, and custom-tailored suit. Each makes their sex appeal obvious by the show of resources available to "play" with the opposite sex, but if they plan more than a quick fling, her appeal for him and his for her will need to get past the overt signs of power and focus on cooperation and tolerance.

Another obvious reason for our urge to get past the "display" stage is the ease with which people can cheat at it. They can wear the fake Rolex and read book reviews instead of the books themselves. They can "kiss up" and "go slumming" to attract mates of a different social class and try to trick them into thinking they have more than they do, in terms of status, health, and youth. In early courtship, there are incentives for signalers—especially males—to cheat, but as time moves on, such signals get more costly. That's because, as proposed by Amotz Zahavi in his work on handicap theory, females who play the courtship and sex game best are probably those who can tell the difference between "honest" and "deceptive" signals. Thus, males who had the "best character"—that is, established themselves as honest signalers—were selected by choosey females who shopped for markers of the best male genotypes.

Zahavi also proposed that having a major handicap increases the risk of male death, at least in animals, and females choose such handicapped males because they have clearly already survived some lethal rite of passage to get to courtship status in the first place. We humans tend to make our deceptions and tests less deadly, but it's tempting to conclude that there are vestiges of such behavior in the most sophisticated urbanites. As Stanford University anthropologist Carl Bergstrom has noted, women *do* appreciate men who court energetically. "It's easier for females to judge males when males are running a marathon than when they are tossing back Buds and watching *Cheers* reruns," he said. Thus we have the courtship rituals of frantic activity, in which males often drag along dates and girlfriends to watch them play touch football or drive fast cars or haul themselves up and down mountains. There also are specific handicaps in which the female desires a trait that *only* high-quality males can even think of having. Rock stars come to mind, and bodybuilders, NBA stars, and billionaires. And there are strategic handicaps, in which each male himself chooses how large of a handicap to produce, taking into consideration his own best qualities and the females he wants.

A long-term study of female olive baboons in Tanzania supports the case for the biological roots of extended courtship signals, at least when it comes to dominance and submissiveness. High-ranking females generally get more rewards and resources and rid themselves more easily of competition for the best males. But the dominance has a price that the alpha males often quickly understand: a higher rate of infertility and miscarriage brought on by their masculinization through somewhat excessive androgens.

According to Dr. Craig Packer of the University of Minnesota, who did the study, these females' fertility problems send a clear signal, at least to some males in the troop, that dominance is a marker for mating failure in women. Analyzing Jane Goodall's decades of information about individual members of baboons, Packer and his Minnesota team analyzed the reproductive status of 138 females with 584 pregnancies over twenty-five years, along with the alpha female miscarriage rates. While the inability to test blood hormone levels for androgens in a direct way weakened this study somewhat, it was stunningly clear that while on the whole, the dominant ladies of the pack achieved sexual maturity earlier because they were better fed, a sizeable number experienced reproductively crippling delayed menarche.

Troops of baboons are matriarchal, with mothers, daughters, sisters, aunts, and nieces bonding in a community of up to fifteen or so females, while the males come and go and viciously compete for female favors. The stress of maintaining her status and passing on her social register address to her daughters makes for fights and screeches and frequent conflicts among the females, and there is huge competition among them to see whose infants get the most attention. Sarah Hrdy argues that the baboon behavior Packer saw evolved because of conditions that forced the females to exploit social dominance for the same reasons males do: to be rulers and make sexual choices that suit them best.

In the end, detecting deception, signaling "gotcha," and accepting the deceiver anyway is the only means animals have for negotiating the conquest of the best genes they can get. To work, courtship must bring females and males the ability to get more of their genes into the next generation, so both must employ strategic moves that involve a certain measure of deception, at first. Because unless a female is completely naive or unable to comprehend the world around her, she knows that some men bring more to the bedroom than others. So she—and he—must constantly, during each courtship, assess their potential mate's worth *and their own* to that candidate, because their self-awareness also means they must accurately assess how much they are worth in the sexual market-place. If a "10" is desired by a "3," then chances are it won't work in a rational courtship. But perhaps the "3" can find a "5" who for his or her own reasons is willing to be tricked long enough to mate "up" or "down" because of circumstances.

The whole process is partly random, partly tactical, partly strategic, and partly circumstantial. The point is that it's the *capacity* for such decisions and negotiations that are biologically driven and derived. While no human is programmed biologically to choose any one particular other human to have sex with, the evidence suggests that what *is* programmed biologically is the motivation and ability to participate in the choosing, in the courtship.

Just as scientists are looking for courtship "genes," they have evidence that the battle of the sexes that undergirds courtship also is played out between sperm and eggs, and among individual sperm and individual eggs. Each gamete uses its own tactics to further the cause of its own DNA at the expense of the other's, or at least not at its own expense. Every man and woman has a maternal genome and paternal genome and those genes must struggle to assert dominance in the offspring and capture more of a mother's intrauterine resources. The egg works to maximize conservation of its nutritionally rich innards, while sperm must battle each other to gain the egg and then in addition give its genes ascendancy over the female genetic code. The egg, hardly passive in this war, may have chemical tactics that not only choose particular sperm from particular mates, but reject some sperm during times of stress or scarcity, or when the woman who carries the egg wants to delay or alter the timing of her use of mating resources. The notion of such biochemical feminine ploys still doesn't sit well with some biologists, not to mention governments and male-protective societies. They prefer the old thinking, that throughout evolution, males, who make all that fast, cheap sperm, spread their seed around to maximize the odds of finding a vehicle for their DNA to survive, while females are passive, focusing their energy not in mate seeking, but nurturing and nourishing the outcome of the fertilized eggs they carried.

The newer thinking is supported by biologist William R. Rice of the University of California, Santa Cruz, who relied on those old standbys, the fruit flies, for the experimental evidence. Rice wondered what would happen if either a sperm or an egg were not able to compete for nutrients in the usual way. He chose to study fruit flies because males produce proteins in their seminal fluid that interfere with the action of sperm deposited in a female by other males. These proteins increase the egg-laying rate of females and send chemical signals that make females less prone to mate again. The fluid also is poisonous to females and if they get too much they die, so females have evolved biochemical defenses against toxicity. If they couldn't, the scientist reasoned, males would soon gain a big edge and overrun females.

For his experiment, Rice made a population of females that he prevented from "evolving" an ability to fight off the male's toxins, and bred them with a select group of males. All the offspring were killed except males that had all of the paternal genes, and these in turn were bred with more of the essentially "devolved" females. After forty-one generations like this, the males were indeed more often able to mate again with females that had already mated with competitors, and they made semen that made 24 percent more male offspring; but the semen was so toxic it decreased survival among females. Rice is considering a reverse experiment giving females the evolutionary edge. But the essential point is that the sexes have coevolved for mutual benefit by *balancing the conflicts* at the biochemical level.

Gerald Borgia, the University of Maryland zoologist, calls such coevolutionary behavior a "biochemical arms race that can evolve when reproductive interests have diverged." However, Borgia emphasizes that many species, including ants, bees, rodents, and songbirds, have also evolved more cooperative strategies, less toxic and less lethal. Indeed, it is likely that for billions of years, sex, or, more accurately, reproduction, was a cooperative effort, a joint venture between gametes or sexes of the same size and stock to repair genes by borrowing from each other's healthy stock. The antagonistic conflicts that exist in modern courtship and sex arose not from sex per se, but took hold when gametes for ever more complex organisms divided up various growth and development tasks more or less equitably. What emerged was an everlasting "armed truce" between two powerful forces, each trying to get the most out of their union.

Working with sea urchins, which reproduce by mixing sperm and eggs in seawater, Don R. Levitan of Florida State University in Tallahassee set up an experiment to see if large eggs were part of a female's activist courtship strategy, the kind usually attributed to males. He squired sear urchin sperm and eggs in Barkely Sound off the coast of British Columbia, then collected the eggs to see which ones got fertilized. Egg size turned out to be a great predictor of success, especially when sperm were rare. Either by their size (making them harder to "miss") or by some chemical that sperm sniffed, bigger eggs were more likely to be fertilized by sperm. Levitan concluded that competition *among eggs* for available sperm drove female evolution just as much as sperm competition is assumed to have done for male sexual evolution.

Do sperm wars occur in humans? Yes, according to Robin Baker and Mark Bellis, British biologists and authors of a book on the subject. The evidence is clearly thin or contradictory for their claims that sperm competition is the principal driver of human evolution, and that sperm from two or more men in the reproductive tract of a woman do battle by organizing themselves by size, shape, and function into "kamikazes," "killers," "blockers," and "eggetters." But their analysis of the interaction of conflict and cooperation among sperm in the female's reproductive tract suggests that in at least some situations, a complex and completely subconscious aspect of human sexuality may indeed be driven by military-like divisions of different men's sperm.

The foundation of their evidence is data from an unscientific yet intriguing national survey of human female sexual behavior in the late 1980s, from which they concluded that about 4 percent of children were conceived while their mother contained within her reproductive tract sperm from two or more men. From their survey sample, they were able to arrange to analyze and count sperm inseminated into and ejected by thirty-four women after intercourse. On average, 35 percent of the sperm were ejected by the female within thirty minutes of insemination. The occurrence and timing of female orgasm in relation to copulation and male ejaculation influenced the number of sperm retained at both the current and next copulation. Orgasms that climaxed at any time between one minute before the male ejaculated and up to forty-five minutes afterward led to a high level of sperm retention. Moreover, sperm from one copulation appeared to hinder retention of sperm at the next copulation for up to eight days.

The Baker-Bellis hypothesis that human sperm competition is the principal driver of human sexuality and evolution is unsupportable based on such a small sample. Moreover, counterhypotheses are bolstered by evidence that because the size of the testes relative to body weight is much larger in promiscuous-female mating systems such as gibbons than in small-testes species such as gorillas, the relatively small size of the human testes means that probably relatively few children are conceived along the lines of the Baker-Bellis 4 percent. Nevertheless, their data do provide some affirmation that a woman's reproductive tract has the capacity to function as a war zone and that sperm do battle there.

Regardless of how many women in the human population have sex with two men during the life span of a sperm, the phenomenon does occur and there must be some evolutionary reason for the fact that it has survived.

Moreover, Baker and Bellis argue, they have some anatomy on their side. In a "penis-eye view" of the vagina during intercourse taken by BBC filmmakers with a camera hooked to a fiber-optic endoscope and strapped to the underside of a man's penis during a "missionary position" encounter, it's clear, they say, that no amount of penile thrusting or forceful ejaculation of sperm can actually "shoot" semen up into a womb. That is because the cervix is blocked with mucus and because when a man thrusts his penis, the cervix retreats and tents up, creating a wall. At the moment of ejaculation, the semen hits the wall and runs down the cervix.

At some point after the penis is withdrawn, the cervix does end its retreat and dip into the seminal pool. "We don't know why the cervix sometimes does one thing, sometimes does another, but either way at some point the cervix dips into this seminal pool like an elephant's trunk," Baker explains.

That's what the camera sees. The subconscious part of this, Baker and Bellis claim, is that a man apparently can adjust what goes into the seminal pool. How do they say they know? By collecting ejaculates and counting sperm, the team has evidence that the number of sperm in this seminal pool can range from tens of millions to nearly a thousand million, and that the number of sperm a man puts into his seminal pool depends in part on the chances of his sex partner already containing sperm from another man. In a few experiments, they say, they found that if it's been three days since a man last had sex with his partner, and he has spent absolutely all of his time with his partner during those three days, he puts about 150 million sperm into the seminal pool, "because there is hardly any chance that she already contains sperm from another male." But if he has spent less than 10 percent of his time with his partner during those three days, and even if he has sex no more frequently or less frequently, then he can put up to 600 million sperm into that seminal pool. "In other words, if there's a chance that she already contains sperm from another male, he is going to put in a lot more sperm to have more chance of winning the sperm competition."

"Nobody is suggesting this is conscious, but there is something very sophisticated going on," Baker says. "Subconsciously a man is adjusting for sperm competition. This is programmed into him. But his is not the last word. *Hers* is, because the woman can adjust how long the cervix is dipped into the seminal pool. What the sperm have to do is swim out of the seminal pool through the cervical opening and up into the womb, and the longer she leaves the cervix dipped in, the more sperm can escape, but at some time the cervix lifts out of the seminal pool and that's it. It's curtains for the rest of them."

Thus, a man may do everything possible in his biological armamentarium, but the call is hers. "The woman does not lie there in the missionary position saying, 'Shall I leave the cervix dipped in for thirty seconds or a minute,'" says Baker. "This is part of our sexual programming."

If the work of Baker and Bellis has even partial validity, it would appear that even if a male "succeeds" at courtship and gets to have sex, he indeed still hasn't won the competition because his sperm may have to compete with sperm from his beloved's other partners to actually fertilize an egg. Says University of California primatologist and anthropologist Meredith Small, "conception, once thought a dance of simple beauty between one female's fertile egg and a solo male's hopeful sperm is now seen as a do or die cellular brawl that determines which males will pass on their genes to the next generation. All other forms of male rivalry, from the youngsters' first bickering over scraps of food to the adults' fierce struggles for dominance, lead up to these sperm-size moments of truth."

While it's true that men and males of most species are the most active, visible pursuers in courtship, at least initially, women and females of most species—just like their eggs—are far from passive in the process. They're just sometimes more subtle about it, a state of affairs that brings high dudgeon to feminist scientists who argue that male scientists, who have dominated biological research, have been professionally and culturally blind to the power of female courtship for centuries.

In any case, as the Hyatt cocktail lounge research described earlier demonstrates, in pursuit of advantage, women produce a variety of non-verbal signals and patterns at *all* stages of courtship.

Women historically have spent huge resources on clothes, lotions, potions, high heels and low-cut dresses, power suits and jewelry to make themselves attractive, while men have hair transplants, face lifts, bespoke clothes, and body armor. Dress and adornments are as old as loincloths, corsets, and codpieces. Turn-of-the-century women wore breathtaking (literally) corsets to squeeze waists to an "ideal" nineteen inches in an effort to advertise fertility (youth), and today aging male actors wear them to look youthful and military. Men of the Huli tribe of New Guinea wear gigantic, ornate penis sheaths while American men pour themselves into butt-outlining Levis. Beauty parlors flourished eight thousand years ago in ancient Egypt and a head of Nefertiti clearly displays the handiwork of some pharoanic Max Factor, down to and including plucked eyebrows, eyeliner, hair dressing, eye shadow, and lipstick. Japanese men tattoo their penises and Middle Eastern nomads stain the whites of their eyes blue. Foot binding to the point of crippling kept Chinese women so "youthful" they were infantile and literally had to be carried around. And of course today, Western women, especially Americans, have breasts, bottoms, noses, and hips enhanced.

In her book *Sexual Strategies: How Females Choose Their Mates,* Mary Batten notes that since women seek mates with resources, men during courtship can be expected to exaggerate their wealth by driving expensive cars, buying gifts, or spending more on dates than they really can afford. Since men prefer young, fertile females, women during courtship tend to do everything possible to enhance their youthfulness and personal appearance. But whether it's teenagers piercing body parts, wearing bikinis to make a sultan blush, shaving their heads, or wearing dreadlocks, or the upscale matrons who enclose themselves in fur and polish their complexions with five hundred dollars' worth of bottled Chanel, all are nonverbal and even subconscious devotions to giving off and receiving signals of courtship, advertisements for oneself, particularly for women.

As Meredith Small says, the active if subtler role of women in courtship became apparent when scientists began to carefully analyze who actually gets to mate and who doesn't. Assuming that only those who actively court get the best pickings, these studies have pretty much put to rest the ideas fomented in such popular books as *The Rules,* which call for women to always demand that males take the lead in courtship.

Careful observations of colonies of animals, DNA analysis of offspring, and other research show that females pursue multiple mates with as much fervor as do males, and for as many different reasons as males do. Right now, *she* perhaps wants sex with a man who will give her a child and the immediate resources to rear it to sexual maturity, because her family circumstances demand that whomever she chooses stay around and create the stability and security she'll need to take care of a baby. But at some point, she may be more interested in casual sex and he in a long-term relationship that gives him the security that it is *his* offspring he is supporting.

It's important to point out here that while Stephen Jay Gould's and Richard Lewontin's "spandrels of San Marco" scenario about the nonpurposeful existence of the clitoris and female orgasm remains unconvincing to many scientists, it's possible that some of the bolder courtship rituals, in both animals and humans, may no longer have much substantive value but have stuck around because at some point in evolution, they *were* honest markers of genetic fitness for mating. Animals that ignored these cues died out, while those who caught on survived, and with them a preference for the traits. Suppose a random mutation once yielded some peacock with a more colorful tail. If for some equally random reason, he got to mate with a preferred female because she happened to like his big tail, then their male offspring would tend to have big tails, too, and the trait would spread until fewer and fewer females want to mate with less colorful birds. In evolutionary terms, if the most reproductively successful female animals happened for whatever reason to prefer the equivalent of guys with gold chains or hairy legs or colorful tartan kilts, they were more likely to get them than other females and to bear daughters who had the same preferences for the same kind of males.

Particularly loud mating calls, aggressive lekking displays, butterfly-wing coloration, chest pounding by gorillas, and even penis size fit into this category of male courtship activities that at some point satisfied female evolutionary whims. And whims may best describe what a woman "wants" in her man's courtship armamentarium, as argued by Alexandra Basolo in her work with the plastic swordtails she affixed to a Central American freshwater fish (see chapter 3). In a similar vein, wearing form-fitting jeans and offering other courtship displays of big penises have nothing to do with the successful ejaculation of sperm, but everything to do with courtship's means of advertising something much more important—compliance with female preferences for the most *competitive sperm.*

The story is subtle but simple. Small penises will ejaculate sperm for fertilization no less well than big ones. Instead, as noted earlier, the biggest or at least most unusually shaped penises evolved in species in which females are promiscuous, mating with more than one male and choosing sperm from among several male competitors. It's the sperm competition among males that led to the array of spines, coiled tubes, pincers, plumes, and other outsized or strange-shaped penises that are found throughout the animal world—what biologist William Eberhard has called "internal courtship devices." That's because these give the male's *sperm* an edge when mixed with a competitor's sperm. In most mammals, including humans, sperm goes in the vagina, but fertilization of an egg goes on in Fallopian tubes far afield of the vagina, so female contractions and movements are needed to push sperm to the point of fertilization via orgasm and other erotic pleasures. Eberhard concludes that large or bizarre penises evolved because they provide some added sexual stimulation to the female, making her more likely to desire males with the most elaborate penises; biologically, such penises have been the most successful in getting certain sperm to her waiting eggs.

So powerful is this incentive, some scientists say, that in many species of animal, the only thing you can tell apart in the sexes is the male genitalia. "Females clearly impel this," says Jerry Coyne, a University of Chicago biologist. They select for male genital display, so the guy with the best-liked penis wins in this aspect of courtship. She chooses the penis and takes the guy wrapped around it.

Where such displays of largesse and other courtship dances take place is also the province of biology. Indeed, "your place or mine," known to animal researchers as philopatry, or place preference, is grounded in genetically controlled behavior in many species. One is the blue-headed wrasse, a common reef fish in the Caribbean. Biologist Robert Warner noticed that male wrasses have particular territories, and that females choose a male in a particular territory. If he were to remove that particular male, however, she'd mate with any male in the territory.

Location also plays a major role in human courtship, of course, in part because particular places offer opportunities for comparisons of a date with other potential mates. The desire for such comparisons has been investigated in animals that lek (from the Swedish word for "to play"), that literally dance and parade around before the females, competing with each other with sounds and ruffling their feathers, even giving the bird equivalent of towel snaps and "noogies" to intimidate the other males.

Biologists believe such behavior evolved principally in these game birds as well as fallow deer, fireflies, bats, and frogs, in which the males have no other courtship edge to advertise.

During courtship and breeding season, the females watch all this and apparently choose mates from the show. Few males actually get chosen, however, and why males go through this beefcake competition with so little chance of winning is not really known. But one compelling theory is that on his own, a guy grouse might not stand much of a chance. In the lek, he gets to compare his abilities to his mates, and if he pulls out all the stops, even a somewhat genetically inferior male can win the lady. There is evidence for this thinking in that females who participate don't always pick the dominant males. If his squeak, whistle, and feather expansion appeal to her eye, he's in, even if he can't deliver the trip to Paris or the beads at dinner. He literally makes the most of a bad thing.

A human counterpart to the lek has been described by Christopher Tilley and Craig Palmer, anthropologists at the University of Colorado at Colorado Springs. In their study of gang membership, they suggest that belonging increases an individual's sexual access to females. The scientists came to that conclusion on the basis of a study of sexually transmitted diseases as markers for sexual activity, and found that male gang members have significantly higher average numbers of sex partners than non-gang males.

So what if your dream boy doesn't roam with gangs but only with another buddy to whom he seems glued at the hip? The process is similar, it seems, when Butch brings Sundance along for the big date, or Butch's girlfriend is asked to watch the guys play football. Behavioral ecologist David McDonald of the Smithsonian Institution and Wayne Potts of the University of Florida have shown that in males belonging to a family of brilliantly colored birds called manakins, the males have given up claw, fang, and brawls to dance with another male. In other words, the males dance in pairs in order to impress the females. Moreover, the dance partnership can last for years, and in the end, only one of the birds gets the girl; the other male almost never.

Previously, it was thought these males were related to each other and that the "loser" or beta male indirectly promoted the transmission of his own genes by helping his male relative woo the mate. But McDonald and Potts have found that the males are rarely related and it is the *females* who force the males into their partnership. The beta bird isn't being altruistic but paying dues, reaching alpha status when the partner dies and building his own dance repertoire to win the ladies, even if it takes years.

This kind of cooperation evolved, the scientists say, because the ladies evolved to prefer the manakin equivalent of tender, sociable Baryshnikovs with great genes. The dancing is a test *she* prefers and that has served her well. The winners tend to be males at least ten years old, able to sing four thousand times a day for months on end, dance really well, and be good partners. If he measures up, she gets the best. To many observers, it's much the same story with women who have developed a preference for professional athletes and wealthy executives who can get lots of other men to cooperate with them.

Dating couples may not cook or dine on coq au vin and champagne, but they often eat special foods in "romantic" places, and work to enhance and extend the physical and emotional "highs" they feel with drugs and alcohol, necking and petting, and closer contact. Ancient hormones, acting on old and new parts of the brain, turn on subliminal, invisible signals and processes to assure our compliance with biology's plan to keep us attentive to our DNA as it readies itself to inhabit our grandchildren.

Few activities related to sex more ably demonstrate the intricate interaction of nature and nurture than courtship. Return again, for a minute, to the female member of our imaginary couple as she anticipates and dresses for her romantic date. Skin flushes, sweaty palms, stomach in knots, every nerve twitching and fingers fumbling as she fastens buttons and earrings. She seesaws from momentary euphoria to a sense of impending doom and she tenses her whole body to keep from wavering. Hormonally speaking, her body "in love" is reacting to an impending event—intimacy with the object of her affection—the same way it would to an opening night on stage or food poisoning, because hormones and their chemical relatives are what connect our bodies and brains *in the moment* to act synchronously and to respond to each other and the world.

Courtship is awash in the same natural drugs that first mold male and female brains and bodies, then propel and shape our sex lives. They carry messages between psyche and soma the instant we want or need them to engineer arousal and intimacy, to send blood vessels pulsing, neurons sizzling, glands oozing. Sex hormones influence hearing, smell, touch, taste, balance, and the response to visual stimuli. They make our hearts race and our digestive systems shut down so that we pay proper attention to not only fight or flight, but the opportunity to mate.

We properly associate adolescence, the sexual coming of age and interest in courtship, with "raging hormones," but the chemistry of sexual desire may begin as early as age six, with hormones secreted not from the undeveloped ovaries and testes but from the adrenal glands that sit atop the kidneys. These hormones influence sexual behavior and development by triggering the first glimpses of sexual *attraction* between ages nine and ten. Getting to the point of sexual desire, or to the even more distant capacity for intercourse, takes more hormones and more experiences. But it looks as if one of these adrenal hormones, a weak androgen called dihydroepiandrosterone, or DHEA, brings on a certain awareness of our full-tilt sexual potential. DHEA surges just before birth, slackens in childhood, but peaks at age ten in a phenomenon known as andrenarche, a sort of hormonal pump priming that leads, in both boys and girls, to a readiness for or sensitivity to later hormonal influences on brain and body. Such things as reasoning ability, flirting and teasing, embarrassment and shyness all may flow from DHEA. (DHEA has recently become the focus of a blitz of books, articles, and testimonials to its reputed role as a sexual and general fountain of youth. Sold as a dietary supplement by hucksters, without benefit of much more than theoretical and anecdotal evidence, DHEA has so far been studied only in mice and rats, which, unlike humans, make very little DHEA naturally at any time in their lives, making them unlikely "models" for the hormone's effects.)

To understand the role of androgens and estrogens, the most powerful sex hormones, it's useful to think of two basic kinds of sex behavior: motivational or "appetitive" (desire, the behavior most associated with dating or courtship) and consummatory (intercourse itself).

Hormones play a role in both kinds of sexual activity, but not to the same degree and not evenly between the sexes. For instance, androgens drive men to find and get females and perform mounting, erection, intromission, ejaculation, and all the components of coitus. Estrogen formed from testosterone induces arousal in many species, but principally desire in women.

In female mice engineered to knock out their cells' ability to respond to estrogen, and surgically deprived of their estrogen-making ovaries, Emily Rissman and her colleagues at the University of Virginia found that under no circumstance would the knockouts go for courtship, although the males found them just as sexually appealing as their normal sisters. After finding that the knockouts had greatly reduced levels of estrogen receptors in their brains, Rissman concluded that activated estrogen is an absolute requirement for sexual willingness, although not for sexual attraction, somewhat the opposite of the DHEA story.

If estrogen is tied to mood and molecular action in the brain, it stands to reason that chemicals made and released directly from neurons would be sexologically important in the activity of steroid hormones as well. And so they are. Working with rhesus monkeys, whose reproductive cycles parallel women's, scientists have discovered that both estrogen and progesterone actually may control the production and expression of serotonin, which regulates a great deal of libido and mood.

Turning to male biochemistry, the principal sex hormone testosterone is a precursor chemical of estrogen and progesterone, and has more than earned its reputation as the father and mother of all hormones of sexuality, active in all aspects of men's and women's sex lives, only more dominantly in men's. Testosterone's effects, including virilization, muscle building, aggression, dominance, and status seeking, have understandably led people to think of it as "man juice," the very essence of masculinity. If men—and the women who want them—find power, status, and physical prowess a form of aphrodisia, then biology is on their side. Psychologists have tested the saliva of men in many jobs and found that corporate presidents, pilots, and sports stars have in general higher levels of testosterone than clergymen and poets.

First isolated from blood in 1931, testosterone is present in its original form in women as well—albeit in less than a third of the total found in men—but the aggressive end of testosterone's effects appears to be modified by estrogen. Men use testosterone fairly simply, and as it is their only real sex hormone, the effects are fairly uniform—erection and libido, along with such "secondary" sex traits as low voices, beards, and muscle mass. For women, testosterone's role is more complex; it's secreted by her glands continuously throughout her menstrual cycle, and beyond its role in influencing and regulating some aspects of fertility, it probably accounts for why this gender, rather than men, have bodies that always are "ready" for sex as long as they are not pregnant. Dr. Adrian Dobs of Johns Hopkins has found that testosterone, even high doses, may make people more sociable and improve their sense of well-being, just as estrogen seems to elevate a woman's mood after menopause.

But although powerful enough to register effects on our behavior and emotions in only billionths of a gram, testosterone, like the other chemicals of courtship, are no more a "love potion" than a nocturnal emission is passionate sex. In isolation, they are without much of their potential because evolution designed animals to account for circumstances, experience, and what's going on in the world outside our bodies before putting hormones into play. Otherwise, we'd be biochemically frazzled to the point of exhaustion with every opposite-sex encounter.

Context counts, and wherever researchers look, that lesson is affirmed. Among male white-crowned sparrows, it's social activity that triggers mating hormones to surge, not the other way around. If a guy doesn't go out and meet the girl, nothing is going to happen. In some species of birds and lizards, male courtship is a must if the ovaries of the female are to grow at all. Aggressive behavior by the male, such as attempted "rape," can inhibit ovarian growth in reptiles, and some studies of humans have shown that rape inhibits conception.

Hormones won their unfortunate reputation as love potions because of the close association scientists uncovered between them and the observed differences—and sexy behaviors—between the sexes.

Over the centuries, and long before the scientific study of biological systems got under way seriously at the end of the nineteenth century, doctors had approximated and intuited the role of sex hormones by prescribing for patients with drooping libidos—or those who just wanted more and better sex—indigestible concoctions of gonads and other glands, along with animals' penises, breast milk, and even menstrual blood. Ancient Egyptians preferred testicles and Pliny the Elder of Rome prescribed hyena penises as sexual fetishes. Around 1000 B.C. the Ayurveda of Susruta of India suggested eating testicles to treat impotence, and in Germany, 250 years ago, doctors talked about horse testicles and whale penises as fountains of sexual youth. Monkey gland transplants in the 1920s earned Franco-Russian surgeon Serge Voronoff a fortune despite no hint of evidence that these worked; and just before World War II in Europe, Adolph Butenandt, the man who first isolated natural testosterone, got backing from the Schering Corporation in Berlin to synthesize the hormone from cholesterol and created a market among men yearning for unflagging erections on demand. He earned both a Nobel Prize (which the Nazis insisted he decline) and the somewhat dubious honor as the founding father of testosterone-replacement fantasies.

Fortunately, investigators turned to more scientific, not to say palatable, experiments to discover the actions and reactions of hormones in the context of courtship's settings and conditions. Notes evolutionary biologist David Crews, "not even rats are hormone driven automatons. When we give animals hormones we [only] change the probability that they will show male or female-like behaviors." R. C. Schiavi and his associates in the department of psychiatry at Mount Sinai School of Medicine in New York demonstrated the truth of Crews's statement when they gave testosterone shots to healthy men who had trouble keeping erections. The men reported more ejaculations and small increases in sexual desire and masturbation. But the added testosterone did nothing to influence penile rigidity, orgasms, or sexual satisfaction overall, nor did their moods or psychological attitudes change. Merely giving testosterone to men who have normal gonads may get their sexual behavior going, but it does nothing to enhance the performance in the absence of the "right" environment for any given man. Further evidence for this point comes from the fact that even castrated men—whose testicles are removed—are capable of masculine sex activity. In the past, in fact, the practice of assigning eunuchs to guard harems was akin to putting the fox in the henhouse. Many eunuchs enjoyed sex, not as frequently or in the same way as ordinary men, but often with quite functional erections and orgasms. And Masters and Johnson, in their landmark studies of human sexuality, showed that even people with very low levels of sex steroids can experience normal arousal, orgasm, and satisfying sex because androgens or estrogens seem to enhance desire and sexual motivation more than erections, vaginal lubrication, and orgasm.

As behavioral neuroendocrinologist Gregory Ball of Johns Hopkins has found in more than a decade of study in songbirds and rodents, courtship can't take place without hormones, but hormones don't directly "cause" it. They do influence "consummatory" behavior in animals—genital rubbing, copulation, ejaculation, and the like, but they don't do much for falling in love or choosing a mate.

His best evidence, says Ball, comes from observations of normal male quail. In the presence of springtime spurts of testosterone and a willing female, these males will stop their mating songs, grab the females' necks, mount, rub, transfer sperm, and jump off; but nothing will motivate the studliest, most testosterone-laden quail unless a female is *present*. "The point most laymen and even many scientists don't understand is that even when sexual behavior is inexorably linked to a sex steroid, it's not a direct response to a simple stimulus. The right way to think about hormones and sex is that if you have a stimulus situation, such as the presence of a female for a male, this situation elicits a response and the hormones influence the intensity and probability of the response, but don't cause it. Even with huge amounts of testosterone, a male quail won't strut unless a female is around."

Ball also looks to so-called 'roid rages, violent behavior in men taking anabolic androgens to build muscle mass, to make his point. Based on what he's seen in birds and other animals, Ball believes that "if a guy on androgens is just doing normal things, he probably won't behave in a violent way just because he has extra androgens coursing through his bloodstream. If he goes into a bar with his girlfriend, however, and another guy makes a pass, he may be capable of murder. It's the combination of the hormones and the particular stimulus that creates the violence." In birds, the story is much the same. "If in the springtime, when seasonal breeding birds have normally raised levels of testosterone, you play a taped male song to a male bird in the presence of a male stuffed bird, the male will attack the stuffed bird because it sees it as an incursion on its territory. In other words, the external stimuli of the song and the bird raised the subject's testosterone levels and aggressive streak. If you do blood samples before and after such a stimulus, you see that the stimulus of that competitor, stuffed though he is, creates a big surge in testosterone. That male stays worked up all day long after this stimulus. He's still singing and ready to fight the male incursion until we scientists are exhausted watching him, because the challenge leads to the transitory huge release of testosterone way above normal levels. But if you do this to the same male in the fall, this same male's response will just shut down. He's downright inhibited. If you play back the song and present that stuffed bird, he doesn't pay much attention to it. And, if you inject testosterone into that male, he still won't react—unless a real female is present. Life is complicated."

Comparable studies in female ringdoves, using laboratory lights to simulate spring and fall and trigger breeding "seasons," found that the *perception* of the male in courting influences the female's hormones and behavior. When the male shows interest, the female coos to herself, and hormonal studies demonstrate that only after *he* signals does *she* coo and only when *she* coos do her own sex hormone levels rise to meet the challenge of breeding. If she doesn't coo at his signal, her hormone levels are less intense and so is her sex response.

John Godwin, a zoologist at North Carolina State University, has studied gender-bending, sex-changing fish, in his case a species known as the Caribbean blue-headed wrasse, which has a large repertoire of unusual sex habits. Along with Bob Warner of the University of California at Santa Barbara, Godwin has been working twenty-three years on these fish, their social system, ecology, mating systems, and of course their major sex hormones. What he's found is that contrary to conventional wisdom, aggressive male sex behavior can and does happen in males without any testes at all, and that a hormone-like chemical produced not in the testes but in the brain may be the reason. The chemical, arginine vasotocin, says Godwin, is present in males and females, and when these fish change sex, "this chemical changes its level right in line with the behavior change." Arginine vasotocin is one of a family of proteins, which also includes oxytocin, a chemical much more associated in animals with the desire for intimacy, the closeness sought out during courtship, than for the sex act itself.

While Godwin's studies don't rule out the role of testosterone or estrogen in the sex changes his fish undergo, the take-home lesson, he says, is that "social stimuli" mediated by oxytocin and its chemical cousins "appear to me to play a major role. . . . The chemicals are highly conserved among vertebrate species."

One reason that scientists have long suspected that neurotransmitters, perhaps more than sex steroids, were deeply involved in the cascade of events that sex steroids facilitate was that the psychological "footprints" of these brain chemicals had "sex" stamped into them. Thoughts, memories, feelings, moods, anger, love—these are the products of the brain, not the body, yet they are part and parcel of courtship, as well as commitment, intimacy, and parenting. Investigators' search for sexy neurotransmitters, therefore, has not been in vain.

Dopamine, for example, probably is involved in the "motivational" components of sex behavior. Male rats release dopamine in the nucleus accumbens of the brain in high concentrations when exposed to an estrous female, even when he can't make contact with her.

J. G. Kohlen, R. K. Rowe, and R. L. Meisel of Purdue have shown that merely stimulating the vaginal and cervical areas of female hamsters increases spurts of dopamine release in the nucleus accumbens of their brains. Their methods were imaginative: They used micro-dialysis to sample dopamine during mating in females with and without vaginal "masks" that prevented penile insertion. Those who had the masks had no increase in dopamine; those who went bare into their love nests had surges of dopamine. They tried the same tests in females who were ready for sex but were only allowed to look at a male, and in females who had oil put on their vagino-cervical areas; they got nowhere. Nor did dopamine rise if the male mounted but was a performance dud. E. T. Larson at the University of Colorado has been able to trigger the process of natural sex reversal in saddleback wrasses by norepinephrine, or stop it by giving dopamine. In the natural state, these fish—relatives of the bluehead wrasses of sex-change fame—do their sex and gender hopping as a result of social necessity. That is, if the ratio of other fish in their group has too many females, a female becomes a male, good team player that she is. This transformation takes two to ten weeks, on average about eight weeks, and during this time, the behaviors are changed immediately and the female starts behaving like a guy as soon as she senses the social cues— that is, *sees* there are fewer males than her society needs. The body then follows with new male gonads, making it clear that the brain's chemistry is driving this, since the female makes her sex change decision based on the fish she can see.

Using fifteen fish, Larson tested two drugs that increase norepineph-rine in the brain. He also used a tranquilizer, haloperidol, which blocks dopamine. Catching the fish from Kaneohe Bay in Oahu, Hawaii, he put them singly in tanks, fed them lots of their favorite chopped shrimp, and let them get used to captivity. Unlike most fish, these wrasses will mature as either male or female, reproduce for a season, then undergo a change in gonads and reproduce and act as the opposite sex. Wrasses begin as females and change to males to improve reproduction. He put time-release implants of the drugs in the fishes' abdomens while the fish were anes-thetized. For control groups, he implanted other fish with placebos. He kept them around for a month, which is the halfway point of sex reversal, then killed them and removed gonads to see what their "sex" was—female or male.

Remember, the test fish had no "social cues" to trigger sex changes. When the results were in, the scientist found that the gonads from the animal getting the drug treatments were, at the halfway point, reversing. Norepinephrine alone did the trick. The control fish produced no changes. By simply increasing norepinephrine and blocking dopamine, he literally made sex changes. In similar experiments, investigators working with mammals—mostly rats and mice—have been able to manipulate dopamine to literally turn the typical "I'm ready for sex" signal of lordosis on and off in rats, like turning on and off a light switch. They've discov-ered that dopamine promotes all kinds of sex and courtship movements as well, including ear wiggling hopping, and darting.

Serotonin's story is equally dramatic. A study at the St. Louis University School of Medicine Department of Psychiatry and Human Behavior by Winston Shen and Jeffrey Hsu evaluated the side effects of psychiatric drugs that inhibit the reuptake of serotonin by special serotonin receptor cells in the brain on women. Thirty percent of the subjects taking these drugs, notably Prozac, complained they had lost interest in sex and could no longer have orgasms. It was clear to Shen and Hsu that imbalances of serotonin, independent of sex steroids, create sex problems. Too much makes people so calm they lose libido; too little creates aggression, violence, and even depression, probably by interfering with the brain's pleasure chemistry.

Hormonally speaking, says Hopkins's Adrian Dobs, interest in and capacity for sex is about equal in men and women, although because of cultural norms, women at times may appear to hide their desires and behavior when compared to men. "Each sex has an equal stake in successful sex and hormones are what help make things happen."

Along with our chemistries, of course, we also have a neocortex and a culture that allows us to modify our environment and adapt our courtship practices to unforeseen circumstances. We are strategic and *thoughtful* self-promoters, and what we do often depends on what's around. When Steven Stills wrote, "If you can't be with the one you love, honey love the one you're with," he was stating the pop-song version of courtship biology unearthed by neuroscientists and social scientists alike.

Just consider our repertoire of courtship behaviors. Men and women can be flirtatious or not. They can turn it on and off at will, they can be seductive, they can proposition the other, they can dress "dirty" or "demure." They can, honestly or dishonestly, tell the object of their attraction that other women or men are bad, smelly, stupid, unavailable, promiscuous, disloyal, or mean. They can lie or tell the truth about their fidelity or others' fidelity. They can show off their wealth or hide it. They can give gifts now or later. They can take or reject gifts. A man can allow a woman that he's interested in to think he's wealthy or that his fraternity brothers are not. He can act macho or tender, she can be nurturing or domineering. She can pretend to love sports or actually hate them. Or vice versa. Men and women can be helpful, nice to kids, friendly, and communicative, or they can avoid the "M" (Marriage) and "L" (Love) words assiduously and mock sensitivity in others. They can enhance their appearance or cover it up. They can tell their best friend that his date's breasts are silicone, then steal her for himself. She can pretend she's pregnant or that she isn't when she is. He can wear a vasectomy button on his lapel that is true or a lie.

Timothy Taylor, in *The Prehistory of Sex,* suggests that it was in fact our ancestors' *consciously* directed sexual choices of mates that in the end led to ever increasing brain size and cognition, which forever separated us from other animals and led to both the consciousness we have about our own behavior—including courtships—as well as sex as the centerpiece of modern human culture.

Modern humans spend more time and energy on dating and mating and the social systems that support that enterprise than we do on any other. And we have done so for eons. (If death and sex are the two primary human realities, and if language is the cornerstone of human society and culture, it's probably no accident that, according to some sources, Spanish has 105 words for the male penis and testicles but no word for condolence.)

Biology, Taylor says, has thus intertwined with culture to create a vast body of practices and products designed to enhance our sex lives, from our sex appeal to courtship and intercourse itself. Taylor sees every ivory phallus extender and aphrodisiac herb, every breast implant, penile augmentation surgery, pornographic magazine, circumcision, and fifty-thousand-dollar wedding as an extension of culturally and biologically determined courtship.

It was in this way, Taylor argues, that "love songs and floral bouquets" became as central as aggression in the life, if not the survival, of our species. Cultural as well as biological artifacts are carried around as part of our personal and group sexual baggage, like traveling museums of our sexual past. Unlike other animals, we picked up the flowers and songs along with the genes and the hormones.

Taylor believes humans have "pulled out of Darwinian evolution" by dint of our big brains. Perhaps or perhaps not. But our big brains have certainly complicated every evolved behavior, rendering courtship an everlasting negotiation for interest, commitment of resources, and emotional attachment. And as in all conflict resolution, whether hot war or cold, there are strategies and tactics, each with their genetic and biochemical components or accomplices.

Scientists have revealed the brain's key role in courtship by marking off the two distinct stages or characteristics of courtship, called appetitive and consummatory. Consummatory, you'll recall, is the label given to those contiguous and actual parts of the sex act, including arousal and penetration, thrusting and orgasm. In courtship, however, it's the appetitive aspects of sex whose biological roots are on parade; the motivation and desire to get sex, the seeking itself.

Studies have shown that animals, including humans, can have motivation and desire even when there is no expectation, at least immediately, of copulation. Similarly, under certain conditions, there is the possibility of sex without preamble. But what is interesting is that scientists have determined that different parts of the brain—certainly in other species—are involved in these two distinct dimensions of sex. In humans as well there are probably brain, mind, and bodily differences. In 1956, Frank Beach was the first to suggest that, at least in males, different hormone and brain mechanisms controlled sexual behavior with respect to courtship and copulation. In the late 1980s, Barry Everitt and his colleagues found that indeed there were different brain centers involved in these two aspects of sexual behavior. Compared to the amount of scientific interest in pelvic thrusting and ejaculation, research on the appetitive or courtship areas has been scarce, but there's enough of it to whet the appetitive behaviors of all of us.

Everitt, for instance, showed that obliterating the effects of male hormones by surgical or chemical castration in male rats markedly decreased or completely obliterated copulation, but lesions in the preoptic area of the hypothalamus, even if they suppressed copulation, had *no* apparent effect on the willingness of a male to work to gain access to a female. In essence, he would work like hell to be near her, but then just stare when he got there. In contrast, damaging the amygdala or ventral striatum specifically disrupted the courtship behavior, such as anogenital sniffing and ultrasonic vocalizations, without affecting at all the rat's ability to copulate.

In Japanese quail, as in rodents, such consummatory male behavior as neck grabs, mounts, and cloacal contact movements can be disrupted by specific damage to a special part of the brain called the medial preoptic nucleus. In an elaborate set of experiments at Hopkins, Ball and his coworkers set out to show that even minor changes in brain chemistry can dramatically alter who goes after whom; to do this, they rigged solid-walled quail houses with battery-powered, spring-mounted window "curtains" designed to measure and test just how long and under what circumstances a male will continue to "court" without being allowed to copulate.

After allowing male quail to copulate once, he found that such males will spend the majority of their time, literally days at a time, standing in front of the opened window and looking through it at the female (talk about yearning!). When Ball, by using the curtain, sharply curtailed male visual access to the female, he could measure the males' interest by recording how long they would literally crook their necks to get a peek at a female, which Ball considers a pretty good measure of appetitive sexual behavior. In each set of tests with each pair of quail, the door between the two areas was ultimately lifted and the feathered Romeo and Juliet allowed to have five minutes together in the bird-flesh, during which the frequency of mounts, cloacal contact, and copulation are recorded. Results? Ball found that males who go through the test many times progressively spend an increasing amount of time in the area in front of the window. They're willing to court again and again in hope of getting the girl again, a possible feathered version of absence making the heart grow fonder and, in courtship at least, practice making perfect.

Not surprisingly, courtship seductions and our senses are closely tied. Caresses and massage that trigger pleasurable biochemicals, foods that tempt our taste buds, odors that stimulate and arouse us, and musical sounds that evoke lullabies or happy memories are hallmarks of many species' dating rituals, including our own, and are universally coevolved behaviors between the sexes.

Gifts of food and the mutual enjoyment of a meal may be the most observed example of the tie between our senses and courtship. Most males in the animal world who want to copulate with females, in fact, first offer a gift of food as a metaphor for his willingness to give her even more. It's likely that prostitution first involved a simple practical exchange of food for sex that rewarded both men and women in times of hardship. The film *Pretty Woman* is an evolutionary biologist's script, and seduction dinners mimic feeding between a mother and her child, evoking a time when women were cared for and protected by their parents.

Food given in exchange for sex is found in humans, birds, fish, and cockroaches. She cooks for him and he buys her elaborate dinners (or, these days, vice versa). University of Chicago fruit fly expert Jerry Coyne notes that in drosophila, males taste a female's waxy body with their forelegs, which have taste buds and can tell right away if she's carrying the right pheromones. If she is not, he takes off for other flies. Male Israeli gray shrikes collect and hoard food for months before the females arrive to breed, and males with the biggest storehouses get the most female attention. And male scorpion flies, tiny, inch-long creatures, risk death to catch bugs in spiderwebs, kill them, release enzymes to break down their bodies into a sort of liquid protein diet, then hang from a stick to release their version of a dinner invitation, a pheromone, to attract the females. A female looks over the menu and even tastes it. She wants a big gift but will not take it if it doesn't taste good. If it suits her palate, his wooing wins. She copulates for up to twenty minutes, but eats throughout the whole enterprise. A note to the wise male: Get extra. If the meal isn't big enough, she'll eat his lunch, then find another male with a bigger gift to mate with right away and push the piker's sperm out of contention.

The *type* of food that works in modern courtship also has roots in our biological preferences. For example, before-dinner drinks and wine with the meal enable alcohol in its fabled job as loosener of inhibitions, sexual and otherwise. And it turns out that our perception of how willing some-one is to drink alcohol in the first place may be hardwired in our brains because it gives us potent signals about how much the object of our affection wants us. In a study based on subjects' interpretation of a vignette recounting a couple drinking, William George, Susan Gournic, and Mary McAfee of the State University of New York at Buffalo found that perceptions by men of sexual looseness of women were significantly enhanced if a man in the story bought the drinks.

Sweet and salty are the two most universal taste preferences, and pre-historic cave paintings show people who went to inordinate lengths to get honey from bees, even when bears were a threat. Harvey Weingarten, a Canadian psychologist, suggests that the energy expended to get craved foods often is so great that people may be driven by pleasure receptors in the brain, forming an actual addiction to certain tastes. In his study of a thousand college students in 1991 at McMaster University, Weingarten found that 97 percent of the women and 67 percent of the men had specific food cravings. Women craved chocolate more than any other food and men longed for high protein and fat. In 1996, a study by the Monell Chemical Senses Center in Philadelphia, which surveyed young adults and those over sixty-five, found that in the younger group, women craved high-fat sweets over entrees two to one, mostly chocolate in any form, while in the young men the reverse was true. In the older group, the older women craved sweets and entrées in equal numbers while the men still wanted the protein-high entrées over sweets two to one.

Chocolate, it is interesting to note, is a very complex food, containing more than four hundred separate compounds, twice as many as any other food, including caffeine and theobromine, stimulants that tend to arouse and elevate blood pressure. It also contains phenylethylamine, a chemical released into the brain during infatuation.

In the dating game, right next to the sensory value of taste and its links to life-sustaining nourishment, the sounds of love may rate almost as high. As Helen Fisher notes in *Anatomy of Love,* the goal of courtship is to build trust, and emotional sound-making—including conversational talk—is a way to reveal a weakness or strength, but in a positive, non-threatening way. Like other animals that produce mating sounds, people produce songs, whispers, poems. In one recent article published in the journal *Science,* researchers put forth stunning evidence that the love and persistence of music in the human experience is universal, essentially instinctive, and as significant in our courtship and mating lives as food and drink. We are the beneficiaries of a love of music that clearly bene-fited those of our ancestors a million years ago whose genes made it to the twenty-first century.

Analyzing evidence from birds, whales, and people, Roger Payne of the Ocean Alliance in Lincoln, Massachusetts, and one of the article's authors, found that elements of the same acoustic, compositional, and artful patterns of sounds are preferred by all the species studied. Whales, capable of far more musical range than humans, stick to shorter intervals in a discipline akin to singing in a single key, and when they vocalize, they favor a theme-and-variation format that Beethoven brought to full bloom. "Music," according to Payne, "is much older than our species."

Because songs, speech, and rhythm are highly conserved appreciations and we evolved an auditory cortex and vocal instruments to make and receive such sounds, scientists have long believed that what we hear and what others want us to hear are critical to courtship. And they have found lots of evidence to support that belief.

In some species, sounds from a male or a female alter hormone output by a receiver or partner. One investigator tape-recorded a variety of songs of male canaries and played them for groups of females. The group of females who heard a long song built nests sooner and laid more eggs than their sisters who only got a relatively few bars. In another study, male parakeet songs were recorded and played daily to females in two-hour sequences for six hours or in one continuous bout. When the tape was of males singing a soft warble, the females' ovaries grew and ovulated.

When the music was of other sounds, described as squawks, chedelees, whedelees, tuks, and loud noises, no dice. Females also did better when the songs were played in the morning.

In *Xenopus laevis,* the clawed frog found in sub-Saharan African lakes, bouts of spawning during breeding are triggered by drops in temperature after which the male makes a metallic trilling sound that excites females. The mate call lets the female locate the male in muddy water after the temperature-cooling rains, and the courtship ends when the males hang onto the females for hours while eggs are released and fertilized. And there is nothing random about this music; it's as rigidly composed as a Bach fugue, its trill made up of single clicks at either 72 Hz or 36 Hz, fast and slow, endlessly repeated. If a female isn't interested, she ticks at a slow 6 Hz. She also produces a slow kind of ticking sound when she has had enough of his love clasp. Scientists have exhaustively dissected the muscles of the larynx, the hormones and the brain structures involved in these animals' vocal systems, and learned that compared to the male, the female larynx has very few muscle fibers and those she has are weak. The adult males produce mate calls only when they have high levels of androgens and if they are castrated, they can't manage it at all. Even giving androgens to females can't produce male-style trills because her laryngeal nerve fibers and muscles won't support the effort. Unreceptive females can tick and respond to a clasping male, but if you give her an androgen, she'll stop ticking. One lesson learned is that if females did have the capability of mate calling, their own androgens needed for this process would also inhibit ovulation and her own fertility. Moreover, it is the brain, not the larynx that generates the mate call or ticking pattern; scientists have also identified particular spots in the frog forebrain (the anterior preoptic areas) and in the thalamus that are involved and are different between males and females.

Mark Jude Tramo, a Harvard neurobiologist and musician whose closet-sized office looks as if it's been tossed by racoons or possibly the macaques he studies, uses synthesizers, computers, and a thorough knowledge of the brain to sort out the music of courtship and sex. In a paper published in the same issue of *Science,* Tramo's neuroimaging studies of people making or hearing music revealed the absence of a specific part of the brain devoted solely to musical processing, and he is principally interested in using music and sound to address the question of how the brain is organized. But he is also convinced that musical sounds are integrated into our emotional memories and shed light on why our memories of a special song evoke powerful emotions.

In infancy, for example, almost all "speech" is musical. Babies coo and babble, their voices rising and falling in singsong cadences and tones. Later, when spoken and written word language dominates, the musical quality of speech is suppressed but still there, "hidden" in grammar (word order, subject-verb constructions) and in the way we end a sentence that is a question with a pitch up.

When experiments suggested that individual neurons in the auditory cortex will respond to pure tones in very rigidly restricted frequency ranges (some will "fire," for example, only between 1,000 and 1,500 Hz), Tramo began to record and analyze animal "music" in the form of mating calls, theorizing that if nature were to reveal the source of the brain's musical preferences in any scenario, it would most likely be one that is critical to an animal's survival—sex.

Specifically, Tramo monitored activity in the auditory cortex of macaques in response to the patterns of sounds or screams made during copulations. As both a musician and a neuroscientist, Tramo knew that sound-wave frequency determines pitch, so if one could figure out how the brain codes frequency, one could also figure out how it perceives and plays pitch. A "note" in music has the same "pitch" regardless of how the tone is made (by piano, guitar, voice) and neurons in the auditory cortex are spaced and arranged very specifically to organize pitch from low to high. Because of that arrangement, animals and humans all have some sense of combinations that sound "good" or "right" and others that sound bad, and the key is in the timing or frequency of neuronal "firing" in response to a sound.

Tramo's recordings of highly distinct male macaque copulation screams show that these shrill, multitoned "songs" are completely unlike aggression or anger screeches. Moreover, brain imaging shows that the response to this sound is different in male and female macaques. The females consider it a sexy love song and subordinate males consider it a warning to stay away. Says Tramo, "This scream is not just a mechanical or acoustical thing, but has strict levels of meaning."

As brain mapping has revealed that musical circuits involve both the thinking brain and the limbic or emotional brain, it's clear that people *feel* as well as learn music, and they both react emotionally to it and perceive its mathematical precision and rhythms. It's no wonder at all that love songs play an important role in courtship and that couples return to "their" song over a lifetime.

In an oft-cited essay, sexologist and anthropologist John Money declared that unlike subhuman animals, people's lovemaps, or idealized sense of the perfect lover or mate, were freed along the evolutionary way from the "biorobotic" preprogrammed restrictions of mating "calls," much as they were freed from other programmed aspects of courtship and sex.

By way of example, Money describes the stereotypical, almost biologically robotic courtship rituals of urban dogs. First the male dribbles urine on some markers and sniffs the ground for the pheromone-laden urine of a bitch in heat. When a willing female appears, he nuzzles and smells her hind parts, and they chase and parry until she agrees to mate.

Human beings were emancipated from rigid ritual first by the ability to walk upright, which freed us from always having the nose and genital organs on the same plane and from rear mounting to copulate. Finally, says Money, humans were emancipated from the stereotypical soundmaps of hoots, yowls, clicks, whistles, squeals, barks, snarls, chatters, screams, grunts, and yells and were free to form more versatile communications such as crooning, cooing, humming, and chanting, in short, musical songmaps. "Like the mating invitation of a songbird, the love song would have been at first a song without words," he writes, "a song of solicitation and allure. If the human species were not a singing species, it may never have evolved into a talking species. The songmap is the evolutionary bridge that connects the lovemap and speechmap, sex and language."

Beyond taste and sound, when it comes to courtship, the evidence for a biological fragrance factory devoted principally to this part of sex couldn't be stronger, although in many cases, the odors in play are "odorless" volatile compounds and the capacity to detect them housed not in our ordinary smell brain and nose, but in a much older part of the brain's sensory system.

Dr. Martha McClintock and her colleague Kathleen Stern of the University of Chicago have shown that exposure to some of these compounds, taken from the underarms of young women during their menstrual cycles, can dramatically extend or compress those cycles in a very predictable way and clearly do so by influencing the intricate release and interaction of female sex hormones. The experiments, published in the journal *Nature,* were among the first solid pieces of evidence that human pheromones—long suspected—actually exist and that like all mammals, people produce a veritable *parfumerie* of chemicals that influence, control, and manipulate the sexual behavior of others.

In 1971, McClintock first documented menstrual "synchrony" itself in women students housed together in dormitories. Such parallel ovulation also occurs in rats and other animals who have periods of "heat" in which they get pregnant and bear and nurse offspring communally. She and others suggest that coordinating the timing of fertility and sex offered animals group protection and backup at vulnerable times. Because reproduction is often risky and always time-consuming, one theory holds that women evolved a means of coordinating ovulation at times when courtship and mating are riskiest or as a means of helping each other breastfeed. If pheromones play the role McClintock and others think they do, they may also help fend off competition by a woman living in close quarters with sisters, daughters, or others who might be vulnerable to a husband or lover living with her.

In any case, the compounds McClintock and Stern soaked up from the armpit secretions of women in the early part of their menstrual cycles, when the ovarian follicles are maturing, shortened the menstrual cycles of women exposed to them. Compounds taken at midcycle, during actual ovulation, lengthened the cycle. Chemicals taken after ovulation, in the luteal phase, had no impact. All of this happened merely by asking nine healthy women to wear cotton pads under their arms for eight-hour stretches, cutting the pads into sections, treating them with alcohol, and wiping them under the noses of twenty other women daily for thirty days.

Unlike most hormones, which principally facilitate "internal" sex messages which then indirectly affect the behavior of the opposite sex, pheromones directly transmit sex signals to others through the air, by touch, smell, or taste. Among pigs, for instance, breakdown products of androgen form the molecules of an attractant secreted in male salivary glands that the boar literally breathes into the face of the sow. Only the sow can "smell" or process this pheromonal perfume. If she is eager for sex, one whiff puts her in lordosis, the spine-raising behavior that exposes her genitals and signals her willingness to let the boar mount. (Pig breeders use an aerosol spray called Boar Mate to prepare sows for artificial insemination.) Female goldfish produce a progestin that when excreted into the water sends the male hormone system into overdrive, making increased concentrations of gonadotrophin and progestins, which stimulate sperm making. Prostaglandins, which bring on spawning in these fish, also trigger male sex behavior when they swim through them in the water. Males have apparently evolved the ability to detect these chemicals, and those whose prostaglandin sensory detectors are the most refined win the copulation race.

It's fitting, perhaps, that the first pheromone was identified in a silkworm moth in 1959 by the same man, Adolph Butenandt, who won a Nobel Prize for isolating testosterone. Since then, scientists have searched for human counterparts, convinced it's unlikely that so powerful a sex chemical and the machinery to detect it would be absent in human mammals.

If pheromones are operating in people as they are in other animals, they probably work the same way as they do in lower animals, not through our "smell" noses but through that knot of tissue in one of the oldest parts of the brain, the VNO. Pheromones aren't smelly in the same way that flowers or rotten eggs are smelly, but are "sensed" through touch or inhalation, after which they trigger sexual arousal. Butenandt looked for his pheromone in moths because for a century, scientists knew that female silkworm moths could arouse male moths miles away by releasing a chemical at a rate of 10 billionths of a gram per second that could be picked up by males on the hairs of their antennae. (The word pheromone is Greek, and means to carry [pher] a hormone.) Animals respond to and produce dozens of these tiny, potent sex molecules, which are regulated by steroid hormones. Female monkeys, for instance, make estrogen-regulated copulin, a small amount of which can excite an entire colony of males. At last count, bees have been found to make thirty pheromones.

Some studies suggest that humans use pheromones and outright odors during courtship and mating to signal genetic fitness. Zoologist Claus Wedekind of Bern University in Switzerland, for example, tested women's responses to armpits and sweaty T-shirts and found that women have strong preferences and are attracted most to the scent of men who are most *unlike* themselves in a very special way—the array of immune system genes known as MHC, or major histocompatibility complex.

Inbreeding tends to concentrate deleterious genes, so clearly we evolved to not breed with sibs, parents, or close relatives. MHC genes are the most diverse of all genes, so different between people that organs transplanted from even close relatives can be rejected and cause graft versus host disease so terrible it is lethal. Biologist Lewis Thomas was one of the first to suggest that MHC differences conferred distinctive odors, and scientists found that inbred mice who were identical except for MHC genes could tell the difference among animals with different MHCs and that young mice tended to prefer the odors of their nest mates until puberty, but not afterward.

Wedekind knew that humans tested could also detect MHC-dependent odors, but he was the first to demonstrate a mating preference component. He recruited forty-nine women and forty-four men with a wide range of MHC genes and asked each man to wear a clean T-shirt for two nights without benefit of deodorants or perfumes. Wedekind put each sweaty T-shirt in a plastic-lined box with a sniff hole and brought in the women, all of whom were at the midpoint of their menstrual cycles, when women's noses are thought to be at their best. Each sniffed seven boxes, three of which contained T-shirts from men with MHC genes closest to their own, three with shirts from men most unlike themselves, and one unworn T-shirt. The women were asked to rate the shirts as pleasant or malodorous. Overall, the women preferred the odor of men with the most *dissimilar* MHC.

When mice in the MHC studies were pregnant, their odor preferences regressed to what they were as pups, preferring close "family" odors, probably because when pregnant mice nest with relatives, they get help with the pups and protection against strange males who want to kill her offspring and copulate with her. Wedekind's women who were on the pill, which elevates estrogen and mimics the hormonal status of pregnancy, also preferred the odor of men whose MHC genes were closer to their own—and MHC-compatible couples are known to be less fertile.

A mouse of course is not a man. Even if it turns out we have pheromones, humans are never going to be slaves to pheromone chemistry in the same manner as boars, lobsters, scorpion flies, or mice. Nevertheless, pheromone scientists at the Monell Center and the Athena Institute, both in Philadelphia, are among those actively seeking to isolate human pheromones, convinced that they may hold the keys to a variety of human behaviors including courtship, violence, and even phobias.

As courtship plays out, the tactics of courtship competition are only successful, and perhaps only fun and rewarding, when they eventually give way to the sexes' equivalent of diplomatic negotiations, with both agreeing to selectively forget or overlook past exploitations and even aggressions in the name of mutual desire for orgasms and offspring, or at least peaceful, companionable coexistence. A successful end to courtship must be a win-win and not a zero-sum game unless rape is the goal, and there is evidence that nature has found a way to keep date rape to a minimum by making forced copulations less likely to achieve fertilization, both in animals and humans.

And so, it seems, nature has given us plenty of physiological and psychological tools to help us get to "yes." And in the next chapter, we'll see what our biologies have in store for us once we agree to risk arousal and intercourse, and to mingle our genes.

7

KISS AND TOUCH

"The first time one performs the act of love, kisses, scratches, bites and other such caresses should be used with moderation, and the whole act of love should not last long. But on following occasions, on the contrary, all moderation should be thrown aside, the act should be prolonged for as long as possible, and to make the fires of desire burn even more brightly, all manner of caresses, cries and other stimulants to love should be used."

—KAMA SUTRA, EDITED BY VATSYAYANA

Trobriand Islanders in the Pacific spend hours at it. The Lepcha of Sikkim only minutes. Bonobo chimps spend more time kissing and fondling each other's genitals than copulating. Female dolphins and whales stroke each other's flippers. Male minks draw blood with their nipping and females won't ovulate without experiencing physical injury. Whether it lasts moments or hours, the animal kingdom's vast repertoire and appetite for arousal is compelling evidence of its evolutionary roots and biological value to sexual reproducers.

In a series of studies designed to examine the nature of the relationship between physiological arousal and attraction and gender, for instance, Douglas Kenrick and his team at Arizona State University found that both men and women who are physiologically aroused behave even more in their biological and evolutionary self-interest than men and women who are not aroused. The scientists figured that if evolutionary theory is right, and the main response a male has to first encountering an attractive female stranger is sexual attraction, then ipso facto males should be more attracted to any attractive woman when he's already aroused. And similarly, if the main response of a woman first encountering a male stranger is to assess danger and the probability he is willing to support any offspring he sires, then an already sexually aroused woman would be even *more* wary of a stranger. The Arizonans asked men and women to rate their feelings toward a stranger and a current dating partner under differing levels of physiological arousal and discovered that indeed men's sexual attraction toward a stranger increased with arousal, while, more interestingly, a woman who is aroused rated an unfamiliar male as less friendly and more threatening when arousal was heightened.

Just as the shape and proportions of the opposite sex lure us to one another, so do touching, dancing, and, quite literally, petting evoke erotic lessons of one of the first physical delights of life, the cuddles and caresses from swaddling parents. As one psychologist has noted, children can be seen as young as age three in "spontaneous erotosexual rehearsal play" such as pelvic rocking or thrusting movements "as [they] lie side by side at nap time." Even fetuses in the womb somehow come upon the means of sexually arousing themselves.

In grown-ups, the keys to arousal are more deliberate and more varied. A long-married woman may begin in a shower sudsing her breasts with scented soap and end up on the bathroom floor rug stroking her clitoris and imagining a romantic interlude with a man not her husband. Whatever the means, intercourse and orgasm may be grand, but getting there is often far more inventive, consuming, and engaging. Mating's warm-up, along with all the biology that attends it, outdraws even copulation in the sex game. For many, in fact, the pleasure of arousal matches or exceeds that of orgasm. Given the central purpose of sex—the survival of species—evolution could have selected for traits that conferred ever-ready status on animals: permanently erect penises, vaginas awash in lubricant. Instead, what we get is all of that kissing, cooing, petting, groping, cuddling, massaging, tickling, rubbing, teasing, smelling, tasting, talking, licking, nipping, pulling, fondling, and fantasizing. The reasons why this is so help answer how the biology of arousal works.

First, it is because nature is fundamentally practical. The need for preliminaries, for arousal, prime the body and mind to facilitate sex and give sexual partners up-to-the-last-minute chances to back off in the face of injury, danger, or illness. The *absence* of arousal, particularly in women, makes penile penetration harder or painful, so one of the benefits of foreplay—for both man and woman—may be to raise emotional barriers to rape. Second, a high capacity for arousal may also have evolved to let us ignore pain when survival of the species was at stake. The late Alfred Kinsey was among the first to popularize the fact that pain subsides during sexual stimulation and that foreplay desensitizes people to even severe injury, reducing the sense of pain in much the way an anesthetic might. Another pioneer of sex research, Theodore Van De Velde, long ago reported that during arousal, when mind and body are focused entirely on the genitals, pain can be blotted out as successfully as it is during orgasm. Sex researchers William Masters and Virginia Johnson discovered that women engage in long bouts of masturbation to relieve lower back-ache, headache, and menstrual cramps. ("Not tonight, honey, I have a headache" may be a cliché, but a premenstrual woman may *need* her sex partner to go away so she can masturbate and get faster and easier relief.) Most of all, like Olympic coaching, arousal allows bodies and minds the time and effort to focus and prepare; we must recruit incredible physio-logical resources such as hormones and muscle strength to reproduce, rebuild resources between reproductive cycles, prepare the body for force-ful sperm release and fertilization, and ultimately muster the energy for "pair-bonding" and parenthood.

What arousal offers is the means to make sex as deliberate and serious for psyche and soma as it is pleasurable. Notwithstanding all of the (mostly) bad jokes about foreplay and who wants it, what's enough and what works, it is both prelude to and echo of intercourse and, unlike the mechanics of copulation itself, an act of will, a declaration of making love as well as sperm war. Because as we've seen, sex takes place as much between the ears as between the legs, foreplay's design is to give each location the opportunity to prepare to the maximum capacity for all of the stress of coitus. As Masters and Johnson first documented, sexual excitement—for men and women alike—follows a pattern of gradual increase in physical *and* emotional sensitivity promoted as much by the mind's stimulation as by physical stimulation.

The principal consequence of all this is that evolution made sure we would not only *pay attention* to who or what arouses us (see chapter 6), but also have separate, redundant, and persistent systems for becoming aroused. For instance, the parts of the brain so far known to be involved in copulation and arousal are *separate:* Even if sperm and egg aren't des-tined to meet, we are quite willing and able to become aroused, and studies described in chapter 4 on the sexual brain showed that animals with damage to the preoptic area, or POA, can be aroused and reach erection but can't copulate. And there is evidence from human studies that when things are working the way they should, it's our brains that must turn on first in order to release chemical messages that harden the penis and wet the vagina. And foreplay is the medium for those messages.

A 1976 survey of foreplay practices worldwide found stunning patterns that underscore their power and their biological underpinnings. Topping the list in almost universal descending order of preference were general body fondling and petting, hugging, and stroking, followed by simple kissing, tongue kissing, and breast fondling. Finally came genital touching, fellatio, cunnilingus, anilingus, and painful stimulation of body parts, such as biting and hitting.

Strong evidence that we have redundancy in our arousal system, or at least multiple routes to arousal, comes from a variety of sources. One is that foreplay without intercourse is well known to produce an orgasm physiologically identical to the mix-your-sperm-and-egg kind, but at times even more intense. Then there is the tingle in a mother's nipple during nursing that is physiologically identical to the sensation her lover can also create during foreplay. It all feels the same by design because it starts and sustains the sexual engines; the biochemical rewards and incentives we have for arousing ourselves and others are the same, even when reproduction remains out of sight or mind.

Further evidence of the importance nature invests in arousal is the fact that the more we *practice* it, the more we want it and the better we get at it. This means that to some extent, at least, arousal is a *learned* behavior, one over which we have some purely psychological control as well as some inborn motivation to generate at will. Jean Cocteau was reported to have entertained party-goers by lying down naked on a bed in front of a gathered crowd and willing himself—"no-handed"—to an erection and orgasm. (A public tour de force, to be sure, but performed more privately by lots of us.) In animals and humans alike, observations of sexual virgins as well as novices and veterans like Cocteau show that the author of the Kama Sutra got it right: Foreplay is not just a means to an end, but, if we put our mind and body to it, also an end in itself.

Before an erect penis can penetrate a lubricated and flexed vagina, foreplay ignites a firestorm of immensely rewarding, blissful physiological and psychological events, fueled by neurotransmitters and hormones. Priming blood vessels and nerves, foreplay sends the initial motivation and desire for sex into overdrive, supercharging, tensing, and focusing all of the sexual body and brain and using up energy at an athletic rate.

Masters and Johnson showed that the human sexual response is a predictable series of *physiological* events that successfully alter the shape and function of the genital organs to get two sets of gametes joined. If arousal is not achieved, conception the old-fashioned way is not possible.

If the penis is flaccid and the vagina tight and dry, coitus, unlike sleeping, eating, and fighting, can't happen. Thus, beyond its own pleasurable ends, foreplay must generate erection (his), lubrication (hers), and a buildup of sexual tension and excitement marked by erotic feelings (theirs).

To that end, chemicals and hormones act on blood vessels feeding the genitals and internal reproductive organs. They expand and constrict muscles, engorging tissues with blood by reflexively dilating blood vessels in the penis and vagina in response to pressure, stroking, and other stimuli. They increase blood pressure, make muscles rigid and hold them hard and fast in advance of spasm, and accelerate breathing and heart rate. A cascade of potent molecules thicken the scrotum, flatten the scrotal sac, elevate the testes, swell the nipples, lift the uterus from the pelvic floor, and create the vasocongestion and organ swelling within the female pelvis.

Arousal, as noted earlier, originates not as feelings of physical bliss, but as erotic-emotional *feelings*, mediated by brain chemicals that act in and on the limbic system. Even at times when one feels simply physically excited—horny, to use the vernacular—emotions and neurotransmitters are brought into play. We've reacted psychologically to some stimulus and we use our mind's focus to either sustain the arousal or help brake it if the time and place are inappropriate. Only when feelings kick in can all this chemistry make blood rush to the spongy tissue in the penis, filling the spaces and expanding the organ, while a clear fluid leaks from the cells of tissues deep in the vagina, lubricating it with super-filtered blood plasma and increasing its width and length.

As arousal advances from the earliest phases of "excitement" to the "plateau" stage just moments before orgasm, chemicals released by ovaries, testes, the adrenal and pituitary glands, and the brain continuously engorge and sensitize the genitals of both genders to the point where release is not only inevitable but painful if it is not achieved. The male's penis is filled and stretched to the limit with blood, its shaft erect and his testicles 50 percent larger than in the non-aroused state. Muscles controlling the spermatic cords are strained, which has the effect of elevating the testicles against the tissues around the anus and sensitizing nerves there as well. In the female, the lips of the vagina swell and turn bright red or burgundy and form a thick plate of tissues around the swollen vagina. The clitoris swells with blood, creating a sort of female erection in this small knot of tissue, which when fully aroused pivots 180 degrees and flattens behind the pubic bone. So engulfing is arousal physiology that even the tissues of chest muscles or pectorals fill with blood.

The stages of physiological arousal through foreplay, whether "real" or imagined, differ in men and women because their hormonal and biological foundations differ. For example, although blood vessel congestion is blood vessel congestion, regardless of where the vessels are, in men it's the corpora cavernosa and spongiosum of the penis that selectively swell to produce erection and in women it's the bulbs of the vaginal vestibule that inflate. And it's the more specialized and complex male vasocongestive response that is particularly vulnerable to distractions that can undo the process. One consequence—and proof of this male vulnerability—is that impotence is a common complaint in men whereas a failure of either lubrication or swelling of the vagina is very rare, even in women who don't experience orgasm. The runaway best-selling impotency drug, Viagra, won't make a penis bigger or an erection harder; it has no impact on men who get normal erections. It *does* enhance a penis's responsiveness to foreplay, however, by blocking the effect of an enzyme that causes erections to *wane*, thereby prolonging them.

The chemicals of arousal, facilitated by what we label foreplay, play a role in all stages, sometimes alone, but usually in combination with testosterone, estrogen, pheromones, and endorphins.

Vasopressin, for example, a hormone made by the pituitary gland at the base of the brain, is an "antidiuretic," helping tissues retain fluid. Its action is probably responsible in part for the achy swelling pressure that increases sexual readiness and tension during foreplay. Recent studies suggest that vasopressin increases or enhances alertness and attention, at least to certain parts of the body, letting us focus on sexual cues and our genitals.

Dopamine, norepinephrine, and serotonin, the brain's principal "chemical messengers," have substantial effects on libido and arousal. Evidence comes from studies of such mind-altering drugs as antidepressants and amphetamines on both laboratory animals and people, which reveal they alter not only brain chemistry but also willingness to mate and responsiveness to arousing stimulation.

These neurotransmitters also are known to regulate such behaviors as impulsivity, anxiety, depression, obsessiveness, and hostility, factors in many so-called paraphilias, or obsessive preferences for unusual forms of arousal. This is why drugs such as Prozac—which ease depression, obsessive-compulsive disorder, and anxiety by enhancing serotonin, the brain's "calm down" chemical, and relatively decreasing the impact of dopamine—also inhibit sexual arousal common to pedophiles, sadomasochists, and exhibitionists. It may also be why chemicals of anxiety, such as cortisol, both inhibit *and* facilitate arousal depending on the level of anxiety men bring to the bedroom. A *little* anxiety, perhaps related to performance with a new partner, seems to be arousing, while a *lot* of anxiety related, say, to having just lost a job, is quite another (wilting) thing. Biologically, this makes good sense, since a man faced with *major* anxieties, such as an animal attack or the risk of starvation, would not be wise to get too easily aroused by a willing female.

Such information also underscores the fact that the *normal brain* is wired to respond to foreplay and the use of fetishes and fantasy to arouse us erotically. In the case of some mental illnesses and extreme paraphilias, some kind of damage or miswiring hijacks the brain chemistry nature gave us to respond easily to erotic stimuli. For example, areas of the brain involved in the regulation of masculine sexual behavior tend to be richer in the receptors for the brain's own opiates as well as dopamine, particularly areas involved in processing such sensory information as smell. Studies suggest that the opiates in our brain can activate the dopamine system and stimulate sexual motivation by helping our minds link environmental stimuli (such as odors, colors, tastes, and sounds) with pleasurable feelings and sexual arousal.

As for kissing and other forms of erotic touching, oxytocin, along with prolactin (a chemical that plays a role in milk production) and other biochemicals, may be part of the reason these actions are so central to foreplay and arousal. Stores of oxytocin rise in the body in response to touch and genital stimulation. It stimulates uterine contractions, increases penile sensitivity, and, as we've seen, draws animals to nuzzle, cuddle, and seek body contact.

Sue Carter, an expert on oxytocin, sex, and neurobiology at the University of Illinois at Chicago, has uncovered some of the almost magical properties of the chemical while working with everyone's favorite, the prairie vole, in particular the monogamous variety.

Carter first observed the female's unusual habit of never ovulating until she meets a strange male, along with the male's devotion to her once they mate. "What this suggested to me was that these animals were using social, rather than hormonal or gonadal factors to run their lives. That meant there were factors mediating their behavior that weren't sex hormones, but other kinds of biochemicals." Intriguingly, Carter says, this may be what is going on with women as well. For women to be aroused, to seek out sex, they must feel safe, while males are capable of sexual activity in conditions that are far more stressful. A female can conceive (with more than normal difficulty) during a rape, but she is not having "fun" or enjoying "sex," because—as with the voles—sex is a social system function that requires a given response to a given set of circumstances.

Looking for the natural chemistry that might regulate such behavior led her to oxytocin in part because it goes way back in evolutionary biology and is conserved broadly across the animal kingdom. "I was led to oxytocin's door time and again. I believe it is at the very central core of what integrates everything involved in reproduction and everything involved in sexual behavior. Oxytocin just always is around at the right place and at the right time, from arousal to orgasm, to pair-bonding."

A chemically close relative of pitocin, the drug used to induce labor by provoking uterine contractions, oxytocin is like other neurotransmitters such as dopamine and serotonin that influence how we respond to environmental challenges. Because the voles in her study were committing to sex only when the environment, or circumstances, were aligned, Carter thinks that women are doing the same thing, and that is why oxytocin may indeed be so heavily present during arousal, sex, birth, and breastfeeding. It also has been found to increase production of nitric oxide, a neurotransmitter necessary for erection in males. As suggested earlier, it once was thought that testosterone and its male sex hormone relatives were the major players in increasing male libido and response to foreplay, but the most recent studies in animals and humans show that nitric oxide synthase, the enzyme that makes nitric oxide, is the key player, although it is dependent on sex hormones for the sensitivity of tissues to its impact. A study looking at how changes in testosterone levels alter erectile function in rats, for instance, used electrodes to stimulate a key erection nerve in animals that had been castrated and therefore made no testosterone. After treating the animals' penile tissue with a stain that shows off nitric oxide synthase, the scientists gave them shots of testosterone. When they measured the rats' ability to maintain an erection and the concentration of the enzyme in nerves, they discovered that testosterone only plays a direct role insofar as it influences the uptake of nitric oxide synthase in the cylinder of tissue that forms the clitoris in the female and the penis in the male.

Although it may seem like a more straightforward process, men's arousal and erection requires a complex interplay of psychologically and physically stimulated supplies of neurotransmitters like nitric oxide, muscle-relaxing chemicals, and hormones. Smooth muscle fibers in the corpus cavernosum and arteries that go to the penis relax in response to the release of chemicals made by certain nerve fibers. Nitric oxide then kicks in and influences the release of noradrenaline by other nerves, which contracts the penile smooth muscle. Clearly, the links between brain signals and the contraction and relaxation of the penis's erectile tissues are subtle, embracing a balance between the relaxing and tensing of fibers and chemicals that control muscles.

Just as nitric oxide, which travels around with other chemicals—one called VIP also dilates blood vessels—is fast emerging as the main chemical of erection, noradrenaline appears to be the main neurotransmitter of flaccidity. In order for a penis to become erect, it must fill with blood through wide-open vessels: while nitric oxide relaxes these vessels, noradrenaline contracts them.

The role of nitric oxide cannot be underestimated. A fourth of men with prostate cancer who have been either surgically or chemically castrated to save their lives are still found to have erections, even in the complete absence of the major supplier of testosterone, probably because nitric oxide offers "redundancy" in the erectile system. Indeed, studies of nitric oxide suggest it may be time to give the "raging testosterone-laden beast" theory of male arousal a rest. Scientists at McGill University, for instance, found that testosterone does not even directly influence how fast men get erections, or how strong the erections are. Studying testosterone levels in men exposed to pornography—a typical "hands-off" style of foreplay for males—indicates that the hormone works by impelling the men to *attend* to the erotic stimuli. In short, the testosterone worked on the *brain,* riveting the men's attention to the stimulation and keeping their minds from wandering to other distractions.

Contrary to conventional wisdom, then, it's not raging hormones that arouse us initially; instead, our arousal is more likely what turns them on, and only subsequently do our minds and chemistries mutually stimulate each other to keep the arousal in play to its end. John Bancroft, the head of the famed Kinsey Institute, and his colleagues once studied 141 women, forty to sixty years old, to investigate which factors determine sexuality and well-being. They interviewed them and studied their sexual function, menopausal status, bone mass, smoking, and sex hormone levels. The findings? None of the hormones' levels, or even their absence, significantly predicted sexuality or interest in it; the most important predictors were things such as sexual attitudes and overall health. And the best single predictor of the sense of well-being that facilitated sexual desire and performance was tiredness, or lack of it.

Anyone still persuaded that, singly or collectively, one's sex hormones are all that move us to desire and arousal couldn't withstand the mountain of evidence accumulated by Arthur Arnold and Roger Gorski of UCLA, who pioneered studies of hormonal influences on sexual development in the brain and body of a wide assortment of birds and other animals. Manipulating sex hormones, they not only demonstrated that male and female brains respond differently to steroids, but also disproved the simplistic idea that females act the way they do sexually because of estrogen and males because of testosterone. No one single manipulation—not even castration—could account for either the presence or the absence of certain sex behaviors, or variations in male and female sexual behavior.

The communication between mind and body during arousal also explains the biochemical success of Viagra, first developed as a drug to increase blood flow to the heart by dilating blood vessels. Test subjects noticed better blood flow to their penises and drug makers soon realized that Viagra suppressed the effect of an enzyme called phosphodiesterase type 5, which breaks down a major energy molecule called cyclic GMP. Cyclic GMP, which is produced only during arousal, triggers the erection, and the enzyme cuts it off after orgasm so that the penis can relax. In impotency, the enzyme keeps the upper hand, producing a relative shortage of cyclic GMP. What Viagra does is muffle the enzyme long enough for a man's own cyclic GMP to work. And because erectile tissue comes with a set number of receptors for cyclic GMP, no amount of additional cyclic GMP will have any effect. That's why men who can get aroused already need not waste their money on Viagra, hoping for a bigger or stiffer performance. Viagra works to increase *sensitivity* to arousal, rather than to directly arouse, and in men who already have sufficient testosterone levels when erection occurs, further arousal will not keep generating more testosterone or harder hard-ons.

It turns out that women's sexual responses also are sensitive to nitric oxide synthase and its fellow travelers. In the vagina, nerves that contain nitric oxide synthase play a role in controlling blood flow in response to foreplay just as they do in men. But it's VIP, which stands for vasoactive intestinal peptide, that may become the newest celebrity chemical of women's arousal. According to its discoverer and chief interpreter, Joseph

Bohlen of the University of Minnesota School of Medicine, VIP increases blood flow in several parts of the body and in the vagina produces the lubricant in a woman's vaginal canal. But its real significance may lie in the fact that it is possibly the big player in arousal responses that seem to come without conscious recruitment or warning. VIP is made by nerves found in the digestive tract, respiratory tract, and urogenital tract that control smooth muscle tone and movement, blood flow and secretions. These so-called non-adrenergic nerves are wholly unlike nerves involved in foreplay and arousal that are activated by drugs such as epinephrine (adrenaline), a stress hormone made by the adrenal glands that increases blood pressure, heart action, and constriction of peripheral blood vessels. The adrenergic system is much more subject to our control, while the VIP-making system may not be. That's why people with high blood pressure who are given drugs like clonidine have trouble getting and maintaining an erection, decreased vaginal blood volumes, and lower arousal responses to erotic films. Clonidine and similar drugs reduce fluid levels, which lowers blood pressure by activating adrenergic nerve cells.

It seems the human vagina is rich with VIP nerve fibers, and Danish researchers have demonstrated that they can dilate genital blood vessels and increase fluid leaks in the vagina. The vagina generally is a moist space with minimal blood flow, but with the first signs of arousal it begins to engorge. Oxygen tension rises around the vaginal opening to squeeze out lubricants that permit painless penile penetration and coital movement. In studies of this process, VIP has been found in nerves closely positioned near blood vessels in the vaginal wall. And giving VIP by injection into the skin of the vaginal wall increases vaginal blood flow and induces vaginal fluid production.

As noted earlier, our capacity for arousal—and orgasm—may have been one of the ways we evolved to keep us interested in sex even when in pain. Beverly Whipple, a New Jersey neurophysiologist and sex researcher who has investigated the notorious "G" spot reputed to activate female orgasm, reported on a woman injured in a car accident whose shoulder and neck pain disappeared during orgasms and for up to eight minutes after.

The relationship between pleasure and pain may be at least as strong in our arousal chemistry. From "love taps" to nips and bites, the biochemical and physiological lines between what hurts and what feels good have always been fuzzy. Pain at its worst blots out every other sensation, no matter how pleasurable. However, the most recent research by a close colleague of Whipple's, Barry Komisaruk, and his colleagues at Rutgers University has now shown that the reverse is also true.

The way they showed this began with the observation that foreplay techniques such as vaginal and cervical stimulation during masturbation, as well as coitus and orgasm, can produce in women and other animals an analgesic effect—a selective attenuation of the sense of pain but not of touch. Komisaruk reasoned that the pain-blocking mechanisms triggered in women with spinal cord injuries were the same ones activated by vaginal stimulation in those with an intact spinal cord. If true, it would mean that physical foreplay and erotic fantasy combined can block or dampen pain independent of vaginal stimulation.

Komisaruk first tested his ideas on rats and found that vaginal stimulation acts through the spinal cord to directly block responses to painful stimulation in the brain's pain centers *and the other way around.* Further neurophysiological studies revealed that in the thalamus, pain neurons could be blocked by vaginocervical stimulation, and vaginal self-stimulation substantially eased the perception of pain caused by sticking the finger with a pin (a standard lab test).

Komisaruk speculates that at least part of the reason we evolved a means of blocking pain by vaginal stimulation is that it could reduce potentially aversive or painful sensory stimulation that may occur during coitus, and thereby increase the positive reinforcing effects of the pleasurable component of the stimulation. This may sound both obvious (that sex feels good enough to blot out pain), and hard to believe (that sex could be painful), but intercourse does have its painful components; at the very least it is hard work that taxes the muscles and sensitive genitals. The chemicals released by stimulation directly into the spinal cord may bathe the body in a countervailing pleasure, and later it's the pleasure, not the pain, that the brain recalls. These pain-attenuating and pleasure-inducing processes would, by acting in concert, facilitate the performance of coitus and thereby promote the process of reproduction. Getting hooked on foreplay and arousal gets us hooked on the analgesia, which reinforces sex behavior from start to finish. Additionally, one reason women may "forget" the pain of childbirth is that the brain at the time is also occupied with the chemical pleasure released by vaginocervical stimulation, which produces a wide array of responses such as increased blood pressure and pupil dilation, and secretion of oxytocin directly into the spinal column. In 1995, Komisaruk and chemist Frank Jordan, also at Rutgers, patented what they believe is one of these chemical pain-blocking peptides related to VIP.

Where the analgesia comes from is just as extraordinary to the researchers as its presence. Komisaruk and his team found "robust pain-blocking responses" through vaginal and cervical self-stimulation even in women with proven, complete spinal cord injury from the chest down, an injury that classical thinking held should block all feelings of genital arousal—and everything else below the belt—en route to the brain. The reason, they speculated, was that "analgesic" messages were traveling along a path that bypasses the spinal cord and enters the brain directly, in this case via the so-called vagal nerve system.

The nerve pathways and mechanisms that underlie this pain blockage were revealed in the cord-injured women, he says, both by analyzing nerves that mediate pain in women and from the surprising finding that these women show the same kind of analgesia responses to vaginal or cervical self-stimulation as women with intact spinal cords. Clearly some other sensory pathway was involved.

Komisaruk first tested this hypothesis by stimulating rats mechanically with a simple tampon-like plastic cylinder mounted on a rigid stick and equipped with a transducer to measure vibrations. The penile-like cylinder itself is not a vibrator and is not agitated or jiggled, just pressed on the vaginal wall to produce the effects of arousal and pressed rhythmically to trigger orgasm. Komisaruk and his colleagues reproduced this effect and identified what they believe is a previously unappreciated feel-good pathway that leads or at least operates from the reproductive tract directly to the brain, bypassing the spinal cord.

The implications of such findings are profound, certainly, for the sexual rehabilitation of spinal-cord-injured patients. But they also underscore the probability that nature programmed us to have great redundancy of sexual pleasure sensors and feelings.

In further experiments with volunteer women, Komisaruk and his team painstakingly plotted and tracked, using electrodes placed on selected nerves and body regions, the sensory path of pain that feeds the vagina, womb, and cervix. They found that the uterus and vagina have different nerve supplies. Pelvic nerves serve the vagina and cervix while hypogastric nerves serve the womb and cervix. Thus, the cervix gets both supplies of nerves. The pudendal nerve is also involved, providing sensation to the skin around the vaginal opening; the pelvic nerves also supply a narrow zone around this skin. The clitoris and skin around it are served by the pudendal nerve.

But the most interesting findings focus on a nerve called the vagus, which helps orchestrate such everyday jobs as breathing, swallowing, and vomiting and supplies major organs; but it also bypasses the spinal column, hooking directly into the base of the brain, never even touching the parts of the body or other nerves involved in the sexual body the way the pudendal, pelvic, and hypogastric nerves do. The vagus nerve, also known to anatomists as one of a group of cranial nerves that carry sensory and motor information to and from the face, eyes, and other parts of the head, is named after the Latin word for "wanderer." And so it does. Its fibers wend through gut, lungs, heart, stomach, intestines, through deep organs, and as part of the "involuntary" or autonomic nervous system, carrying sensory information directly to and from the brain and the viscera. Until recently, scientists believed the vagus nerve responses stopped at the intestines, never reaching as far as the genitals.

When women with complete spinal cord injury underwent vagino-cervical stimulation, they were found to experience all of the physiological hallmarks of arousal and orgasm—skin flushing, pupil dilation, elevated heart rate and blood pressure, and elevated pain tolerance. And the reason was the vagus nerve, which carried the stimulatory messages where they needed to go, even when spinal cords were severed at about the middle of the chest, making them essentially paraplegics.

Measuring a variety of responses to vaginal and cervical self-stimulation in these women, Komisaruk and company found that these women not only had analgesia but also orgasm. John Money had reported such orgasms, which he called "phantom," in 1960, but it took thirty-five years more to find the physiological evidence of how they occurred, and to demonstrate that the orgasmic pleasure was as intensely "psychological" as the ordinary kind, suggesting that somehow the sensations are occurring in the brain even in the absence of a direct link—the spinal cord—between the brain and the genital nerves.

The part of this work of key importance to the story of foreplay and arousal is that of the sixteen women in Komisaruk's study, thirteen produced analgesia without orgasm, suggesting that arousal alone produces the analgesia. Just stimulating the vagino-cervical area produced the anti-pain mechanism. Further tests in rodents that selectively knocked out

vagus nerves below the diaphragm found that such interference significantly reduced the magnitude of pupil dilation and pain relief but left the
heart rate alone, providing further evidence that the vagus nerve conveys
the vagino-cervical nerve activity form the reproductive tract to the brain.

"Taken together, these findings indicate the existence of multiple sensory pathways from the reproductive system to the brain," says Komisaruk. "There is convergence between the pain and sexual systems in the
brain. Only when I cut the vagus nerves was the analgesia response abolished." In one study in animals, he was also able to cut the vagus nerve
but electrically stimulate the end still connected to the brain and get the
analgesic response. "This is something we were intended to have in our
sexual arsenal," he concludes, "or it wouldn't be so persistent a force."
Komisaruk has also reported that PET brain scans of a few women during
cervical self-stimulation show high metabolic activity—energy use—in a
special cluster of nerve cells called the nucleus of the solitary tract. What's
exciting about this finding is that the vagus nerve has "projection zones"
in the medulla oblongata, just below the cerebellum, the most ancient
part of the brain, and ends its wandering in the nucleus of the solitary
tract of this structure.

Whether the practice of "intentional application of pain"—as in sado-
masochistic foreplay—is related to the analgesic effect of vaginal and cervical stimulation is an unanswered question. Even in plain vanilla sex,
people seem to prefer—or at least need—harder and harder stimulation
as arousal mounts. Tugging at rather than caressing nipples may elicit an
"ouch!" from most women at the start of foreplay but an "aah!" later on.

Another piece of evidence for the link between pain and pleasure is
the effect found by Komisaruk and Whipple in a test of hot chili pepper
diets, vagino-cervical stimulation, and analgesia. The fiery ingredient of
pepper that causes a burning sensation on the skin is a peptide called
capsaicin that can block out other pain-transmitting messages. Its characteristics make it a useful drug test in lab rats, where scientists have
known for decades that injecting capsaicin into newborn rats permanently
destroys certain fibers, called C fibers, in pelvic sensory nerves in the
spinal cord, brain, and peripheral nerves that bypass the spinal cord.
When Komisaruk tested these injections on the effects of vagino-cervical
stimulation in adult rats, he found that if the pups had been injected as
newborns, they no longer got the analgesic effect of such stimulation, nor
could they fully function sexually. This finding suggests that analgesia-
triggering neurotransmitters are contained in the C fibers (probably by

reducing the release of the amino acid glycine). Giving newborn rat pups capsaicin also has been reported to block the automatic release of brain hormones during copulation in female rats that is usually triggered by vaginal stimulation. In an extension of his studies on chili pepper protein, Komisaruk and his colleagues conducted vagino-cervical stimulation tests on volunteers in Mexico who had eaten a diet high in hot chili peppers since childhood. They "showed a significantly lower magnitude" of pain blockage produced by vaginal self-stimulation than did women who did not eat such a diet. This would seem to suggest that all those stories about "hot-blooded" pepper-eating men and women being the world's best lovers are just that—stories.

Among the implications of Komisaruk's work is that while the end points of foreplay are always in the brain and genitals, the origins of the stimulation we crave are much more numerous. We can stimulate ourselves, or each other. We can even imagine our way to arousal, with erotic images, with romance novels, with obsessive concentrated thoughts of someone or something that our minds have previously—or even perversely—linked to sexual pleasure.

A kiss is still a kiss, but it is possibly a great deal more—an echo, perhaps, of intercourse when the tongue penetrates the lips and lips suck the tongue. And the widespread existence of kissing certainly supports the long-held notion that tongues and mouths, full of soft skin, nerves, and fluids, are echoes of genitals. The mouth kiss, even when relatively chaste, satisfies the almost irresistible urge to touch and join exterior membranes to interior ones. The pursed lips, even in a Hollywood-style "air" kiss, evokes erotic orifices, especially when colored red. The first real kiss in adolescence is a marked event, remembered as vividly as first intercourse; and studies have concluded that people with aversions to kissing are seriously hung up.

While some blanch at serious (French) kissing in public, it is one of the few socially approved intimacies. It is so intimate that, as with Julia Roberts's prostitute character in *Pretty Woman*, a single kiss cannot be bought without love or some form of emotional attachment, while intercourse can. And that may be kissing's most erotic characteristic.

Kiss, taste, and smell are intricately entwined as well. Babies' early attempts to reproduce kissing, for instance, wind up as licking—with their tongues on any handy face or set of lips—or a nose plant against a mother's cheek, inhaling her scent.

When our mouths open and close to those we want to arouse us or seek us for sex, smell goes along for the ride, our noses aware of our own breath and our lovers'. Smell is the oldest of the senses, managed by the primitive or old reptilian brain that controls memory, appetites, and the emotional states of joy, love, fear, anger, and sexual arousal. When combined with touch, the sense of smell is not only potent but individual as our fingerprints. In fact, scientists have discovered that many primitive peoples such as the Yanamamo of the Amazon and the Kanum Irebe of New Guinea touch each other in greeting and in courtship, then smell their fingers to get a scent print that tells them if they are, at some level, compatible.

Indeed, scientists believe kissing is derived from a form of smell checking called face licking and face smelling: Animals lick those they like, including their own offspring. Moreover, the mucous membranes of the lips are very sensitive to touch and the food we eat with our lips and mouth is "tasted" in great measure by the smells we experience at the same time as the flavors.

French kissing has been explained by anthropologist Desmond Morris and others as an offshoot of the earliest methods of infant feeding by mothers. Just as mother birds chew and digest food, then deposit it in their fledglings' mouths, so did human mothers develop their version of Gerber's strained carrots. Adult "erotic kissing," says Morris, "is almost certainly a Relic Gesture stemming from these origins." Indeed, if he's right, much of lovemaking's preludes are "neotonous," or a throwback to the joy and playfulness we experienced as children—nakedness, body contact, laughing, licking, rolling around, and tongue kissing may be yet another echo or memory of the time when our parents mouth-fed us, recalling the trust, security, and love we felt as babies.

As we grow from infancy, we learn to associate some smells with love, pleasure, and trust and others with danger and pain. Whatever it is that makes an odor "male" or "female" is what makes a particular male or female odor important or repugnant to us. Body hair traps the odors the apocrine glands put out. Saliva and urine, too, deliver odorant hormones; and musky steroids, such as androstenol, androstenone, and androsterone, live in the armpits. Although there are cultural differences in what smells "good" or sexy, much of the difference stems from the number of glands that emit odors, so that Japanese and Koreans, for instance, who have few armpit scent glands compared to northern Europeans, emit different kinds of odor-arousing signals. There is even evidence, from Duke University's Susan Schiffman, a medical psychologist, that two people who don't like each other's smell, consciously or subconsciously, will never be lovers.

Kissing, like all touching, promotes sensations in nerve endings that are perceived in our minds as well as our brains, and women of all cultures adorn, puff up, or enhance their mouths, usually with softening agents and "sexual" coloring. Like the baboons reddened rear ends, reddened lips, full and fleshy, mimic the blood-engorged labia of the female genitals, so mouth kissing, especially tongue kissing and the exchange of saliva, also evokes the most intimate "kiss" of all, intercourse. Having watched the scene in the film *Tom Jones* depicting the hero and the object of his latest affection slobbering their way through a feast—all juice and flesh and tongues and shiny grease-covered fingers, forks, and food plunging into open mouths—who could fail to understand its erotic underpinnings?

Among children and adolescents, another form of touch—tickling—may seem an innocent enterprise, but like kissing, it is both more and less than that. "Solo tickle is even emptier than solo sex," says Robert Provine of the University of Maryland at Baltimore County. In his study of the tickle (a "touch-based stimulus of laughter"), he concludes that it is an intricate and intimate part of a social relationship, one reason why it's impossible to really tickle yourself.

Another reason we can't self-tickle, says Provine, the pioneer researcher of the matter, is the way our brains are wired for the sensation. Observing that the intensity of the tickle sensation varies inversely with a person's control over the predictability of the touch that tickles, he concludes that a touch is only a tickle when it's unpredictable at the level of what we consciously feel and at the level of what our brain perceives as plain old touch. Without such a distinction in our brain wiring and skin nerve circuits, we'd be constantly tickling ourselves to helpless laughter every time we touched ourselves, and turning ourselves "on" at inopportune times. "The world," he says, "would be filled with a lot of goosey people." By the same circuitry, the more predictable a tickle is, the more ticklish it is. And it's no accident, he says, that the most ticklish parts of our bodies—underarms, throats, necks, elbow crooks, genitals, and body orifices—are precisely those regions that have lots of nerve endings and are very vulnerable to both arousal stimuli and predatory assaults. If a bug crawls up your neck, you feel the tickle and react! This means that tickle is both a defensive adaptation and sociosexual affair.

In a questionnaire study of 293 subjects between eight and eighty-six years of age, asked fifty-two questions about who tickles, who is tickled, and what they think about tickling, Provine found that contrary to common wisdom, most people *enjoy* being tickled, but they enjoy tickling others even more. Science would say that tickle is a central component of an intimate and emotionally charged medium of touch, critical in establishing and maintaining social bonds between mothers and infants, parents and kids, and of course lovers. Ordinary people would say that like sex, tickle is a measure of who is trusted. You tickle and *allow* tickle most by those closest to you, by those you let come close. Alone or combined with strokes, hugs, fondles, cuddles, and gropes, tickles are most unwelcome when unwanted. When asked who tickled them the most, more than six times as many subjects said members of the opposite sex as said the same sex, a margin that grew to almost twelve times when subjects were asked who they would most like to have tickle them. Tickle was mentioned often as a kind of foreplay, "an innocent (?) ploy that may lead to physical intimacy and perhaps to sex."

In perhaps an uncharacteristic fit of scientific speculation, Provine even suggests that the tickle-challenged are undersexed. The capacity to sense tickle, he explains, may be the key to sexual arousal because it develops so early in life, during the first four months, and may be the primary stimulus for the development of laughter, another critical social and sexual cue.

Babies laugh when touched by their mothers, a vocalization that leads to more maternal touching and more baby laughter. She who overtickles and overstimulates pays the price in a hyperactive insomnolence, but after language develops, tickle and its remembered association with laughter stay an important channel of communication among family members. How important tickle is in stimulating social bonding and arousal can be seen in the fact that tickle is a prime-time piece of entertainment among chimps, gorillas, and orangutans, most occurring during roughhouse play. Animal watchers consider it a kind of "permissible," nonthreatening touch, similar to a poke that could be a threat, but that in a shared moment of tickle glee is no threat at all.

While it may be impossible to tickle oneself, it is certainly not difficult to self-stimulate. Masturbation, along with all erotic touch, is widely used to delay, extend, and enhance arousal and orgasm. And as with many aspects of sex, such triggers of arousal affect men and women differently because of the different ways in which the brains and bodies of the two sexes develop under the influence of sex hormones and culture.

Thus, when women or other female animals masturbate, it plays into their brains and bodies by blending fantasies, lovemaps, brain chemicals, hormones, and nerve endings in a way that is different from when males masturbate. According to surveys, men masturbate far more than women do, but women who do masturbate frequently report significantly more orgasms, greater sexual desire, and the ability to get aroused much faster than those who don't. At the end of the act, men and women may get to the same place, but the roads leading there are pink and blue. Moreover, while masturbation is often an accompaniment to sexual intercourse, there is evidence that nature intended it to have its own stand-alone value—for both sexes—as well.

For one thing, the orgasms triggered by masturbation have been described by both sexes as more intense than orgasms generated by intercourse. For another, studies show that in women, masturbation increases the flow of mucus from the cervix into the vagina, increases the acidity of the cervical mucus to fight germs, and breaks down old sperm. It also makes sperm less able to swim through mucus challenges and may be a way not only of providing and prolonging pleasure, but of producing a form of selective contraception by strengthening cervical "filters."

In men, masturbation's bad reputation, at least in the West, undoubtedly emerged from the biblical story of Onan, who in an effort to avoid siring children withdrew during coitus with his brother's wife and "spilled his seed" on the ground. According to sexologist John Money, masturbation as a term entered the English language in a 1776 translation of Tissot's *Treatise of the Diseases Produced by Onanism*. Tissot was an authority on so-called semen conservation theory from ancient times, which held that wasting or withholding semen, mediated by masturbation, led to nothing good. Science and common sense have eliminated most of the negative medical consequences ascribed to masturbation, such as "spermatorrhea" (wet dreams) and cereal king John Kellogg's belief that eating grains would save the world's children from the evils of masturbation. The irony unveiled by less moralistic and more scientific studies of masturbation is that the solo practice of sexual pleasure may have evolved in direct contrast to Onan's legacy of avoiding conception.

Mark Bellis and Robin Baker at the University of Manchester in England came to that conclusion after observing species where monogamy is not universal (that applies to most species). Among these animals, where the tendency of females is to mate with more than one male in a short period of time, a male's sperm will sometimes need to compete with other sperm in the same vagina or cervix, explaining in part why mammals produce sperm in such huge numbers. But sperm wars aren't just a question of numbers. Some sperm may not be made to fertilize at all, but to just altruistically get rid of any *other* sperm that is already around or to block access to yet another male's sperm coming in afterward. With the help of volunteer couples, Baker and Bellis examined sperm samples from their sessions of both intercourse and masturbation and asked a lot of questions about the circumstances that preceded each "contribution."

The findings suggest that masturbation is indeed linked to fertility rates. For instance, if a couple has sex twice, the woman retains very little of the second ejaculate, but if a condom is worn the first time, the second ejaculate is greatly conserved in her vagina, suggesting that the first sperm in is exercising some kind of control. Bellis and Baker also found that sperm count at copulation gets lower the more time a couple spends together, regardless of how often they have sex. The biological mechanism for this is unclear, but it supports the notion that if competition is the reason males make copious amounts of sperm, then a guy who's fairly sure he has little or no competition doesn't have as great a need to claim his territory.

In masturbation, the case was reversed. The sperm count depended mostly on the time that elapsed since the previous ejaculation. Moreover, the more a man masturbated, the more able he was to get rid of older, presumably less fertile sperm and make way for the new stuff. As sperm age, they become less motile and more sticky, and sperm first and freshly made in the genitals with each arousal and erection must get in line behind sperm already there to await ejaculation. If sex is infrequent, the backup of older sperm can seriously reduce fertility. Men in a stable relationship don't masturbate that often, unless, for example, they are away from their sex partner for long periods of time, but even *in* a relationship, most young men masturbate alone, with intercourse following, at least once every ten or twelve days. The main factor in this is not how long a man has waited for sex, but how soon he can expect the next opportunity for intercourse. More than half of all masturbation events took place within forty-eight hours of the next copulation, suggesting, from an evolutionary standpoint, that masturbation "primes the pump" and is a good way to get rid of old, less fertile sperm.

Of course, today's young men might say just the opposite is true, that masturbation slows things down. Critics of Bellis-Baker-style psychobiology might suggest that men may masturbate within forty-eight hours of the next sex act because it keeps them from climaxing too soon, and in fact such a hypothesis might be tested as an alternative explanation for what the British researchers report. Nevertheless, if a hallmark of good hypothesis is that it explains something in context, it's hard to entirely dismiss what Bellis and Baker are suggesting.

As for women, Bellis and Baker say that while men are ditzing around trying to make fertile sperm, women have control over how many and whose sperm get retained after copulation. When they have orgasm, especially if it's close to his ejaculation, they hold more sperm in their vaginas. But an earlier orgasm during foreplay can significantly reduce the number of sperm at a subsequent copulation, giving women a way to use masturbation to regulate their conceptions. Women masturbate more in stable relationships than they do when they are single and even more during times when they can't conceive. Other studies show they are more likely to have an orgasm during coitus with a lover than with their husbands, suggesting that women may in fact seek greater fertility with a lover than with a longtime mate if the relationship with the mate gets as old and tired as some of his sperm.

Finally, given a choice between foreplay, intercourse, and afterplay, women found foreplay the most important part of a sexual encounter while men (surprise!) found intercourse the most important. This may also be true in part because women are more likely to have orgasm from foreplay that only indirectly stimulates the most sensitive clitoral nerves, rather than during intercourse, which may either clobber or miss the clitoris entirely. The direct manipulation of the clitoris is sometimes painful because it is so sensitive, and although some women achieve arousal by placing fingers or other objects into the vagina, most masturbate by touching the clitoris lightly and rhythmically or pulling on the skin around it to tug indirectly at its sensory nerve supply. In any case, as a form of mutual masturbation, foreplay gives a woman control over her reproductive destiny, even if she's unaware of the sperm competition that may be going on.

Far more than masturbation, however, oral and anal sex continue to fill the top ranks of sex behavior people don't talk about easily. Still, surveys report that while they may not talk about it, most men and many, if not most, women like cunnilingus. But fellatio—the "blow job" (with or without handling the testicles, with or without helping hands, with or without whipped cream or Jell-O) is The Sex Skill Men Really Love (*Redbook* magazine proclaims) and women hate. It's the rare woman, popular wisdom has it, who likes it or gives it and rarer still are those who don't engage in "choking, biting, slobbering" when they do.

What's wrong with this picture, of course, is that it's highly unlikely that women really would have *evolved* to dislike fellatio given that men are so desirous of it. As we've seen, most sexual traits and behaviors have coevolved to work cooperatively rather than divisively. If women *really* were programmed to dislike what men like, it's unlikely the desire for it would be so universal and persistent.

Sydney Biddle Barrows, the infamous Mayflower Madam of the 1980s, was probably closer to the biological truth when she wrote that "most men believe that oral sex is an integral part of any sexual relationship" and suggested that when cultural sensitivities lead wives and lovers away from it, prostitution flourishes. The situation seems to be that a man, in this case very much like a woman, is turned on mostly by the fact that his partner is aroused by giving oral sex, and that her willingness and eagerness to do so are signals that she trusts him and is aroused by him. It's just that men have been more willing, perhaps, to acknowledge that despite their legendary disinterest in lengthy bouts of foreplay, cunnilingus is a turn-on for them as much as for their partners.

In this, humans are not alone. Dogs, apes, monkeys, and many other animals sniff, bat, and touch each other's genitals. Even spectacled flying foxes enjoy foreplay marked by raucous barks, rough play, and prolonged periods in which males groom female genitals with their tongues. One of the most comprehensive sex surveys undertaken in the United States, the National Health and Social Life Survey, randomly sampled 3,432 men and women eighteen to fifty-nine and reported that vaginal intercourse is the overwhelming favorite: 83 percent for men and 78 percent for women find it "very appealing." Receiving oral sex, however, appeals to 50 percent of men and 33 percent of women, and giving oral sex, 37 percent of men and 19 percent of women. A fourth have tried anal sex, too.

But the numbers may be larger. As many sex scientists have reported, no sex survey gives a true picture. "It's not," said one, "that people won't tell you the truth. It's that they won't tell you all of the truth." It's likely that the results are either an overestimate or an underestimate, but experts are betting it's an underestimate.

Like other forms of kiss and touch, orogenital and anogenital stimulation are tied to sex hormone function and brain sex. One piece of evidence for this comes from observations of the anogenital licking observed by mothers with their male rat pups, whose echo may be found in our love of foreplay and the smell of each other's genitals and urinary tracts, which become perceivable when we have our mouths close by. Mother rats lick the anogenital regions of the sons and daughters and even drink the pups' urine. But they lick male pups more than females, and males who aren't licked enough later show less ability in their willingness to mount females and they take longer to reach ejaculation. There is also evidence that the licking increases testosterone.

As neuroscientist Simon LeVay notes in *The Sexual Brain*, "male rat pups know what sex they are" without this motherly help, because their bodies and brains are putting out "male" odors that tip moms off in the first place. So why does she do it? It's possible that the licking communicates other information to a pup, such as clues to *her* body chemistry reflective of the lack of food, for instance, or disease. The victims of Kallmann syndrome can't smell, and doctors have shown them to have delayed onset of puberty, two conditions that at first blush would not seem related. But as LeVay notes, delayed puberty is caused in this case by the absence of certain hormone-producing neurons of the hypothalamus, and recent research in animals has discovered that these neurons originate not in the lining of the ventricles like other brain neurons, but in the outer surface of the tissue that later gives rise to the olfactory mucosa at the back of the nose. Only later in development do these neurons reach out their filaments and migrate along the olfactory nerves to land in the hypothalamus and operate there.

It is not just marketing but biology that accounts for the pornography industry's almost nonexistent sales among women and the romance novel industry's almost nonexistent sales among men. These are not just assumptions: Numerous studies have affirmed not only the roles of fantasy and erotic images in arousal, but also the sex differences in the brain that prime us for—though don't necessarily drive us to—those roles. Typical of the studies is one involving eight men and eight women, all heterosexuals. When comparisons were made of their physiological and subjective arousal, what their bodies did and how their minds behaved during exposure to erotic stimuli, results showed that men and women differed in the direction and magnitude of their arousal, and that there was a much stronger connection between subjective and physiological measures of arousal for males than for females. In another study, during an erotic film presentation, measures of genital fluid buildup and groin skin temperatures found that males were aroused more than females. Hardly a surprise.

A Dutch psychologist, Ellen Laan, and her coworkers at Amsterdam University have shown that stimulation from erotic images, being "in the mood" for sex, and genital arousal are closely linked in the minds of women. Using a "musical mood-induction" procedure, they randomly assigned fifty-one women to one of several experiments—such as being shown an erotic film preceded by mood music as well as without; and being asked to conjure an erotic fantasy preceded by mood music and without. When they measured "vaginal pressure pulse force"—a physiological marker for arousal—and recorded the women's own subjective reports of sexual arousal and emotional reactions, they found that a positive sexual mood did not facilitate immediate genital sexual arousal, but having a positive sexual mood did "inhibit negative emotional experiences to erotic film." They concluded that if women are sad to begin with, mood music and erotic, romantic films might help, but just because a woman's in a good sexy mood doesn't mean foreplay is not needed.

Psychological doctrine for years held that sex fantasies were far more predominant in men than women and in a National Survey on Sex published by Dr. Edward Laumann at the University of Chicago, 54 percent of men but only 19 percent of women said they thought about sex at least once a day. A 1990 study found that college men had almost twice as many sex fantasies a day as did women. But the problem with these numbers is in the concept of what anyone calls a sex fantasy. If you add in the kind of fantasies that predominate in women—the romantic kind— then the gender gap closes fast and tight.

Using an anonymous student questionnaire, Kenneth J. Davidson of the University of Wisconsin found that among 144 never-married men and 166 never-married women who had experienced sexual intercourse, men were more likely than women to report sexual fantasies about being the recipient of sexual activity. Women were more likely to fantasize about past sexual experiences and romantic settings for sex. Dr. Donald Symons, an evolutionary psychologist at the University of California at Santa Barbara, has found that women's sexual fantasies emphasize touch, feelings and emotions, her mood, and the context of her sexual expectations, while men's fantasies tend to be more directly involved in visual erotic imagery or intercourse. Women's fantasies, in his study of them, emerged more slowly, in the romance-novel meanders of description, tease, and flirt, with more time spent imagining kissing than intercourse. Men's fantasies were more graphic, often involving multiple partners and kinky positions. Symons calls male fantasy "pornotopia." It goes directly to the genitals, all lust, libido, and hormones "devoid of encumbering relationships . . . courtship and extended foreplay, where all the women are easily aroused and willing," and undoubtedly, in their erotic abilities, like Garrison Keillor's mythical population, all above average.

Among fantasies, however, there is no "right" or "wrong," only what works and what doesn't in each sex. Women do certain brain tasks differently than men, with women slower than men to even recognize explicit sex words, according to one study. Erotic stories are remembered differently, too. James Geer, a Louisiana State University psychologist, played tapes of erotic stories to subjects and the women could remember the romance ("they looked deeply into one another's eyes") far better than the men. Maybe that's why romantic encounters that leave lifelong impressions on a woman aren't even remembered by her mate. This difference was immortalized in Maurice Chevalier's and Hermione Gingold's rendition of "I Remember It Well," wherein, to her bemusement and his embarrassment, he does not remember any of the details of their first meeting.

For most couples in a mutually satisfying sexual liaison, arousal by foreplay is the standard operating procedure, but many individuals and couples routinely use sex aids in the warm-up to coitus or even to masturbate to orgasm. Again, the purpose of the aids is not the orgasm itself but the arousal, and the more drawn out, in some cases, the better. Pornography is clearly the leader in this field; among pubescent boys, it is the by now almost traditional *Playboy* centerfold, sticky with semen and stuck under their mattresses, while among adults it may range from *Playboys* to films to peep shows, offering depictions of everything from "normal" sex to sex filled with violence and paraphilias.

While one person's erotica is another's "filth," scientists are interested in the effects of pornography, fetishes, and "perversions" for a number of reasons. First, because it is so widespread, pornography's effects are especially interesting. Also, men's sexual lives are indeed more driven by visual cues, and understanding how that operates can shed light on what is "normal," what is kinky, and what is to be either feared or promoted as healthy.

Because the images associated with pornography are easy to reproduce in drawings, videos, and photographs, their value in tests of arousal is also unmatched. It's really hard to avoid a response to what nature intended us to respond to. For example, researchers at the University of Utah asked forty-eight men to suppress erection while viewing a pornographic video and to encourage arousal while viewing a neutral one. In each case, the experimenters hitched the subjects' penises to a plethysmographic device that measures increased penile volume and made sure the guys were paying attention to the video by asking them to give a running description of what they saw (to keep them, presumably, from suppressing arousal by simply not watching). While the men were able to effect some control over arousal, the degree of success varied and in no case could a subject "fake" it when trying to encourage an erection during a neutral video.

By varying the kinds of pornography used—woman-made, man-made, female-focused, male-focused, and so on, behavioral and biological scientists can also explore the differences between the sexes and between individuals within a sex. The results? In study after study that used erotic videotapes to measure sexual arousal and response, women just as much as men are *physiologically* aroused by explicit sex image—genital blood flow and congestion give them away. While they may enjoy romantic stories, neither women nor men get sexually aroused—physiologically speaking—if the explicit sexual imagery is missing, even though *subjectively*, at the level of rational thought, women rate the romantic stories and scenes as arousing and sometimes more arousing than the sexual ones.

As we've seen consistently, evolutionary theory holds that women more than men tie sex to love and romance for very adaptive reasons having to do with their need to get substantial investment from a mate to pay the high costs of pregnancy and motherhood. Genital arousal is not the issue for such between-the-ears decision making, and nature made sure that arousal and intercourse were unbundled. In a series of studies, for example, neither Laan, the Dutch scientist, nor Julia Heiman at the University of Washington in Seattle could find any cause and effect association between measures of markers of genital arousal, such as increased blood flow and hormone output, and feelings of sexual arousal. Women enjoyed the buildup to sex as much as sex itself; one was not dependent on the other. The scientists concluded that this capacity for arousal alone—with or without intercourse or even orgasm—is, like "love at first sight," a part of our biological program. As noted earlier, one reason for this may be that the capacity to enjoy the arousal phase of sex without going all the way may give a woman more time to assess the man arousing her and sort out Mr. Genetic Wrong from Mr. Right, without losing interest in case she has indeed found Mr. Right.

In a second study, Laan wondered whether erotic films made by women are more arousing for women than erotic films made by men. She exposed forty-seven women to a woman-made, woman-centered erotic film and a man-made, man-centered erotic film. Vaginal pulse amplitude, or force, was recorded continuously along with the women's subjective reports of arousal and emotional reactions to each film. Contrary to what they expected, *genital arousal* did not differ between films, although genital response to both films was substantial. But the women's *subjective experience* of sexual arousal was much higher during the woman-made film. The man-made film evoked more feelings of shame, guilt, and aversion, despite its genitally arousing impact.

Laan and her colleagues concluded that the largest contribution to female sexual excitement might result from a woman's processing of both the content and emotional meaning of erotica, and not from any feedback her brain gets from genital congestion. Sexual arousal, for a woman, at least, is mainly in the brain.

In another study, this time on the effects of "performance demand" on sexual arousal in women, Laan asked forty-eight volunteers to become fully sexually aroused within two minutes during sexual fantasy and while watching an erotic film excerpt. She monitored vaginal pulse amplitude and the women's self-reports of sexual arousal and emotional reactions after each erotic condition. Results showed that pressure to perform sexually resulted in higher genital response and was most effective in the fantasy condition, much the way it has been found to be in men. She also found that women who masturbate an average of four times a month had higher subjective sexual arousal during "performance demand" conditions compared to women who masturbate more or don't masturbate at all, suggesting there may be an optimum level that saves women from exhaustion but reinforces the practice.

At the University of Washington in Seattle, behavioral endocrinologists explored the possibility that a sexually induced hormone response might "prime" further sexual arousal in women who respond to erotic movies. Healthy, premenopausal, heterosexual women in the ovulating phase of their menstrual cycle were randomly assigned to be exposed to either a sexually explicit video or a nonerotic movie. Ninety minutes after watching either film, both groups saw another sexually explicit video.

Vaginal vasocongestion and hormones such as cortisol, prolactin, luteinizing hormone, and testosterone were measured continuously and subjective responses to the films were recorded every twenty minutes. Compared to those who didn't get the first round of erotic video viewing, experimental subjects who did showed a greater amplitude and longer duration of vaginal response to the second video; and the subjective reports among the group of women who saw the first round of erotic videos claim greater sexual response to the second erotic tape compared to the first, an effect that was not matched by hormone measurements.

In just about every species tested, scientists have found particular "arousal" touches that are absolute hallmarks of courtship and sex. This appears to be true, whether the contact is chin rubbing and tongue flickings that deliver pheromones from male red-sided garter snakes, or bites by female brook sticklebacks inflicted on other females on spawning days. Moreover, touch, at least in nonhuman mammals and probably humans as well, is required by the body and brain before prolactin and oxytocin can be released to "mature" the breasts and stimulate nursing.

In fact, the pressure of massage and touch alters hormone levels at all sexual stages, and even moderate pressure stimulation of the vagina and cervix can stimulate the output of hormones needed for ovulation. In cats and rabbits, the mere act of physical mounting and penile insertion triggers ovulation, and experiments have found that it's the touch during copulation that causes a surge of luteinizing hormone from the brain's pituitary gland, a "reflex" that prepares the body for ovulation. In still other experiments, scientists have cut or anesthetized nerves leading from the genital area of female mammals so that the nerve impulses caused by touch don't reach the brain; they found that in such animals, male intromission fails to trigger these hormone surges and ovulation. Any doubt that there's something in male sperm, rather than the tactile stimulation, that's key to this phenomenon has been put to rest by similar experiments with intromission by male mice after vasectomy. The vasectomized males have no sperm, but their penile insertions alone in intact females led to pseudopregnancies. The touch to her genitals triggered ovulation and a surge of pregnancy hormones—quite without sperm meeting egg or a true conception.

Studies of sexual "first touches" also reveal the pattern and purpose of male versus female arousal and reproductive strategies. Frank Willis and Leon Briggs of the University of Missouri in Kansas City recorded and described so-called touch initiation among five hundred couples in a variety of public areas and then asked the couples to identify their relationship. Results showed that men were more likely to "hit on" a woman first—literally, to initiate touch during a courtship date. Women were more likely to do the first touching in public only after marriage. Couples married for a year or more were less likely to touch each other at all. The investigators believe such touches reflect the fact that in general men have more to gain from casual flings than women, and that for men, a casual fling generally requires someone very attractive. It's likely that women interested in one-night stands will have developed special signals that tell a man she's interested, too.

When touch is involved, arousal is sometimes devoid of words, dominated by body language and gesture that convey meaning by themselves. More often, however, terms of endearment as well as frank sexual language are used to enhance arousal. The words may be the "sweet nothings" and tender coos of romance, or those that evoke more pungent images of sex.

"Dirty words," social scientists report, exist in every human language and culture, and the famous four-letter variety like fuck, cunt, cock, twat, and slit are in themselves semantic dittos of pelvic thrusts and onomatopoeic sounds of copulation. "Narratophilia" is recognized by sexologists worldwide as a paraphilia, or bizarre erotic practice that becomes part of distorted lovemaps. John Money, who named the condition, describes it as "being dependent on reading or listening to erotic narratives in order to initiate and maintain sexual arousal and facilitate or achieve orgasms."

Our interest in dirty language may also reflect the links some cultures have made between the ear and the words. The Greeks had a word for dirty language, of course: erotalalia; and one collection of English words for copulation ran to almost thirty pages, with many containing slurpy sexual "l" sounds such as libidinous, voluptuous, lusty, lewd, and lick. (The word "sex" itself, from the Latin *secare*, "to cut," first emerged in English usage of any widespread nature in a fourteenth-century translation of the Bible and was first used in its modern sense by John Donne in 1631 in a poem.)

Much has been learned about erotic language by looking at what happens when the normal brain is damaged, and by examining how the brain is organized to deal with symbolic representations of sexual bodies and relationships. Some stroke patients, for example, lose their capacity to speak, but not to use "foul" language, suggesting that evolution separated—and perhaps protected—the capacity for sexual sound making. Tourette syndrome, a brain-wiring disorder, compels some victims to discharge explosive bouts of cursing and obscene yelling, a symptom called coprolalia. Afflicting men much more than women, Tourette tends to run in families and is commonly linked to other learning disorders.

In his widely acclaimed essay, "A Surgeon's Life," physician and medical writer Oliver Sacks describes the case of a Canadian doctor and amateur pilot with Tourette syndrome (TS), details the involuntary barks, twitches, and other tics that convulse these patients, and notes that the syndrome, named for French neurologist George Gilles de la Tourette, also often includes impulsive, provocative behavior, "a constant testing of physical and social boundaries, a restless reacting to the environment." People with TS "are driven to do this, to do that, against one's will." Long thought to be a "moral disease," scientists now know that it is organic, a discovery first made when its symptoms eased under the influence of psychoactive drugs such as haloperidol, which dampens dopamine, the brain's "alertness" chemical.

But Sacks and others see TS as not just a biological but a psychological and social disorder, just as sex is a biological, psychological, and social phenomenon. Tourette, the experts say, releases primitive behaviors of *many* kinds and in that sense is a disease of "disinhibition."

In such a way, dirty language may be a release, a "disinhibition" of what organized societies' rules say is inappropriate for public consumption. Just as societies have put restrictions on sexual behavior in public, so have we adapted to the restrictions on language that imitates sex. When we let loose, it's a form of foreplay. What is forbidden is ever enticing.

A rising tide of anger and opposition against pornography, among women's advocates, religious organizations, and other groups, has drawn attention to sexually explicit media, but rarely to evolutionary adaptations that may be going on in humans' reaction to it. According to most scientists who study arousal, even if every last pornographic magazine, book, or movie were to vanish tomorrow, pornographic imagery and fantasy would be sought, depicted, and recalled at will when the mood struck. One of these scientists, Neil Malamuth, points out that "both media and fantasies reveal mechanisms of the mind that developed as a result of evolutionary adaptation." There is no way to "ban" them without banning the mind that creates and responds to them.

A good example of this comes from research by Terri Fisher, Robert Pollack, and Victor Malatesta at Ohio State University in Mansfield, who a decade ago used sexually explicit films to try to understand sex differences in arousal and response to psychosexual stimulation. They recruited twenty-four men and eighteen women to participate in sessions during which the subjects were asked to masturbate to orgasm while viewing pornography and then rate the film for its arousal value. In their results, published in the *Journal of Sex Research,* they found that for men, the amount of time it took to go from arousal to orgasm had little connection to how they rated the film's arousal value. Some took longer, some got there sooner, but there was no predicting from this "latency to orgasm" how they rated the films. For women, however, films perceived as being "average" in arousal quality resulted in the longest latencies, suggesting that they were much more sensitive to the images, or perhaps the story content, of the films than were the men.

Numerous studies have suggested that for women, the most sexually stimulating films show romantic heterosexual behavior. Next in line was mild group sex involving two men and a woman, followed by heterosexual sex behavior that showed some genitals but nothing kinky, explicit group sex involving three men and three women, mild sadomasochism, brutal explicit sadomasochism, and male homosexuality. Females preferred and were significantly more sexually stimulated by films in which men and women had relationships, even if the man was treating the woman badly, than by homosexual stimuli.

It's also the case, of course, that people can and do overdose on erotica and our minds and bodies have evolved means of coping with that, too. Constant arousal, as noted earlier, is a sure road to exhaustion, not sex.

Years ago, medical students at Johns Hopkins were offered a course on human sexuality that began with the students sitting in a darkened auditorium watching XXX-rated movies. At the start of the "show," for which the students were unprepared, they sat almost still, riveted to the images and undoubtedly parsing the effect of it all on the course teacher's expectations, on their minds, and on their bodies. Half an hour later, the film still rolling, the students began to visibly tire of it all. They squirmed and fidgeted, began to talk to each other and look away from the screen, not in disgust but out of sheer boredom. And that, of course, was the point of the lesson.

Only when these twenty-something, presumably sexy students "got over" the general discomfort most people share at naked sexuality and a lifetime of taboos could they begin to use plain English and explicit data to talk about human sexuality and how it can become dysfunctional. But what the students also learned is that people become refractory, normally fairly quickly if they get a steady diet of sexual stimulation.

Michael Hammon, a sociologist at the University of Toronto, has pointed out one possible evolutionary reason for what he calls "emotional habituation," the decline of emotional arousal that occurs with the repetition. In general, anti-habituation is "evolution's safety valve," limiting the pursuit of the same action even when it offers positive effects. In humans, he says, it is an "unconscious physiological constraint on expanding needs. If basic needs are met, a declining habituation curve means that for any individual, additional action will take on the average more and more effort to have a significant impact." Psychologists and most parents learn quickly that the best teacher is reward, but not *constant* reward. Like the child who is *always* given a chocolate for saying "please," recipients of constant rewards given long after the lesson is learned soon tire of either performing or the reward. Thus, our brains have evolved to recognize that the best reinforcers of any behavior are those that are intermittent. Addicted to arousal, we honor it as much in the breach as in the experience.

Notwithstanding any social or commonsense value in limiting access to pornography and sex aids, it's fair to say that what once was considered truly shocking is now more likely to generate ennui. What was hidden under the Victorian bed did nothing to necessarily stop what happened in that bed, but these days, sexually explicit media are so intrusive that even libertines might wonder what's left to really arouse us. The answer: the same things that always did. It may be on the Internet now, but it was always available somehow, if only in the imagination or the Bible. (Pay close attention to the Song of Solomon.) The caves at Lascaux and other parts of the Dordogne are full of paintings and carvings so explicitly erotic that reproductions made for tourist consumption have apparently been sanitized. In one image, described by Timothy Taylor in his book *The Prehistory of Sex*, a lioness appears to be licking "the opening of a gigantic human penis which is hanging down from between the legs of a rather confusingly engraved man. Others have seen a vulva somewhere in this picture . . . an anus and buttocks. Whatever the truth may be, neither the explicit sexual imagery nor the connection of male human and animal sexuality is in doubt."

Also depicted in the caves were batons in the unmistakable shape of a phallus with testicles attached. Perhaps they were used as "sex toys" and perhaps not. But as Taylor notes, "ritual defloration of virgins using such implements is still commonplace in many societies and dildos used as objects for sexual pleasure have been observed among primates in the wild."

Pornography is always with us. During the Civil War, pornographic daguerreotypes described by modern standards as pure *Hustler* were taken into battle. In 1986, the U.S. commission on pornography reviewed 3,050 separate magazine and book titles sold in "adults only" outlets in Washington, D.C., Baltimore, Miami, Philadelphia, New York, and Boston. Of these, 746 could be assigned to a paraphilia category listed in the *Diagnostic Symptoms Manual* of the American Psychiatric Association. Incest and pedophilia were included with sadomasochism, by far the most common. And what's on the Internet is orders of magnitude more accessible.

Paraphilia means literally love (philia) beyond the usual (para), and in several books and essays, John Money has described the existence of more than thirty that are frequently apparent. How paraphilias develop is still a matter of debate within the scientific community, but one prevailing idea is that sometime in early childhood or early adolescence, just as the sexual body is becoming the match of the sexual brain, a boy or girl will become aroused by an idea, object, or image and perhaps achieve orgasm through masturbation at the same time. If the arousal is particularly intense, the boy or girl will seek to repeat its effects, and at some point the idea, image, or object becomes *a requirement* for either copulation or copulation fantasies. Without this particular trigger, the penis will not easily, or at all, get erect, the vagina will not easily, or at all, lubricate, and orgasm will not occur. Paraphilias also are remarkably stable. "Once hung up, always hung up, hung up always on the same thing," says Money, and rarely do those hung up on a paraphiliac lovemap have more than one. "A man whose erotosexual arousal is governed by a primary compulsion as a peeping Tom may be completely incapable of . . . sadism, homosexuality or pedophilia. A lewd telephone caller may panic if the woman who responds invites him to a bondage and discipline sex party."

Far more males than females become paraphiliac, according to Money, an "inequality that may derive from the fact that nature has designed males more than females to be dependent on their eyes for erotic turn-on and females to be more dependent than males on skin feelings."

While some paraphilias are benign, others are not, and most are at least socially destructive if they go too far. Thus, a man who likes his women in black, silky nightgowns is clearly using images and touch as part of his foreplay, but a fetishist addicted to silk, leather, feathers, or wearing diapers or rubber pants has crossed the line for many. Smell fetishism may be an overreliance in some ways on vaginal pheromones or crotch, armpit, or genital odors. Stroking, scratching, nibbling, licking, and puffing of warm breath may be related to the "grooming phase" of animal estrus or massage and the licking, preening, and cleaning that is prelude to copulation in most mammals. Such paraphilias as inflicting pain and receiving pain may be acceptable if they are in the form of spanking or nipping, but in extreme cases cause serious injury and death. Coroners have reported mini-epidemics of autoerotic asphyxia, in which mostly teenage boys hang themselves by their necks to simulate strangling and choking as an enhancement to arousal and masturbatory orgasm. Pedophilia, in which the arousal object is a child, is a particularly loathed and feared paraphilia, and for good reason. It involves nonconsenting children who don't understand the purpose of the arousal and haven't the strength or wiles to avoid the pedophile. They get hurt because they don't share the pedophile's lovemap but have no choice in what they wish to do.

Among other paraphilias identified by Money are gerontophilia, the fantasy or actual use of a much older partner to initiate or maintain arousal; kleptophilia (stealing); homicidophilia (murder), very rare; apotemnophilia (amputees); rapism; scoptophilia (watching coitus); zoophilia (animals); coprophilia (feces); and urophilia (urine).

While rapism is rare, rape fantasies—especially among women—are not and some feminist sex scientists speculate that for most women, the turn-on is not the rape but the female's controlling capacity to *make* a man lose all control. In a rape fantasy, the object of a woman's lust is so overcome by her desirability that he literally cannot stop his desire. She wills him to "take her," and he obeys her command. She chooses it, he loses it. In another scenario, however, rough-and-tumble foreplay that mimics rape can be a mutually agreeable piece of foreplay, and in yet another interpretation of rape fantasy and paraphilia, men with the physical potential to rape a woman are perceived by women as more able than less aggressive men to have more fertile sperm. Presumably their aggression has also brought supremacy over weaker males. Though not wishing to be really raped, a woman can still "test" this kind of power by initiating or allowing certain kinds of roughhouse sex or verbal abuse.

The use of clothes or other objects for arousal and foreplay might be eccentricities or outright paraphilias. Valerie Steels's book *Fetish—Fashion, Sex, and Power* says the line between what is normal and what is perversion when it comes to sex and its expressions is getting more and more blurred. Ultra-high heels (catch-me-fuck-me shoes), along with Madonna's pointy bras, frilly lingerie, cat suits, nipple rings, and tattoos, are as much the province of the sexually upright as the "pervert."

Foot fetishes especially are common and commonly the butt of jokes. All but 1 percent of foot fetishists, according to some estimates, are men and they are a varied lot with highly specific preferences. Some like boots and others galoshes, some leather and some cloth, and always there is a story that "plays" with the fetish. Girdle fetishists like stories about "bad" girls put in "chastity belts that resemble corsets. The high heel fetishists generally want to have their genitals crushed by such a shoe and can even get turned on by seeing depressions in grassy areas made by a crushing piece of shoe leather."

However we pursue it, sensual stimulation has become as essential a part of our lives as nourishment and intercourse. Aside from health, good nutrition, adequate sleep, and a normal upbringing, we humans manage arousal quite efficiently. That has not, of course, stopped us from abetting it.

8

APHRODITE'S DRUGSTORE

Nature made it easy to get our engines going. Our sexual exuberance has all it needs to get us aroused, focused, mated, and orgiastic. With the hypothalamus linking the endocrine system's hormones to the nervous system, and the limbic brain orchestrating the neurotransmitters of emotion, the sexual arc is figuratively and literally electrified. As the brain's pleasure centers light up, they bring not only joy to the moment of sex but also the appetite for more and more sex. The all-time best aphrodisiac for sex remains sex itself. The more we have it the more we want it. It is the ultimate addiction.

Far from inhibiting the search for outside help, however, this insight has paradoxically whet the appetite for love potions—to put you and your lover in the mood. Named for Aphrodite, the Greek goddess of love, beauty, and fertility, aphrodisiacs today range from the ancient to the high tech. Sexual adventurers and venture capital alike are fueling the search for pills, potions, and perfumes that will either turn up the user's libido or put stars in the eyes and lust in the loins of a preferred partner.

The interest is understandable. There have always been those who were less efficient in their sexual chemistries. And those who, like the garlic-loving chef, calculate that if some is good, more is better and enough is never enough. Indeed, sales of sildenafil (Viagra), which stifles an enzyme that prevents muscle tissue from swelling with blood, are so explosive—among men and even women—that one wonders if there's an impotency epidemic threatening the survival of the human race.

Ironically, perhaps, the trade in sex enhancers has been fueled recently by modern revelations that basic sex hormones—estrogen, testosterone, other androgens—have relatively little effect on libidinous trajectories in healthy, normal men and women and other animals. To be sure, testosterone is the "bad boy," "go get it" sex hormone and estrogen is the "want it" companion chemical, rendering a woman's genitals moist and producing skin and nipples and soft curves that welcome sexual advances. (In a nice evolutionary twist, sensitivity to estrogen in male brains—in mice and presumably men—is also in play among men who do the advancing.) Testosterone shots have been shown in some to increase libido, and estrogen replacement therapy not only promotes oxytocin but also appears to be a natural dopamine booster and antidepressant for many women. Androgens, made by the testes and the adrenal cortex, naturally prime the brain's sex centers in both men and women and promote genital growth.

But most of our sex hormones' aphrodisiac impact has been documented in men and women with *abnormal* gonads and hormone supplies. The popularity of estrogen replacement therapy and other hormone treatments marketed to meet the demand for sexual fitness notwithstanding, sex hormones simply don't function as aphrodisiacs in those whose sexual biology is intact. The hitch is that where there is no clear deficiency, the effects of androgens including testosterone are variable and usually weak, a state of affairs known in technical lingo as "initial rate" variable. In popular terms, it's a situation in which the lower the levels at which you start, the higher you go; but starting with normal or high levels, you don't go higher. No amount of sex hormone replacements will restore one's full natural capacity for arousal, erection, and orgasm. And in any event, even in the relative absence of these hormones, people can perform quite well in the arousal stage of things.

Rational or not, all manner of risky, reasonable, and goofy attempts to alter, prolong, intensify, or delay arousal and orgasm have been part of the human sexual repertoire as long as history has recorded reproduction.

It's unclear exactly what the ancients intuited about the natural chemistry of sex in animals and people. They surely knew nothing about testosterone, estrogen, DHEA, oxytocin, vasotocin, vasopressin, pheromones, endorphins, serotonin, dopamine, or any of the brain's or body's means of recruiting these chemicals for sex. Nevertheless, they must have deduced the ebb and flow of desire and their links to the sex organs, health, and youth. And it must have made sense to them as it does to us that if one could improve on whatever built-in sexual drug and device shop existed, the payoff might be spectacularly pleasurable, not to mention lucrative.

The trade in mandrake root, whose shape reminded ancients of the human body, was so serious that mythmakers made it the source of the magical aphrodisiac used by Circe to lure Odysseus to a watery grave. Farther to the east, the ancient Hindu Caraka Samhita offered recipes for fifty different aphrodisiac potions of herbs and nutrients. By the thirteenth century, their neighbors the Mongols were tying the cartilaginous eyelid of the goat, with lashes attached, to the erect penis to woo the ladies. This French or rather Chinese "tickler" has been copied in a more permanent fashion by men who scar or embed objects into the penis to sexually arouse and stimulate a woman's vagina or clitoris. Finally, substances that resemble sexual organs have long been imbued with substantial powers, giving support to the market in oysters, eggs, onions, artichokes, and clams, which resemble the pudenda. (Rhinoceros horn, bananas, antlers, sausages, asparagus, and celery stalks leave less to the imagination.)

Body modifications are another historic form of aphrodisia, although most were probably calculated to attract mates rather than assist in the follow-through. Practicing centuries-old tribal customs, Sumatrans today still cut holes in their penises and embed pieces of gravel, resulting in a knobby tickler. Raven Rowanchilde of the University of Toronto, a sociologist and specialist in urban and primitive body art and male genital modification, published a paper in 1996 almost calculated to make male readers—and some females—cross their legs reflexively.

For *her*, there's the plastic surgery for breast enlargement or breast reduction, liposuction for waist-to-hip ratio sculpting, and vaginal tightening procedures we're all familiar with. Then there's the traditional Nama of southern Africa, who prefer hugely elongated vulvar lips, so mothers massage their infant daughters' genitals to get them to hang down by adolescence. Other women enlarge their facial lips with dinner-plate-sized appliances.

But most of the sexual-enhancement industry was and still is directed at men. Rowanchilde, for example, has documented the widespread use of penis sheaths, scarification, body piercing, and other genital "treatments" that "encode and transmit messages about age, sex, social status, health and attractants," but also, in their extreme form, "not only honestly advertise status, but also sexual potency and ability to provide sexual satisfaction." If aphrodisiacs function as sexual turn-ons, these qualify as all-out advertisements—and honest ones at that—of a man's capacity for and willingness to provide sexual pleasure to his mate.

The honesty of the advertisement, of course, is evident in the obvious endurance of pain and risk that the procedures—adult circumcision, infibulation, piercings, penile inserts, and incisions—inflict on the men who seek them out. Like body painting and scars and marks on such erogenous zones as breasts, thighs, and buttocks, genital modifications underscore the patience and pain that is "proof positive" of unselfish devotion to peaks of pleasure. In common with celibacy and abstinence and similar self-deprivations, these "high cost" genital procedures are clearly aphrodisiacal to some whole cultures and some individuals because of what they stand for: discipline, endurance, and wealth. They give new meaning to the seductive lure of the moneyed leisure class.

A modern compendium of penile operations still reads like a torture manual, but in the search for better sex, it seems, anything goes. American plastic surgeons are not infrequently called upon by men to elongate their penises to cure "locker room phobia." Urological surgeons cut the suspensory ligament located in the pubic area just above the base of the penis, a ligament that holds the erect penis in an upright angle. Once cut, it releases an inch or two of the shaft tucked inside the body. Then, one-pound weights are attached during healing of up to six months to keep the penis stretched and prevent withdrawal back into the body. This little item costs about four thousand dollars, plus pain and suffering, and results in a permanently downward-pointing erection.

Other procedures add penile circumference or girth. In one, the surgeon cuts or suctions fat from a man's abdomen and then injects it into the shaft. (The result can be an asymmetrical blob when fat cells migrate into uneven lumps.) Then there is subincision, practiced in Australia and parts of New Guinea, involving deep cuts on the underside of the penis that bare the urethra from the glans to the base of the scrotum. The purpose of this is to increase girth and indeed the penis looks double the width. During erection it is very wide, and aboriginal women consider subincised men superior pleasure givers.

Women, of course, contribute to men's interest in such enhancements.

More than twenty groups in Southeast Asia favor men who insert objects into their penises, including bells, balls, pins, rings, spheres, and pellets, all sewn under the skin of the organ along its length. For a price they get prostitutes to sew in gold, silver, or bronze bells. Palangs, or rods and pins pierced through the glans, are penis modifications almost universally practiced among the Dyaks of Borneo. Women claim that sex without a palang is like rice without spice, and men who want lusty females are willing to provide them with insertions like these because they apparently stimulate and extend the inner walls of the vagina. One social worker quoted by Rowanchilde said that getting a palang is like "turbo charging your penis." It is probably not coincidence that Borneo natives who favor palangs also value the rhinoceros penis for its natural palang, a rigid cross-bar located four inches behind the tip of the penis and projecting two inches on either side. It is considered sacred.

In the West, devices that enhance arousal, intercourse, and orgasm these days are used principally by the sexually disabled, but bigger markets show promise among the aging baby boomers and the sexually adventurous in search of better or more sexual stamina. While penis extenders aren't high on any postfeminist's list of how she wants her partner to dress for a date these days, consider the robust sale of ribbed condoms, vibrators, and dildos. "So much of the interest in penis extenders and enhancers emerges from the evolutionary pressures on men to pay attention to women's interest in and choice for orgasm," says Randy Thornhill, who has studied sexual attraction in numerous cultures and concluded that a substantial amount of what people do about sex today evolved between 2 million and ten thousand years ago as both bodies and minds were reshaped by early humans trying to solve survival problems in a changing and threatening world.

Today, what we're left with isn't necessarily still adaptive or necessary. Bigger penises or hour-long erections don't necessarily deliver better reproductive fitness; small ones will get a sperm to an egg. But at some point in our sexual history, our ancestors found it useful in some way to enrich the sex drive and sex organs. Given the evidence that, as we'll see in chapter 9, orgasm in women can increase the likelihood of delivering the sperm of a woman's chosen mates to her egg, men are programmed in some ways to want to produce that effect in women. And penises that are more extravagant, whose owners have more staying power and endurance, may indeed deliver the goods better. As most orgasmically satisfied women have attested, there are ways around a small penis, some even more interesting and pleasing than accommodating a large one, but biologically, many men haven't yet gotten the message.

Safe or not, penile-enhancing devices are probably here to stay, at least for a while—along with an alternative pharmacy full of herbs and chemicals, most of which are of questionable value and safety. One example is an extract made from the bark of the double-trunked African yohimbine tree, which faintly resembles a humanoid shape. Still used to restore erections for impotency and marketed under the trade names Yocon or Yohimex, yohimbine is an alkaloid substance chemically related to the hallucinogen LSD. It earned its reputation as an aphrodisiac probably because in the brain, it causes the release of greater levels of the "excitement" neurotransmitter norepinephrine. Confusingly for yohimbine fans, however, scientists at Valparaiso University in Indiana in 1997 conducted a double-blind, placebo-controlled study on a group of men with problems getting erections and a comparison group of men who had no trouble. They gave the test group 30 mg a day of yohimbine and measured erectile response in a psychophysiological laboratory where the men were exposed to pornographic videos. They also assessed sexual desire, arousal, and ejaculation. Results indicated no effect on most aspects of sexual response in the men who already were in good shape. "Mixed effects" were found on measures for the erection-challenged, with three of the eleven reporting strong positive effects. With yohimbine, the frequency of sex activities went up as did the men's own assessment of their erections during masturbation, but not intercourse. Together with androgen and strychnine in small doses, it has been used medically, but its adverse effects—abnormal heart rhythms and panic attacks—have taken it out of the formularies approved by the FDA. A synthetic form of yohimbine that increased blood pressure and stores of free noradrenaline in the brain is said to be undergoing clinical studies in Europe.

Another example is nutmeg (*Myristica fragrans*), a spice native to Indonesia and related to a group of chemicals similar in molecular structure to amphetamines. In the quantities needed for an aphrodisiac effect, users will experience red eyes, headache, nausea, agitation, sleeplessness, dry mouth, and slurred speech and gait, along with euphoria and hallucinations. If that isn't enough to dissuade, there is the depression, fatigue, and aching muscles afterward—anything but a prelude to love.

Predominantly, however, the search for the perfect love potion has focused on the search for a pill that, as one observer put it, creates instant "on" and instant "up."

Exhibit A in recent history is Viagra. While only a minority of adult men have medically definable and treatable impotence, Viagra's reputation for erection building made it the best-selling drug in history—worldwide—in less than six weeks. Never mind that it won't do a thing for men who aren't actually impotent. Exhibit B: Rohypnol, an amnesiac sleeping pill also known as "roofies," "Stupefi," and "Rope." Popular with date rapists who drop one into a target's drink to overcome a woman's objections and block memories of what happens next, it has actually become the drug of choice for some misguided "inhibited" couples, despite the risks of brain damage and habituation that it carries.

Perennial news stories also tout marijuana as a way to spark the sex drive by raising testosterone levels in the blood; tranquilizers as a way to prolong male erections and thrusts; substances isolated from human armpit sweat as a way to enhance desire; room deodorants as a way to intensify orgasm; neurotransmitter boosters as a way to extend arousal for hours; and antidepressants as a way to drive women to nymphomania. The fact that use of these drugs is promoted without a grain of scientific evidence for effectiveness or safety does little to dampen their persistence, because historically the road to aphrodisia was paved with imagination as much as with intentions. Whether early travelers along the way sought to activate interest in themselves or in someone else, whether they wanted to seduce or be seduced, whether they wanted to speed things up or slow them down, the means to the end only required a willing buyer and a plausible product.

Perhaps the most important thing that can be said about aphrodisiacs at the start of the twenty-first century is that the idea of them is endlessly alluring. But the question hangs out there: Do any products really work safely or predictably? Do genuine aphrodisiacs exist?

The good news is that the answer is an unqualified "sort of." If by aphrodisiac you mean "some substance that makes the object of your affection fall in a swoon into your arms, the answer is no," says one researcher. "But if you mean something in small amounts that enhances the intensity of sexual pleasure sometimes in some circumstances for some individuals—perhaps."

The bad news is that for healthy, normal men and women, shopping in Aphrodite's drugstore for new sexual thrills may present serious, even lethal emotional and physical risks. Comprised of opiates, irritants, sedatives, mood elevators, disinhibitors, poisons, hormones, or hallucinogens, available aphrodisiacs are variously scary and illegal. Moreover, most of the evidence in support for a sexual effect of some substance is largely, if not wholly, anecdotal, inferential, spotty, and unconfirmed.

Consider just one case, that of L-tryptophan. A naturally occurring chemical that sometimes acts as a sedative, L-tryptophan used with antidepressants and tranquilizers has been reported to create some aphrodisiacal effects by limiting the effects of serotonin (the calm-down, inhibiting neurotransmitter) and by default raising levels of dopamine (the take-charge, get-happy neurotransmitter). In the early eighties, experiments by Gian Luigi Jessa, a pharmacologist in Sardinia, created a sensation with reports that a diet *free* of tryptophan, a nutrient found in milk, cheese, and other dairy products, caused lab animals to become oversexed, apparently by depleting serotonin and, again by default, boosting relative levels of dopamine in the brain. No one since has reproduced the effect he claimed.

Although research on potential or reported aphrodisiacs is staffed by some notable psychiatrists, pharmacologists, endocrinologists, chemists, and neuroscientists, the fact is that precious few controlled clinical trials have ever been done, even with drugs that offered good leads. And because the scientific establishment generally condemns or ignores research on healthy people and especially in healthy people who just want to increase sexual prowess, such studies are likely to be limited if they're done at all. "Practically nothing is known about dosages, tolerance or the influence of age or gender or personality on the effects of any reported aphrodisiac," says John Money, "and there is serious question about whether we'll ever be allowed to find out." As a result, much of what credible scientific evidence there is on the sex-sparking influence of drugs and hormones has been drawn largely from reported side effects of necessary drug therapy. And even that evidence is weakened by the nature of its sources—studies among people who are ill or emotionally distressed, addicts, prostitutes, transsexuals, or hormonally abnormal persons.

Despite all of the drawbacks, there remains enough tantalizing research to keep interest high. Some examples:

Bromocriptine, a fertility drug that acts on the pituitary to influence prolactin, has been found to create sexual desire in women and also in a few men with abnormal levels of prolactin, a fertility hormone. Prescribed for women with pituitary tumors that cause infertility, it is unlikely to be widely sought on the open market, but its mechanisms are informing other studies. Another infertility drug called clominiphene citrate improved sexual function in a man given the drug to treat liver disease, probably by correcting a sex hormone imbalance.

L-dopa, a chemical building block of dopamine used to treat the tremors of Parkinson disease, was found many years ago to give older patients, but not younger ones, a serious libido boost, probably by temporarily restoring dopamine to a brain long deprived of the right amounts of this neurotransmitter. Already tried on a few healthy people with mixed and mild results, it seems to have the most effects on those who biologically and physiologically need it least, but again, its effects on the brain's mood centers make further study of possibly safer analogs of L-dopa likely.

PCPA, like L-dopa, is a serotonin antagonist, and studies in rats show it causes increased frequency of mounting and copulation. In its inhibition of serotonin, it is a useful treatment for migraine, and some studies in women with migraine show increased sexual arousal in a few patients who take it.

Papaverine was found in 1984 by French scientists to induce erections after it was given to men with coronary artery spasms and stroke, although probably by sabotaging biochemical receptors in brain and muscle cells.

Naloxone, used to treat narcotic addiction by blocking the brain cell receptors that activate opiates, also stops the poisonous effects of narcotics, and along the way reverses the suppression of sex drive that long-term narcotics addicts are likely to experience. One test of naloxone on eight healthy men found that three got spontaneous erection; and in a study on four women, three claimed enhanced sexual pleasure.

Dehydroepiandrosterone, or DHEA, enjoyed its fifteen minutes of fame a few years ago. A steroid hormone that men and women have in nearly the same amounts except in the brain, its potential to arouse has been suggested by a few cases in which excess amounts of the body's own supply of the biochemical led to hypersexuality. But beyond all the book-length and naked hype over DHEA, so far there is only theoretical value in taking supplements of DHEA. And even that value would probably rest in the hormone's impact on overall energy, not our sexual arousal system. Studies of DHEA, particularly with respect to its levels in the elderly, are, however, continuing.

In addition to these, some old-time aphrodisiacs have clearly been found to work, albeit with a hefty price tag. Shortly before the American Civil War, French doctors treating soldiers in North Africa witnessed an epidemic of severe priapism—prolonged and painful erections. Some gastronomic investigations (they were French, after all) soon revealed that all of the men were fond of a local variety of frogs' legs. When the frogs were dissected, their stomachs were found to contain the carcasses of myeloid beetles, which are now known to contain cantharidin, a urinary tract irritant also known as Spanish fly. The Marquis de Sade was alleged to take it often, but biologically, cantharidin is a potent poison that can be fatal even in relatively small doses. The beetle has it to murder his predators.

A few years ago, entomologists at Cornell studied another beetle called *Neophyrochroa flagellata*, and in one of the more interesting aphrodisiacal twists found that these bugs have a natural love for cantharidin. They are attracted to Spanish fly bait set in fields and will feed on it in the lab. Males that eat it secrete cantharidin from a gland in their heads, which females sample during courtship; and females will mate preferentially with males fed on this poison. Like Spanish fly or cantharides, which alter the local blood vessel pressure and irritate the genital organs by irritating the bladder and urethra, amyl nitrite or "poppers" dilate blood vessels and when "popped" during the height of sexual arousal may increase the vascular response of genital organs to the muscle spasms of orgasm. However, because amyl nitrate is used to relieve the spasms of angina pectoris patients, it can, in normal people, cause heart attacks.

Many other compounds—raven gall and pumpkin seeds to gold and silver shavings and urine—have enjoyed popularity at some point in history. One, animal odorants, has come by its reputation more honestly than the rest. In fact, the most expensive perfumes and colognes have what olfaction experts call "animal base notes." The most notable of these is musk, a bitter, sharp-sour substance from a gland near the genitals of the musk deer, whose stench bears little resemblance to the thick, warm scents that boast of its use. The perfume industry, whose product names—Seduction, Opium, Joy, White Shoulders—are odes to their purpose, also accounts for most of the market in civet, a rotten-smelling fatty acid taken from the anal glands of an Ethiopian wildcat. There is also castoreum, secreted by beavers, and ambergris, the ash-colored blob excreted from male whale intestines that costs four times the price of jasmine, which is itself more than twice the price of gold.

Sex odor experts point out that "musky" concoctions that smell like steamy barnyards are still peddled in a thinly veiled attempt to mimic their attraction to the creatures who live with the originals. Androstenol and other metabolites of human sex steroids are molecularly similar to musk and civet, and it is probably no accident that as early as the Middle Ages, animal and "natural" scents were far more popular than floral odors as sexual attractants. More recently, a resurgence in the sale of truffle oil might be a tribute to the truffles' taste, but also to the fact that this vegetable contains a chemical somewhat like the principal male pig (and human) sex hormone, one reason that lady pigs are willing to trot at the end of a leash and dig them up. The human body, too, emits strong acid odorants that when exposed to bacteria form chemicals that are smelly but potentially sexy to some. Steroid hormones such as androsterone have been experimentally sprayed on theater seats, and in one set of studies it seemed to attract women at higher rates than usual.

And then there is vanilla—currently enjoying a revival in the U.S. perfume market. A form of the Spanish word for vagina, vanilla somewhat resembles the scent of warm baby, sort of sweet and powdery. To whatever degree people associate babies with lovemaking, it's not hard to understand why the vanilla smells of fresh-baked goods and why vanilla skin lotions are not only in vogue but rumored to be aphrodisiacs.

Ginseng, believed to trigger desire, has some properties in common with caffeine as an overall energy booster; and chocolate, made from the seeds of the cacao bean, has a couple of things going for it as well. First, it is rich in phenylethylamine (PEA), the natural amphetamine described earlier as a possible source of arousal. Second, it often is confected with vanilla, a double whammy. Montezuma reportedly drank fifty cups of chocolate a day to boost his virility before he visited one or more of the six hundred women reported to be in his harem. (It is one of the great ironies that the inventors of cappuccino were celibate Capuchin monks; it, too, was supposed to be a sexy drink.)

Pharaonic Egyptians used radishes; Ovid liked onions and especially liquamen, made from spoiled fish guts and poured like sauce on everything from meat to mushrooms. Foodwise, the wealthy Roman could do nothing more lusty than prepare and eat truffles dipped in liquamen.

Peppers, beans, garlic, and other "heat inducing" foods also have been prized as love lotions. Ted McIlvenna, president of the Institute for the Advanced Study of Human Sexuality in San Francisco, has been studying botanical oat and barley extracts that seem, he says, to protect testosterone in a woman's body and increase libido. And in her book *The Nature of Love,* Diane Ackerman quotes a medieval recipe listed under "venereal pastimes" that contained burdock seeds, brandy, the left testicle of a three-year-old goat, burned hairs from same, and four drops of crocodile semen. After three weeks (to gain "astral influence"), this concoction was to be rubbed on the genitals and results awaited. About the only thing in it with any chemical jolt was the brandy.

Which brings us to alcohol and other mood-altering substances: barbiturates, marijuana, cocaine, amyl nitrate (which we've encountered previously), LSD and other hallucinogens, amphetamines, MDMA (so-called Ecstasy), and methaqualone (quaaludes), all of which have been reported to enhance arousal or orgasm. They do indeed produce intense feelings, sensations, and sensitivity to the environment. They seem to lower inhibitions, relax muscles, enhance the senses, and produce a sense of well-being and power. They certainly can distort unpleasant realities and block out guilt, qualifying all of them as aphrodisiacs by definition. Alcohol, especially, in every form and in small doses fuels the desire for sex.

Less well appreciated, of course—beyond the obvious risk of arrest or addiction for using most of them—is the fact they also produce *counter-erotic* effects with regular use or even moderate overdose. Alcohol is the classic case: a small amount, as Shakespeare said, "provokes the desire" but any more "takes away the performance." It's not a stimulant, of course, but a central nervous system depressant, a classic sedative. Chronic doses of amphetamine, including speed, break down blood vessels and brain cells that affect dopamine uptake, resulting in a burnout of the very sexual system first turned on and turned up by the stimulants. Long-term use of cocaine and narcotics completely suppresses sexual interest. (Cocaine, as Cole Porter's original lyrics noted, inititally delivers a sexual kick. It does so by prolonging the release of norepinephrine and enhancing the dopamine action, altering heart rate, blood pressure, and muscle contraction. But it soon wears out the brain's sensitivity and response.) Barbiturates and other powerful sedatives initially produce drowsy euphoria, excitement, and emotional swings that may temporarily enhance a sexual experience, but they eventually bring addiction and psychotic behavior; when withdrawn, they induce the very anxiety, panic, and loss of sexual interest they were intended to counter in the first place.

However much sensible people and the medical profession may balk, interest in aphrodisia shows no signs of diminishing. So what is the outlook for true aphrodisiacs in the future, for safe, effective means of initiating or enhancing erotic pleasure?

In a review for the *Archives of Sexual Behavior* in 1993, scientists at the University of Medicine and Dentistry of New Jersey Department of Psychiatry described several promising "new aphrodisiacs," including drugs that affect dopamine, serotonin, and other neurotransmitters that act as mood dampeners or enhancers. Used to treat a variety of ailments, their prosexual side effects have been studied in both animals and humans. No one, including the review authors, seriously suggests deliberate tampering with the brain's neurotransmitter system unless there is a good medical reason for doing so, such as Parkinson disease, Alzheimer disease, mental retardation, or the like. But there may be ways to boost a little the natural chemicals that fuel or maintain our sexual desire and performance, perhaps with foods or other substances that mildly alter the ratio of these compounds.

A couple of these deserve a few more words of explanation because they are the target of relatively intense research. Oxytocin, for one, secreted by the rear end of the pituitary gland, influences in still largely unclear ways dopamine, estrogen, serotonin, prostaglandins, testosterone, and other blood vessel and muscle relaxers and constrictors. Because oxytocin rises during intercourse and spikes at orgasm, there's no question that this chemical is part of our natural aphrodisia armamentarium and a reasonably potential candidate for both treating sexual dysfunction and enhancing sex in the healthy. So far, the best and only proven way to get some is to have sex, but that could change.

Prozac and other compounds that lower stores of serotonin should, theoretically, enhance and promote sexual arousal and performance, since serotonin is the brain's "stay in control" chemical. Like its fellow neurotransmitters, serotonin has some seemingly paradoxical effects, both because of the way the brain regulates its uptake and distribution, and because brain sensitivity to the chemical is widely distributed. Thus, serotonin in *some* doses can put the brakes on people whose wacky brain chemistries create panicky, obsessive, or anxious behavior, while in other doses and in other brains, it relieves the depression that comes with exhausting and frustrating efforts to control our behavior or our circumstances. Prozac is one of a class of drugs called a serotonin reuptake inhibitor, and its net effect is not to block the manufacture of serotonin but to slow down its reuptake by various receptors in the brain. The Prozac path to aphrodisia, if it exists, is therefore likely to be extremely complex, because serotonin itself can either facilitate or inhibit sexual activity, depending on which serotonin receptor type is involved.

Similarly, dopamine's potential as a sexual stimulant has been fueled by observations of its activity in the brain during sexual arousal and orgasm, but increasing dopamine's availability in our sexual brains is going to prove just as difficult as blocking serotonin or balancing the two chemicals. What naturally increases dopamine can be scary, notably schizophrenia and other serious mental illness.

Finally, the hunt for safe and effective aphrodisiacs may have some real payoffs if preliminary reports of powerful human pheromones turn out to be valid, and substantial research is under way in animals and humans to identify and understand the role of the "smell brain" and other organs that detect these sexually activating chemicals.

John Dwight Harder, an authority on pheromones who investigates sexual desire and what is known in the biology trade as "sexual receptivity," reports that for now, the search will probably—and properly—stay focused on the roles these chemicals play naturally and not on efforts to jump-start or enhance them. But even at these preliminary stages, the possibilities are seductive.

For example, Harder argues that unlike most historically recorded searches for aphrodisia—those driven by men to increase their performance or to seduce women—the best clues to true aphrodisiacs emerging from studies of pheromones will probably be found by investigating the complex sexual repertoire of females, not males. After all, females are the ones who are multiorgasmic; they are the ones whose moods and minds are so much more central to sexual pleasure. "The human female has a greater range of sexual receptivity than most mammals, she doesn't have 'heat,' " says Harder. Like studying only men who had heart attacks to find out why they had them, studying only erection-causing love potions misses what may be the more informative strategies: how it is that women *don't* have as many heart attacks and how it is that they turn on their sexual bodies for the longer, more enhanced, more enduring experience.

While pheromones and all of the compounds under investigation have—or have the potential to generate—a sex-enhancing effect, even those that may appear to be natural or benign have potentially dangerous effects. Given the inherent risks in tweaking or goosing a complicated set of behaviors that has evolved over millions of years, we ought to think at least twice about using even those that carry very little known risk. Before we try to fool Mother or Father Nature, we'd better be sure the last laugh isn't on us. It's also worth noting that typically it's still the male who wants an aphrodisiac, not for himself but for his partner. In this sense, aphrodisia may eventually fall victim to political correctness in the same way that other means of manipulating women have.

On the other hand, perhaps not. Our bodies, after all, are full of chemical manipulators that both sexes use to woo and win orgasm, the subject of the next chapter.

9

ORGASM: HIS, HERS, AND THEIRS

*"C'est une des superstitions de l'espirit humain d'avoir imagine
que la virginite pouvait etre une vertu."*
*(It is one of the superstitions of the human mind to have imagined
that virginity could be a virtue.)*

—VOLTAIRE

Hearts beat as fast as they would during wind sprints. Blood pressure soars and breath explodes in grunting bursts. Skin and genitals flush and swell red and blue with blood. Central nervous systems, electrified by the firing of millions of cells, expel bursts of mind-altering biochemicals, while sex hormones play in and out of receptor feedback loops and tiny proteins play their own roles and glands goad gametes along fluid rivulets in lubricated passages. Extra energy is recruited to cells and fibers. Faces contort, muscles strain to the breakpoint and convulse. And finally, bodies and brains give up and give over to the warm, sharp, rhythmic pulsations, contractions, and spasms that tear us however briefly from consciousness with fragmentary hallucinations. Then, with no effort at all, the cascades of sensation ease. Breathing slows. Sex steroids and adrenal hormones dissipate slowly and pleasant fatigue settles.

The impact of an orgasm's buildup and denouement is indeed mighty, but, ironically, matched by the mystery and misinformation that shadow its course. Here is an experience intensely sought after, enjoyed, and powerful in our lives, yet whose origins and significance are obscure and enigmatic.

Consider the popular wisdom that orgasm evolved in every mammal to reinforce the desire for sex and bonds of love that keep a couple together at least long enough to rear the consequences to reproductive age. It may sound logical at first, but the hypothesis is weak. To begin with, it's unclear whether most other animals have orgasm or, if they do, what emphasis they place on it. (Only one nonhuman primate, the female stump-tailed macaque monkey, has been definitively shown to have orgasm, and only when stimulated by other females.) By contrast, *all* animals must be wooed and aroused for mating to be successful, suggesting that it's the preliminaries, not the act itself, that serve as the "reinforcer."

There's also the fact that women don't need an orgasm to become pregnant or even desire motherhood, and that, in general, women climax most easily not during penile thrusting or ejaculation but during foreplay or postcoital stimulation. If orgasm itself were so central to desire and reproduction, moreover, why would orgasm be so physiologically automatic in men while rarely so in women?

Oxford University scientists Marion Petrie and Fiona Hunter suggest that if orgasm via intercourse were truly the evolved motivation for having sex, then it would follow that animals that had the highest *frequency* of intercourse would have the greatest pleasure and thus the greatest reproductive success. But that doesn't seem to be the case at all, because the evidence shows that frequency of copulation is neither universal nor stable, even in the same animal over time. The huge variability may suggest that frequency of mating isn't driven by orgasm's *pleasure* at all, but by the need to reduce a partner's availability to other potential partners. In other words, if you keep your partner very busy in the bed or burrow hole, there's less time for him or her to seek out others.

In a similar vein, if orgasm were so central to sexual reproduction or intercourse, why would nature have rendered the process of getting one so relatively fragile? The lust for sex and its progress are fierce and urgent, but the ultimate pleasure of it is easily dampened or altogether undermined by a child's small voice or a fleeting anxious thought.

The list goes on. Why would our brains and bodies be "wired" to permit an orgasm even when there is no chance of reproduction? Transsexuals who have undergone surgery to fashion cosmetic but nonfunctioning genitals have been reported to enjoy sex and "perceive" orgasm. Masturbation, by many accounts, produces the most intense orgasm and both men and women can fantasize to orgasm without even touching themselves or each other. Moreover, from the physiological standpoint of generating orgasm, the anus may be as much a "sexual" organ as the penis, clitoris, vagina, and womb, and during menstruation, when pregnancy is highly unlikely, orgasm is often intense and sought by women to ease the congestion and discomfort. Some men and women achieve orgasm when their nipples are sucked. (So do some nursing mothers; and some women during childbirth even experience orgasm when the clitoral hood is firmly covering the nerves.) And the famous castrati of harem and religious-order fame, contrary to popular belief, were often still very capable of orgasm, although not fertilization.

Some have suggested that orgasm is an "accident" of evolution and that in women little more than an evolutionary "leftover," an incidental by-product of male orgasm, which *is* adaptive because it promotes insemination, a favorite theory of the famed biologist Stephen Jay Gould.

Yet there is also compelling evidence that suggests a woman's orgasm does have purpose: it facilitates sperm penetration of an egg, or, as studies in mice and other mammals have shown, the success of one male's sperm over another. And the almost universal physiological markers of orgasm suggest that even if orgasm *evolved* for no particular "reason," it would not have persisted so powerfully unless it conferred advantages to our bodies, minds, and behavior. Whether sex takes place in a mirrored bordello or the backseat of a Chevy, in the missionary position or hanging from chandeliers, the biological programs have made orgasm central to our *perception* of a "successful" act of sex, so much so that women frequently fake it and men consider themselves failures if they don't produce one in their partners.

However or why ever orgasm originated and then persisted, its awesome power to transform our bodies and minds with pleasure stands in contrast to its power to worry and even frighten us. Perhaps only in common with death, orgasm (the "little death," as it is sometimes described in France) is so compelling a phenomenon that the most advanced societies and the most advanced brains in the animal kingdom consider the whole subject "dirty," surround it with taboo, and relegate those who study it to the research fringe. (The first international conference devoted to orgasm wasn't held until 1991.)

Other animals just go at it, but human social and psychological squeamishness about coitus and orgasm abides, rooted, in part perhaps, in the sweaty, often awkward physical enterprise it takes to accomplish them. Is there anything most people feel *less* comfortable displaying than their bodies on a coital ride?

The external contortions of sexual intercourse as prelude to orgasm do seem wild, but all of that frenetic activity echoes and facilitates an equally energetic biochemical mystery dance. Given the need to design a sexual system for optimal reproductive capability under all circumstances, perhaps it should only be surprising that intercourse and orgasm aren't even wilder affairs.

Consider the challenges of such a process to the natural designers. They had to attend to such differences between the sexes' bodies and physiologies as the relative size of sperm and egg and the competing strategies of selfish genes. They had to assure that a man and woman would find the time-consuming hard work of penis-in-vagina sex at least as rewarding as arousal and courtship. And they had to assure that each sex would come to the task with enough mutual trust to make their mutual vulnerability tolerable anywhere, anytime, and to be willing to have sex again and again, even when one or neither is optimally fertile.

Both coitus and orgasm as we know them are the solutions nature evolved to such challenges, a complicated approach to sex that accommodates a variety of environmental and biological circumstances, checks, and balances.

Thus, studies show, women who have regular sex are more fertile than women who have less intercourse, and men and women have different—but not necessarily less intense—sexual appetites because of differences in their thinking brains as well as their bodies' hormones. We wound up with sexual "specs" that make us like sex more, in general, the more we get it for the most important of reasons. In a study published in *The New England Journal of Medicine,* Allen Wilcox of the National Institutes of Health shows that chances of conception are related to frequency of intercourse during the fertile period, with conception most likely when intercourse has been daily or every other day. This certainly fits our biological design, given that men and women can't easily "tell" with any accuracy when fertility exists or peaks. To play it safe, our mental and physical mechanisms for sex tell us to have it as often as possible.

Although in humans, frequency of intercourse may not *greatly* influence fertility—we've learned to detect the generally most fertile periods in women—we may have conserved some part of what is clearly a strategy of *intercourse-dependent* fertility in animals. Remember that in rats, for instance, it is the male's insertion of the penis into the female opening, and his thrusts, that trigger the hormone release that lets her ovulate and advance implantation of a fertilized egg. The male's copulatory efforts also contract the female's uterus, a requirement for sperm passage to occur.

As we'll see later in the chapter, there is even evidence for some of this in human animals as well.

Along the evolutionary way, we also got a sexual system designed to encourage us to cuddle and bond as much as to seek orgasms, perhaps explaining why women increase the chance of conception if they stay horizontal after intercourse and why men tend to fall asleep after intercourse. We achieved a coital and orgasmic plan that seems to align biological peak performance to *sustained* mutual contact, to "sleeping together," and sharing not only that pleasant postcoital fatigue but also a commitment to accumulate resources and parent any offspring that result from mating. This is enormously important in view of the fact that our Stone Age biology puts the peak of male and female sex drive in the late teens, when the urge to roam and experiment are high. Modern studies show that boys who undergo early puberty and begin sex activity earlier have the highest rate of sex activity throughout their lives. So do girls. Far from the "family values" conviction that it's unnatural for the very young to want or have sex, young sex is highly adaptive in the game of survival of our DNA—but less adaptive for offspring unless there is something in coitus and orgasm that also works to keep the young and sexy together for a while. This may not be what we want to hear—whether extremely conservative or equally liberal in our social and political convictions—but it is an incontrovertible biological fact. And while it's true that men "peak" in terms of actual sex interest in their twenties and women in their thirties, males, some experts report, are likely to describe such a peak as the "desire" for sex, while women describe it as "satisfaction with" sex. So for men, frequency of intercourse in their youth fits with their Stone Age strategies to spread their seed as far and wide as possible, while for women, the sex-is-best-in-a-committed-relationship sense of satisfaction—more likely to occur when females are older and more experienced—fits best with *their* reproductive self-interests.

If the process isn't as streamlined as it might be, it's understandable. Because what happens in intercourse, orgasm, and afterglow reflects strategies that nature constructed, by trial and error, by random hits and misses, to form biological drive and performance. Thus, the mental and physical mechanisms that make us the sexiest animals in the world took millions of years to become what our ancestors bequeathed to us a mere forty thousand years ago.

What does actually *happen* during intercourse, atop the sheets and beneath the skin? It is far from completely described, much less understood, but what is known is a dazzling mechanical and biological feat.

In a man, reflex-driven arousal and therefore erection is regulated by the parasympathetic nervous system and therefore is technically beyond a man's "will." However, erection is within his control to the degree that he can keep stimulation that causes arousal away from himself, if it is physical, or from dominating his thoughts, if it is psychological. He can also, to some degree, control ejaculation of semen, since that particular function is governed by the sympathetic part of the autonomic nervous system, which is used to engineer and manage genital contact. At the conscious level, a man can manipulate the degree and rapidity of genital contact, while unaware that the process of emission is getting under way during the final stages of intercourse—with contractions in the vas deferens, the prostate, the seminal vesicles, and the abdominal part of the urethra that concentrates and collects sperm and prostatic fluid from storage depots in the seminal vesicles. These contractions shut down the part of the urethra's valves that would let urine flow and push the fertile prostatic fluid to the powerful striated bulbar muscles that control ejaculation. The bulbar muscles are attached to the nearby muscles of the perineum, and during ejaculation they contract involuntarily at 0.8-second intervals, squeezing the base of the penis and the urethra to force the semen out in spurts and creating the intense pleasure of orgasm. A man's orgasm puts semen deep in the vagina near the mouth of the womb, in a small pool, from where the journey toward an egg is most optimal.

In a woman, local tissue congestion triggered by arousal produces a clear fluid on the walls of the vagina to ease insertion of the penis. As excitement mounts, the walls of the womb become engorged with blood, and this organ actually enlarges and literally rises from the floor of the pelvis to position the mouth of the uterus nearer the pool of seminal fluid that her partner will eventually ejaculate.

The thrusts, movements, and rotational touching of the penis inside the vagina further stimulate the congested and engorged vaginal vault, which converts from an empty, flaccid cavity to an inflated, firm balloon, while the clitoris becomes erect and an inexorable sense of wetness and pelvic congestion builds. Photographs by sex researchers William Masters and Virginia Johnson of the internal female organs during sex, made with a camera-equipped clear plastic phallus, show that once an erect penis is inserted, the local congestion and smooth muscle reactions that occur with arousal keep the vagina from collapsing, and maintain it as an erect yet flexible, tight tubal platform along which the penis can glide. During intercourse, this same tube—capable of expanding enough to expel a five- to ten-pound fetus, let's not forget—conforms to and grips the much smaller erect phallus. (Masters and Johnson long ago demonstrated that the size of a flaccid male penis varies greatly, but the variations in diameter of erect ones are incidental.) One set of studies of seventeen healthy women in their mid-thirties, whose vaginas were distended with air in a condom tied to a catheter, also suggest that during coitus the levator ani muscles (which lift muscles around the anus) contract more as the vagina balloons, so that penile thrusting enhances sexual performance and plea- sure for both parties.

During intercourse, and along the route to orgasm, a woman's lower external genital tract, the vulva and labia, engorge and swell further, along with tissues near the anus, becoming thicker and brightly colored with blood. Unlike a male, in which vasocongestion is restricted to the tissues of the penis and is limited by the shape of the shaft, a woman's congestive response is more diffuse. She experiences a general swelling of internal and external genitals and pelvic area all around the vaginal barrel. When she climaxes, she may sense a wide dispersal of her orgasm as well. Mas- ters and Johnson clocked the contractions of women's orgasms at three to fifteen of them about 0.8 seconds apart in the outer third of the vagina. Some orgasms have been timed to last as long as 19.9 seconds and a few have been clocked for minutes.

The genitals and brains of both men and women exhibit much the same signs of climax: a series of reflexive, automatic rhythmic contractions of the structures that make up the orgasmic platform, including a variety of muscles and engorged tissues around the base of the penis or lower vagina. In women, the perineal, bulbar, and pubococcygeal muscles (pri- mary muscles in the lower pelvic region) contract at the same 0.8-second intervals against the engorged vaginal tissues. Deep nerve receptors in the perineal and vaginal muscles and other internal organs transmit the orgas- tic sensation to the brain. The uterus also contracts, although most women are not very aware of it and women without uteri have reported no real differences in their awareness of orgasm, although they do report some vague spreading of pleasure. As with the glans penis, direct or indirect stimulation of the clitoris is always necessary to produce orgasm, because this tissue alone holds within it the nerve distributions that relay pleasure signals to the brain. During intercourse, when there is no direct stimulation of the clitoris, stimulation occurs indirectly by tugs on the outer fold of skin or clitoral hood connected to the vaginal lips, with the tugs created by the back and forth motion of the penis, which exerts rhythmic mechanical traction. Meanwhile, of course, a variety of hormones and hormone-like substances are not only priming us for pleasure but also stimulating the internal reproductive duct system in men. Oxytocin, for instance, has been found to promote sperm transport during ejaculation. And French researchers have found evidence that the release of endorphins, the brain's own opiates, accompany the release during orgasm of certain enzymes into a woman's vagina that may break down cervical mucus blockades, permitting sperm easier entry and access to eggs.

Whether or not there are particularly erogenous zones on the vaginal wall, known as "G" spots (for Grafenberg, the fellow who first identified their possible existence), is still an open question. From Shere Hite to molecular biologists, investigators have failed to prove or disprove any case pro or con.

Despite all the information now in place, uncertainty about whether or not a man and a woman experience orgasm's pleasure in *exactly* the same way persists, and certainty is likely to prove elusive. For one thing, we have no vocabulary equal to the task. Language has hundreds of words for coitus (intercourse, banging, fucking, screwing, bonking, humping, bumping uglies, etc.) but only one word for why most people say they want sex. *Roget's Thesaurus* has no listing for "orgasm" and the best that the *American Dictionary of Slang* can come up with—"cum" and "coming"—have more of a linguistic reference to ejaculate than to the experience of orgasm. A standard dictionary is not much help either. It calls orgasm—the word itself is drawn from the Latin for "swelling"—"the physical and emotional sensation experienced at the peak of sexual excitement," a peculiarly unsatisfying explanation. A medical dictionary is perhaps less helpful with "an apex and culmination" of sexual excitement. Uh-huh.

Alfred Kinsey, William Masters, Virginia Johnson, and a handful of other sex researchers tried the taxonomic approach to defining orgasm. But with all of their electrodes, clear plastic tampons, and questionnaire surveys of masturbating and copulating volunteers, all they unmasked and described were the anatomical constructions of orgasm, elevating the clitoris to center stage, measuring vaginal congestion, calculating angles of erection, and putting masturbatory and coital orgasms on an equal footing. They essentially got the human race no further in describing what is actually *happening* to us when orgasm happens than did the most primitive, inarticulate cave-dwelling lover.

John Money, whose penchant for turns of scientific and social phrase popularized "orgasmology," took a stab at defining the word also: "The zenith of sexuoerotic experience that men and women characterize subjectively as voluptuous rapture of ecstasy. It occurs simultaneously in the brain/mind and the pelvic genitalia. Irrespective of its locus of onset, the occurrence of orgasm is contingent upon reciprocal intercommunication between neural networks in the brain, above, and the pelvic genitalia, below, and it does not survive their deconnection by the severance of the spinal cord. However, it is able to survive even extensive trauma at either end."

His entry is arguably more instructive because it embraces a key dimension of orgasm—its uniform expression in the wake of an almost endless assortment of manipulations, from autoerotic asphyxiation to conventional missionary positions. For whether the arousal that gets you there begins with fantasy or pornography, thoughts or touches, the object of your desire or second best, a foot fetish or a photoplay, with music and flowers or with procreative lust, the same confluence of muscles, nerves, hormones, neurotransmitters, and emotions *happens*. "Metaphorically," Money says, "orgasm is the Rome to which all roads lead, irrespective of the induction site at which they begin. Each road must, however, feed into the two-lane highway of the spinal cord that connects the sexual neural networks of the brain with the peripheral neural networks of the pelvic genitalia and vice versa. Orgasm occurs synchronously in the brain and in the genitalia."

Thanks to the work of many sociobiologists and psychobiologists—and despite the difficulties of studying sex in general, human sex in particular, and orgasm most of all—orgasm is recognized incontrovertibly as both a physiological and a psychological phenomenon, an event that takes place not only in the genitals and brain cells but also in the mind. It consists of what we came wired with and what we learn. It is what Money calls a "phylism," an element of human existence that belongs to individuals, as members of the human race, just the way sleeping, yawning, and urinating belong; but it is an element that also occurs under so many different individual circumstances and as the result of so many different individual experiences that at the end of the day, we are dazzled with the fact that we can only, finally, understand the experience of our own orgasms. Ultimately, they are a defining quality as unique as our fingerprints.

"Try to compare orgasms," exhorts Money, and you wind up instantly constrained by "solipsism," in which you are stuck with the conviction that the "only orgasm you can ever know and talk about in its entirety is your own. Its self-generated sensations and feelings are yours alone. The [arousals and triggers] of orgasm are personal and private." We are left asking each other, as did Pilar in *For Whom the Bell Tolls,* "but did thee feel the earth move?" and left assuming that whatever it is we feel is what the others do, too.

Such has been our ignorance of the geography and biology of orgasm that what scientists *have* managed to uncover in the last century is a testament to perseverance and some of the cleverest experimental strategies yet devised.

In 1919, John B. Watson of Johns Hopkins University was apparently the first to develop a number of somewhat crude instruments to record the physiology of orgasm with the help of his student and lover, Rosalie Rayner, and another student, Curt Richter, who became the pioneer researcher in biological clocks and biorhythms. His apparatus, according to some accounts, is all that survived of his career after a scandalous divorce, wedding to Rayner, and firing from Hopkins on account of the "moral delinquency" of his orgasm research. All of Watson's research findings disappeared or were destroyed and apparently died with him. Adolph Meyer, the head of the Department of Psychiatry at the time, demanded that the university president fire him as a matter of "principle."

It took decades before Alfred Kinsey used "direct observation" of orgasm as a response to pornography to get a fix on the physiology involved, but the state of the art in his era in the 1930s, 1940s, and 1950s was light on both biochemistry and the means of tracking the hormones and neurotransmitters fueling the process. In the 1970s, Masters and Johnson directly observed some of that physiology in copulating and climaxing couples, documenting arousal, "plateaus," and the vibratory rate of orgasm. They also reported that clitoral orgasm is the only orgasm for women. We can only be grateful that their work was not censored, nor were they subjected to the fate of Wilhelm Reich, the famed German sexologist who was sent to prison for his ideas about orgasm and died there.

It may sound surprising today, but since Masters and Johnson, the tables have turned again, and sex research, particularly orgasmology, has been permissible mostly when framed in the context of diseases (infertility, for instance, spinal cord injury, prostate disease) or psychiatric disorders (schizophrenia, epilepsy, pedophilia). Even now, those with the simple desire to know and understand what orgasm is depend mostly on getting information by approaching their topic sideways.

Until recently, efforts were also hamstrung by the lack of technology to get at biological processes in the body and brain. The big advances had to await more modern imaging and monitoring technologies, the kind that record functions such as the use of energy and nerve-fiber activation. Scientists needed not just snapshots of what the brain looked like, but moving pictures, in a sense, of what energy the brain was using and what nerves it was triggering and what brain chemicals and hormones it was transmitting during orgasm. Once the technologies came on-line, orgasm began to come unveiled, although care is always taken to cast the research in nonsexual terms. (Efforts to separate sex from orgasm, particularly women's orgasm, are, of course, an old tradition among researchers. In her 1999 gem of a book on the historic development and widespread use of the vibrator by physicians and midwives in the nineteenth century, Rachel P. Maines points out that the Western "androcentric model of sexuality," which defines coital orgasm as the only "legitimate" orgasm because it was the norm for males, led to the concept of inorgasmic women as disordered "hysterics" whose "treatment" was the achievement of orgasm in a clinical setting, by doctors or midwives employing vibrators.)

Among the more sensible and creative users of technology is Barry Komisaruk, whose work at Rutgers on sensory pathways to the brain forged by the vagus nerve was discussed in chapter 4 and chapter 7. Komisaruk has managed to investigate how the sense of pleasure is released during orgasm by crafting his studies and his publications around discussions of improving the quality of life of people with spinal cord injury rather than in terms of orgasmic physiology. (For example, his experiments showing that vaginal stimulation produces substantial pain blockage in women, and that rectal stimulation does the same in male rats, led him to conclude that there are chemicals released during orgasm that need to be identified and investigated because they could lead to treatments that restore pain relief capacity or directly provide pain relief itself.) "The fact that they may also be able to restore or amplify sexual response, is a conclusion just as profound, but less welcome by the research community," he admits.

Komisaruk and Whipple have, in fact, reported the presence in rodents during orgasm of a neuropeptide they call vasoactive intestinal peptide, which, when released into the spinal fluid, causes the sensation of orgasm in the brain. A neurotransmitter, it also has been identified in women and can, according to the Rutgers team, increase the general pain threshold of women up to 107 times the normal baseline. Studies are under way to determine if VIP has clinical value in the treatment of pain, but the two say that to the contrary, there is no evidence that VIP "orgasm pills" are in development. "To link research with laboratory rats to the human female is not only inappropriate, it is simply wrong," Whipple has said.

Komisaruk has taken his research on the links between orgasm and its deadening effects on pain to logical evolutionary conclusions. It's just possible, he says, that the central evolutionary purpose of orgasm is to reduce pain. In rat experiments, Komisaruk first established a consistent response to a precise electrical shock given to females through electrodes affixed to their tails. He increased the intensity of the shock until the animals would squeak each time the current was sent to the tail and he could easily hear and count the number of squeaks per shock. Komisaruk set the experiment up this way because in rats, mating behavior is so fast—each penile intromission takes a fragment of second. As the female darts in front of him, the male mounts and thrusts and inserts his penis for less than a second, dismounts, grooms awhile, and does this all over again. After about eight such episodes, the male ejaculates once, then waits half an hour before restarting the entire process. Thus, if Komisaruk wanted to test whether there is something like a copulatory orgasm that blocks pain in the animals, he needed to find a very fast way to inflict a consistent level of pain (a squeak) each brief time the male ejaculates. Komisaruk learned that during mating, as measured by a decrease in squeaks, the most powerful pain blockage effects occurred in the female during the time a male ejaculates. In step with the surmises of some evolutionary biologists, he speculates that the intense pleasure of orgasm may well serve to overcome all other distraction or the stress that comes with rigorous sex.

Adding to the mystery, he says, is the fact that female rats appear to require those eight intromissions to occur before ejaculation in order to conceive. In the studies done by Norman Adler at the University of Pennsylvania, Adler put a male rat in the company of a fertile female, but took the male out of the cage after only six intromissions. Then he put the same (presumably unhappy) male in with a new female; the male performed two more times and ejaculated. The female who got only the two intromissions never got pregnant, despite the presence of sperm. Adler concluded that precisely eight intromissions were necessary to stimulate release of the hormone progesterone in the female, which prepares the uterus for implantation of a fertilized egg. But Komisaruk says studies also suggest that such frequent intromissions are painful, or in some other way aversive to the female, so pain blocking is one way perhaps that a female becomes willing to accept multiple intromissions.

Other studies by Komisaruk and his colleagues have been conducted on spine-injured women and animals who are well-suited to illuminate the nature of orgasm because there is a disconnect between the nerves that connect genitals to the brain. The results offer tantalizing clues to the nature of the orgasmic brain and body.

The human spinal cord has thirty-three concentric layers stacked around the cylinder of the cord, beginning with seven in the neck and the rest in the thoracic or chest area (layers 8 to 19), lower back (20 to 24), pelvic areas (25 to 29), and coccygeal area (30 to 33). Research has made clear that if the flow of information along the spinal cord is cut off above the thoracic (T) layer 11, then all of the nerve reflex feedback loops between the genitals and spinal cord required for release of ejaculation in the male stay intact, and vibrating the penis will produce semen. But such a man will not experience "brain" orgasm or even the sense of ejaculating without looking. If the cord is severed below T11, erection can be achieved but not ejaculation. In women, the experience is similar with respect to orgasm. (It may seem confusing that *where* one injures a spinal cord would affect sensation, since many assume that a cut cord eliminates all motor and nerve activity. But that is not the case. Some muscles and other structures continue to function in a variety of ways because of peripheral and vagus nerve activity, as well as reflexive drives.)

In any case, however, with training and motivation, paraplegic men and women, those paralyzed from the chest or waist down, can experience sensual arousal of the nipples, breasts, ears, mouth, and other parts of the body. And on occasion, they experience a sensation not unlike the buildup to orgasm, although it is never "felt" or experienced in the genital area. Concludes Komisaruk, although it could be that there are in such patients remnants of nerve pathways that provide sensitivity in the vaginal or cervical areas, it's more likely that "the vagus nerve may be a primitive pathway to orgasm that really can shed light on how orgasm occurs and where exactly it is controlled in the brain and even on why it feels so good."

The vagus nerve is the tenth cranial nerve of twelve in the head and face, and contains pathways that contribute to the regulation of the heart and other deep internal organs. The nerve fibers of the vagus do not originate in a common brainstem structure that feeds as a single cable to the spinal cord and brain, but have two sources instead. Stephen Porges, an authority on childhood vagal nerve system and the development of sexual and other relationships, speculates that mammals, but not reptiles, have ancient brain organizations characterized by such a two-source vagal complex, a system that forges key links between attention, motion, emotion, communication, and language.

In Porges's model of the vagal complex, moreover, there is competition between the two pathway sources. Sometimes, the competition leads to disease or even death. For example, sudden infant death syndrome and asthma may be related to the competition between the two sources that upsets the regulatory balance responsible for breathing (involuntary action) and holding our breath (voluntary action) and for not letting stress overwhelm out brain's automatic breathing rules. But if he is correct, it's also probable that the vagal system could be responsible for linking lots of other things our bodies do unconsciously—such as erection, lubrication, and orgasm—with the emotional states that so heavily influence when, how, under what circumstances, and how successfully we give our bodies and brains their "head" to do these things.

One way to test Porges's hypothesis is to do what Komisaruk and others have: study people who have damage to the central nervous system circuits and see how they act and respond sexually. In one such study of men with complete spinal cord injuries, investigators at the Kessler Institute said 38 percent reported they still had the ability to have an orgasm and ejaculation.

This was not the first time, of course, that orgasm has been reported among those with cord injuries. Money, for example, described it in some paraplegics, who said that during dreaming and sleep, orgasm is experienced as it was before the spinal cord injury. "This dream orgasm is properly characterized as a phantom orgasm for it occurs only in the imagery and ideation of the dream, deconnected from the pelvic genitalia." It happens infrequently and never more than two years after injury.

Perhaps the landmark paper in this field came from Komisaruk, Beverly Whipple, and Carolyn Gerdens, also of Rutgers. In 1996, they reported in the *Journal of Sex Research* that there was strong evidence from women with complete spinal cord injury that these may not be "phantom" orgasms at all, but psychophysical ones that recruit vagal and other peripheral nerves that bypass the two-way highways Money described in his patients. They obtained objective measures of the subjective reports of women during vaginal self-stimulation with a specially constructed tampon adapted from one used in animal experiments described in chapter 7.

All of the women—sixteen with spinal cord injuries at T6 or below who were at least one year post-injury, and five controls—were aged twenty-one or older and had a cervix and uterus. Of the sixteen in the injured group, thirteen said they had experienced orgasm before injury.

The tampon the investigators developed is a vaginal and cervical "self-stimulator" made from super-absorbency Playtex Ultimate tampons, modified so that the cardboard interlocking cylinders were fastened together tightly on the inside with a nontoxic adhesive to prevent the tampon's parts from moving individually. The entire tampon construction, lubricated with K-Y jelly, was fit into a specially designed holder with a handle, and a pressure transducer was embedded in the holder behind an aluminum head to provide a digital readout of the pressure being self-applied to the vaginal wall or cervix by the women using it. During stimulation, the women were instructed to observe the readout and not allow the reading to go higher than 500, which is about 440 grams of pressure. For cervical stimulation tests, they used the same handle with a small Velcro disc attached to the center of a diaphragm fitted individually to each woman, with a matching disc attached to the tip of the tampon. Pressure of no more than 180 grams was encouraged.

As in the experiments on rats showing that muscle spasm–measured pain of foot pinches was completely blocked by stimulating the rodents' cervixes, these results showed that vaginal and cervical self-stimulation, as well as stimulation to some other sensitive part of the body, such as the breasts, produced significant increase in blood pressure in the control group, but also in the women with complete spinal cord injury below T10. Astonishingly, three women with injuries as high as T7 reported orgasms during the study, supporting the idea that women with virtually complete spinal cord injury experience orgasm from genital and nongenital self-stimulation.

Their experiments further found that cervical stimulation in women stopped pain but not gentle touch, a point confirmed by women's own recollections in conversations with Whipple. Her experiments also found that the threshold for pain detection increased by a mean of 47 percent when the stimulus was applied to the front vaginal wall, and among women who experienced orgasm it increased nearly 107 percent, conditions that might also reflect, she says, the body's easing of labor pain.

The Rutgers team concluded that "the orgasm is not exclusively a physiological phenomenon, nor is it exclusively a disembodied psychological or transcendental phenomenon. It is par excellence, a phenomenon of mind/body unity. We also believe there is a genital sensory role for vagus nerves," Komisaruk says. They carry stimuli directly to the brain, bypassing the spinal cord. When he severs the vagus nerves in rats, the effect of cervical stimulation disappears entirely. And in positron emission tomography (PET) and magnetic resonance imaging (MRI) in two women, Whipple and Komisaruk found increased chemical activity in the medulla oblongata during genital stimulation, structures in the brain linked to the vagus nerve.

So powerful does the vagal pathway seem to be for sex that even in the complete absence of sex hormones, vaginal stimulation produces sexual responses in experimental animals via this pathway around the spinal cord. In another set of rodent experiments, Komisaruk found that cutting pairs of pelvic or hypogastric nerves reduced but did not abolish the painkilling effect of cervical stimulating. When he sliced in two all of the direct "sex" connecting nerves in the body and the vagus nerves as well, he abolished the entire sexual response. But when he stimulated electronically the "end" of the severed nerve connected to the brain, he was able to produce what he describes as "pure brain sex"—neural evidence of orgasm—without any other part of the body or its nerve or chemical supply involved at all. (Animals can't tell us if the "earth moves" for them, but on the other hand, that limitation of theirs eliminates a lot of human subjectivity that can muck up the physiology.)

In a related but separate series of studies, Porges conducted experiments on vagal fitness and the regulation of emotion. He and his coworkers propose that areas in the brain stem that operate the vagal circuit independently of the limbic system and spinal cord are also involved in integrating emotional expression. If those brain stem areas are sensitive to oxytocin and other hormones associated with cuddling, attraction, arousal, and love, Porges suggests, then our ancestral nervous system had a mind of its own, a primitive emotional circuitry now integrated into the thinking and emotional brain, but one that can still at times function on its own.

As with all aspects of sex, the *capacity* for orgasm "happens" in the brain. We are primed even before birth to make the neural connections between genital stimulation and arousal. While it's impossible to know if a fetal brain experiences the intense pleasure we know as "orgasm" the same way we do, the organization and distribution of nerves and muscles and the way they contract and "fire" during intercourse and orgasm make clear that we have a built-in "sex hard drive" that includes orgasmic know-how. One source of evidence is work by Finnish investigators at the University of Kuopio who have clocked an increase in cerebral blood flow in the right prefrontal cortex during orgasm in men. This part of the brain is the area in men that is highly specialized from birth for emotional and visual-spatial perceptions and for abstract "big picture" problem solving. Using PET scans, they studied eight healthy, right-handed heterosexual males and further discovered a *decrease* of cerebral blood flow during orgasm in all other cortical areas. Nature seems to have worked it out so that during the height of sexual pleasure, the emotional brain gets preferential energy supply.

Our minds also seem to experience orgasm *separately* from our bodies, or at least longer. In a study of women, Gorm Wagner of the Panum Institute in Copenhagen found that in 85 percent of them, genital and physiological measurements showed their orgasms to be about twice as long as the women *thought* they were, suggesting the absence of any significant link between the intensity of a physical orgasm and the intensity of what the mind experiences as orgasm. In a similar study in men, the Danish scientists found no link between ejaculation intensity and what a man said he "felt" or experienced.

Joseph Bohlen of the University of Minnesota Medical School in the 1970s and 1980s also made use of modern equipment and federal grants to study orgasm during intercourse and masturbation, measuring heart rate, blood pressure, oxygen exchange, blood flow, muscle contractions, penis circumference, and skin temperature. His subjects could activate monitors the minute they felt orgasms start and end. Combined with subjective questionnaires, his data led to a stunning conclusion that many did not believe: that there was very little association or correlation between what people said they were feeling as an orgasm and what the measurements of the physiology were finding. There was no quick formula for describing the orgasm. No this-number-of-muscle-spasms equals perception or intensity of orgasmic pleasure. There was no correlation between when contractions began and when a person said he or she was having an orgasm "now." And those who had the strongest contractions or the weakest did not necessarily perceive them any differently. A third of the women in Bohlen's study said they had orgasms but had *no* contractions. Imagining orgasm? Probably. But their perceptions were as consistent and persuasive as those who had contractions. As one writer put it at the time, "he couldn't measure the pleasure."

Taken together, the evidence so far suggests that no matter the hormones and genitals and arousal and erections and lubrications and all of the preparations for orgasm, unless we *literally think* ourselves to orgasm, it can't happen. The limbic brain, which integrates and regulates not only feelings of pleasure and emotions but also brain stem–derived primitive urges, is in charge.

Further evidence of the brain's preeminence in achieving orgasm comes from doctors who, in the course of conducting electrical studies of the brain to diagnose epilepsy and other disorders, inadvertently (at least at first) stimulated areas of the brain in the limbic cortex and elicited, incidentally, orgasm. In the 1980s, Tulane University neurologist Robert Health said of such an experience that it was "the most dramatic kind of electrical discharge in the septal region of the limbic cortex . . . equivalent to what the whole brain experiences during epileptic seizure." The brain literally shudders with orgasm.

With orgasm as with sensing danger, the body prepares for everything with an increase in heart rate, shortness of breath, flushing, tightened muscles, fierce concentration, the release of adrenocorticosteroid hormones, and other stress chemicals. Whether the stressor is heat, cold, joy, grief, excitement, or drugs, the physiological reaction of the body is identical, too. We sweat, we shiver, and oxygen transfer in our body speeds up. We perspire and the blood sugar levels rise, the heart pounds, and blood pressure rises to dilate blood vessels and increase oxygen output. The body is on full alert, first alarmed, then rising to the occasion to keep itself going, prepared for fight, flight—or sex—until finally it exhausts itself, expending all of the available and accessible energy it has. Hans Selye, the pioneer stress researcher, worked out most of this response decades ago, concluding that the body and brain always seek balance, or homeostasis, but when healthy respond brilliantly to both good stress and bad stress. "Presumably," he wrote, "in the course of evolution, living beings have learned to defend themselves against all kinds of assaults, whether arising in the body or coming from the environment through two basic mechanisms which help us put up with aggressors or destroy them."

In controlled experiments with a variety of subjects, brain hormone specialists such as Julian Davidson of Stanford have painted a portrait of orgasm that is similar in its basic outline to Selye's reflexive stress response outline. If you tickle your nose, for example, you sneeze reflexively. If the doctor taps your patellar nerve with a hammer, your foreleg jerks erect. And if you stimulate the right neurons and brain chemicals—particularly that cuddle hormone, oxytocin—long enough, you will have an orgasm.

Davidson and his colleagues investigated the links between heart muscle and oxytocin during sex by matching up the physiological and psychological activities that hormones trigger en route to orgasm. Working with twenty-three volunteers, thirteen women and ten men, each subject masturbated to orgasm two or more times while they were hooked up to monitors that recorded their systolic blood pressure and anal muscle and electrical activity, and their blood was sampled for oxytocin levels. In both men and women, they wrote in the *Archives of Sexual Behavior*, "very high positive correlations were observed between the percentage change in levels from baseline through orgasm of oxytocin release, blood pressure, and anal contractions." The subjective experience of orgasm for multiorgasmic women was so aligned with increased levels of oxytocin that this chemical may need to be renamed.

Women's capacity for multiple orgasm is among the most enigmatic of all issues related to the experience. One useful perspective is to see the phenomenon not as an isolated one, but part of an overall evolutionary strategy for survival based on variety that helps females level the reproductive playing field. Consider: If *his* ejaculation always brings orgasm, what's in it for women who don't enjoy that outcome? What benefits would accrue to women—and ultimately to men—if women in any given situation have orgasms, don't have them, have several, have them during intercourse, before, or after?

"Women," says Robin Baker in *Sperm Wars*, "differ far more than men over when, how and how often they climax." Starting with that fact, Baker and an increasing number of sex scholars believe they have figured out the strategic context. Mainly, he argues, this pattern fits the notion that orgasm's variations and even the absence or avoidance of orgasm "can enhance a woman's reproductive successes." The pleasure of orgasm, for a woman particularly, is a by-product of its real role, not the function itself. "Basically," says Baker, "whenever the body is intent upon a particular course of action, it generates an urge to perform that action. As that urge is gratified, the sensation generated is pleasure. The female orgasm is pleasurable *because* it has a function." And the function reasonably may follow from the idea that a woman *should* want or have orgasm whenever her body judges that will enhance her reproductive success. If her sexual brain and body judge that orgasm would *reduce* her success, she feels no such urge. The beauty of this scenario is that it also fits the fact that women can and do have orgasm when they masturbate and from oral sex.

How? First, surveys repeatedly show that the vast majority of women, perhaps 80 percent or more, masturbate to orgasm and many do so routinely, perhaps once a week and more as they age. Some female mammals, especially primates, also stimulate their clitorises themselves, by rubbing their vulvas on the ground or on branches; chimps have been seen inserting twigs. Moreover, a female chimp will climax if someone simply massages her clitoris, as will a cow. Thus, it's likely that something so widely present in animals has some adaptive function.

Second, in masturbation and the orgasm that comes with it, the cervix gapes open and moves, a motion called "tenting." Combined with what happens inside the cervix during orgasms, this temporarily increases the flow of mucus from the cervix into the vagina, increases the acidity of the cervical mucus, and strengthens the cervical filter. Physiologists have discovered that although mucus moves from the cervix all the time, orgasm hastens the trip and masturbation's arousal phase triggers glands at the top of the cervix to speed up mucus making. The tented cervix is better able to handle not just the increased velocity, but also the increased volume of mucus that is made when there is orgasm, moving the old mucus out and lubricating the walls of the vagina for intercourse. Out with the mucus comes old blocking sperm and even bacteria. (Many women report the urge to masturbate when they have urinary tract infections. They have learned that this can help them feel better in two ways: relief of pain and also of disease organisms.) During tenting, mucus, like taffy being pulled, stretches sideways, forming channels that direct and pull fluids into the tip of the cervix. Fluids at the top of the vagina are acidic, and as the cervix tents during orgasm, this acidic fluid spreads and is ejected after orgasm back into the vagina during flow-back. Some will stay in the cervix, where neither sperm nor bacteria can do well, as the acidity retards them.

Finally, orgasm in masturbation triggers about half of the cervical channels to unload sperm from the last insemination, because the sperm collected in the mucus flows through new channels. The more channels that already contain sperm when a woman has orgasm, the more sperm go into the channels and through the cervical filter. Thus, masturbation twenty-four hours after intercourse, when there are lots of sperm, increases the activity of the filter more than masturbation later. Reports that women feel like getting an orgasm via masturbation most during ovulation may be their bodies' physiological signal that they need to prepare the vagina and cervical filter for insemination. Such a biological strategy may also help favored males win the sperm race. Along with Baker's research with colleague Mark Bellis, the University of New Mexico's Randy Thornhill and his team found in 1995 that female orgasm occurring near the time of male ejaculation results in "upsucking" of sperm for up to forty-five minutes after ejaculation, and may hinder retention of sperm from *other, subsequent* copulations for up to eight days.

Like men and some animals, some women have spontaneous nocturnal orgasms, or "wet dreams," while they sleep, which contract their genitals and lubricate them, and, according to Baker, function in much the same way as masturbatory orgasms. "Both . . . help a woman's body in its battle against infection. Both prepare her vagina for its next intercourse by depositing lubricant. And both increase the action of her cervical filter, as long as there are sperm in her cervical crypts. In fact, no physiological differences can be detected between the two types of orgasm. Not surprisingly, therefore, nocturnals show a link with the menstrual cycle (and hormonal cycles) similar to that shown by masturbatory orgasms." That is, they occur generally during a woman's most fertile phases.

These kinds of orgasmic strategies, which benefit both men and women, did not, of course, happen overnight, in a straight progression or in absolute "yes-no" fashion. What the human race wound up with is enormous variety and variability, between individuals and within the same individual. Studies show that some men, as well as women, for instance, experience multiple orgasm. According to Marian E. Dunn and Jan Trost of the State University of New York at Brooklyn, their studies of twenty-one multiply orgasmic Scandinavian and American men twenty-five to sixty-nine years old found that quick penile collapse to the flaccid state did not always follow these men's orgasms and that they often climaxed *before* ejaculation. This suggests that biologically, men may have evolved to have a whole range of orgasmic experiences, but lost most of them somewhere along the evolutionary way.

In their 1995 study of "upsucking" noted earlier, Thornhill, Steven Gangestad, and Randall Comer of the University of New Mexico in Albuquerque affirm and expand such a view of the purpose and origins of female orgasm. They think that female orgasm evolved not only because it serves to help a couple stay in love and bond, but also because it may very specifically help a woman choose whose sperm will be more likely to result in a pregnancy. This is particularly true, they say, when there is more or less simultaneous orgasm during intercourse or when a woman climaxes just before or after her partner's ejaculatory orgasm. If this turns out to be the case in further studies—and Thornhill is firmly convinced that it will be—then female arousal and orgasm are a classic Darwinian adaptation to prevent the backflow of preferred sperm-bearing semen and increase the backflow dump of less favored sperm. Such a strategy is apparent in other species. In rats, males create a plug with their semen to keep their sperm in and the next guy's out. With humans and other species, such as red jungle terns, the females are in control; they dump out inferior male sperm, suggesting that traits can evolve both ways. But in either case, "Female orgasm is not necessary for conception," Thornhill said in an interview, "and it's our belief that it is an adaptation for manipulating the outcome of sperm competition that results because our female ancestors pursued a strategy of polyandry, or multiple mates."

Building on their research on fluctuating asymmetry (see chapter 5), in which they demonstrated that men and women favor mating with other men and women who are more symmetrical in their facial and body features, the study by Thornhill, Gangestad, and Comer found evidence that symmetry is a marker for genetic fitness at both the visible level and at the microscopic level of the sperm. In a survey of eighty-six University of New Mexico couples, they tested the idea that if symmetrical men truly have a mating advantage, then these men's sexual partners should be more likely to reach orgasm *in synchrony* with them than with other men. Otherwise, what would be the advantage to the symmetry?

Their results found that not only did the men who were better looking (i.e., symmetrical) have an easier time getting sex partners, but also that they were more likely to have partners *whose orgasms occurred when theirs did*. "As predicted, when other potentially relevant male features, female features and relationship characteristics were controlled for, women with men who possessed low fluctuating asymmetry (i.e., were more symmetrical and thus better looking) self-reported and were reported by their partners to have had more orgasms during copulation than did women with men having (higher asymmetry)." The lucky women did not simply have more orgasms in general; they were also more orgasmic during copulation itself.

Thornhill says their results also were consistent with findings by Thomas Insel that oxytocin mediates both the retention of sperm of males of high reproductive fitness and the promotion of pair-bonding. Animal studies suggest, for example, that this hormone not only induces erection but also may be responsible for the "refractive" period after orgasm and ejaculation (when the penis becomes flaccid and remains so for some time afterward) that contributes to a man's willingness to cuddle and relax after intercourse. It could also be directly involved in ejaculate "choice" by women, because, as University of Maryland's Sue Carter has shown, oxytocin also increases during the female sexual response, peaking at orgasm, and may abet orgasmic contractions of the vagina and uterus that lead to vaginal sperm retention. Oxytocin also has been shown by Carter and others to calibrate animals' willingness to stick around after sex to nurture their babies. "If oxytocin," says Thornhill, "really promotes conditional and selective affiliation with sexual partners, presence or absence of orgasm could be an important 'in the moment' cause of female mate choice."

The presence of oxytocin and other chemicals that facilitate a man's tendency to snooze after sex, and stick around to support any children that a particular act of sex produces, also is consistent with Thornhill's findings that oxytocin and female orgasm have a lot more to do with the practical need to bind a man to her than it does with any interest in a romantic relationship. "We found no evidence that oxytocin and female orgasm were associated with greater investment by women in their romantic relationships or even greater love for the male partner. Instead, if these things facilitate differential bonding with men of high genetic quality, as we think they do, the bonding basically helps women get and keep the genetically fit man of the moment, to retain sperm of men of high developmental stability and perhaps the sperm of men who are facially attractive and large in body size, which correlates with symmetry." There are certainly other ways nature could have handled this female need, but it seems to have chosen this one.

The idea that human female orgasm influences *rates* of conception, as Baker's theory holds, may be highly supportable, Thornhill says. Studies show that women retain more sperm—about 70 percent more than usual—if they have an orgasm. Study after study also demonstrates that peaks and valleys in a woman's fertility parallel her interest in intercourse and orgasm.

But, Thornhill notes, sperm choice, not conception, probably is the main reason orgasm survived in modern woman. That's because there also is tremendous biological motivation on the part of *men* to conceive, too. Evolution would have favored male adaptations designed for high rates of conception, while encouraging female traits that use, manipulate, and encourage sperm from males she is most likely to consider fit. Female orgasm does just that. If women pick mates who will invest the time and energy to help her achieve orgasm, particularly during intercourse or shortly after, the woman benefits also by increasing the potential for her begetting sons, who will be reproductively fit enough to do the same for *their* mates, thereby winning the best ladies in the next generation. And a man who so pleases his partner gets the benefit by getting the edge in the sperm wars. But he has no edge at all if she does not choose him.

Multiple orgasm has an ever greater effect, as noted already, on the number of sperm that are sucked into the cervical space and kept there. How that happens is instructive. When a woman climaxes during intercourse, the cervix gapes open, the mucus splits, more channels are made for more and more sperm, and the cervix dips up and down into the seminal pool at the top of the vagina, helping more sperm swim for the egg. This is a race for time and space, since studies show that the estimated average lifetime of a sperm is about 1.4 days once it is ejaculated, and less than a day for an egg. The rippling muscles in the womb and vagina during orgasm suck semen and neutralize acidity in the mucus, increasing survival of mucus and getting rid of old sperm.

Some critics point out that optimal fertilization—a bypass of her cervical filter—could then occur only if a woman climaxes *after* a man ejaculates. But in fact, biologists have found that even if a woman climaxes before ejaculation, the orgasm can still operate to get sperm past the filter because the feeling of orgasm may continue for many minutes. As long as the seminal pool is in place before the peak, her filter will be bypassed. She doesn't need the man's penis in her, nor does she need him to be the one to confer the orgasm. Even if he leaves the room and she masturbates to orgasm, the filter is bypassed and more sperm can get in because the seminal pool is filled. "The best moment for a woman to climax or not," says Baker, "varies considerably from occasion to occasion."

For the man, of course, the best time may not dovetail with her best time. He too has anticipated intercourse, at a subconscious physiological level as well as consciously, and may have "tailored" his ejaculate with masturbation, or not, to suit his genes' interests. (Frequent precoital masturbation can alter such things as sperm counts.) In overall evolutionary terms, a man is best served usually by a woman who has an orgasm during intercourse, to get her to retain as many sperm as possible. But he may fine-tune his efforts to bring her to climax during intercourse, depending on how receptive she is. If she doesn't want to climax, she won't no matter how hard he tries. Viewed from the vantage of our prehistoric biology, if a female has few or no sperm in her from any other copulation recently, and if he's *sure or pretty sure* she hasn't been with any other man or masturbated, a man might be less interested in her pleasure and just please himself. He may not have understood *why* he felt less inclined to bring climax to a woman whose sexual activity he controlled closely, but some combination of her interest, her receptivity, and her cervical filters communicated to his sperm, if not his conscious brain. Nature, it seems, gave us some sexual strategies that balance our conceptual interests even if we're unaware of them.

If much of this sounds as if the woman is in complete or at least in most control of intercourse and its outcome, the facts are that men can almost always physically overpower women to get access to them. Thus, it is more of a question of nature giving a female some wiggle room. Moreover, each time a man ejaculates, he is neither a passive player nor an aggressor, but a strategist. With the same partner, he introduces no more sperm than he needs to "top her up" or replenish the sperm supply. His body tends to increase or decrease the amount of sperm he needs to inseminate, but it does not add to the cervical filter much or block the egg-carrying channels.

Studies also show that if more than a week has passed since the last intercourse—and she has not been with another man—a woman's sperm supply will be on empty and the male will deposit up to 400 million sperm to get a million or so in her reservoirs. If the gap is far fewer days, he will produce less. This happens so precisely because testes are always growing and producing sperm, more than a thousand a second. When a man ejaculates, the sperm farthest forward in "line" is sent out of the penis and the others move up in line according to when they were made. First in, first out. It takes about two to three months or so for a sperm to mature and get from the testes to the front of the line and out. Notes Baker, depending on how many of his loading muscles work and how strongly, the man can shunt some length of each of his two queues of sperm into his urethra. "Even after loading, his body can change its mind. By varying the number of spurts, usually from between three and eight, he can ejaculate a different proportion of the loaded sperm. Any sperm and seminal fluid left in his urethra after ejaculation can be flushed out at the next urination."

Baker presumes there is a link between the man's brain, which somehow "knows" when he last inseminated his partner, and his genitals. This clearly is not consciously done. "In the midst of thrusting and at the point of loading his urethra, he does not ask himself is this a hundred million occasion or a four hundred million," quips Baker. But his biochemistry, he believes, apparently does.

In this way of looking at things, according to Baker, a sexual act between a man and a woman—except on rare occasions when the same outcome, bypassing the filter, suits them both—is a case of each trying to always outmaneuver the other into doing something that may be against the body's interests. After all, a guy's safest play is to ejaculate a few seconds after the woman has climaxed. But a woman may insist on having an orgasm during foreplay, then delay his ejaculation. Or she can climax so fast that his ejaculation can't come fast enough. Or she may stop herself from climaxing or cooperate in a postcoital climax. Or not. She can help herself with masturbation. Or not. It's a contest worthy of the human race.

If Thornhill and his supporters are right—that female orgasm is an adaptation for female choice, now complete with its own biochemical supports—two other well-known orgasmic phenomena also make some sense. First, males have learned—and passed on the lessons—to help provide orgasms for their partners; second, faking orgasm is a uniformly female trait that is the source of much male *angst*—and cultural mirth.

Let's consider faked orgasms first. In Thornhill's and Baker's views, for the most part, there are evolutionary pressures on women to "resist" orgasm, or to be capable of resisting orgasm as a matter of increasing their right to "choose" particular sperm for fertilization purposes. In this way, "faking" orgasms is also highly adaptive, and women are very good at it precisely because it is a deception that gives them some leverage during intercourse, even if they are not consciously thinking about or aware of the need for such deception. She may also "fake" *not* having an orgasm (perhaps by tightening down on internal muscles or twisting away to distract her partner's awareness of her unconscious spasms) as a way of keeping an alpha male "in play" until sperm from a less favored male have made their way out of her, for example. Or she may want to fake having one to keep a beta male around until someone better comes along on the theory that some resources are better than none. Females' ancestral psyches, their genetic fitness, their stress levels, and their physiologies are on alert for the optimal conception, even if they are not. Think about it: A woman fakes orgasm by mimicking the way she signals *real* orgasm, with sounds and pants and heavy breathing. "Sometimes," says Robin Baker, "the sounds really do give an accurate indication of what is happening. On those occasions when a virtually simultaneous climax is in both people's interests, the sounds are the method use to achieve this mutual aim. But it is precisely because (these sound signals evolution selected) are sometimes honest that on occasion both men [yes, men can and do fake it] and women can use them to trick the partner."

By hiding her orgasm or lack of it, by masturbating in secret or in the open, a woman increases her weapons in the sperm wars. Helping a woman climax during foreplay is a disadvantage, biologically, to a man. The longer he waits to penetrate and ejaculate, the bigger the risk that he'll lose the opportunity to impregnate, and thus most men would prefer to begin intercourse without too much foreplay, while women have much to gain from a foreplay orgasm. By withholding and faking orgasm, a woman reserves the right and power, from a strictly biological standpoint, to alter her sexual strategies to fit the needs of the moment. There may be times, for example, when it's in her interests to mate with less genetically fit men who are unlikely to produce an orgasm in her, in order to keep her DNA alive when no better men are available, and to fake orgasm to keep him close and still permit fitter sperm to impregnate her.

Many biologists are coming to believe that overall, orgasm and intercourse reflect the long continuing negotiations between the sexes that starts with flirting, continues with sperm and gene wars, and concludes with bonding and parenting. Clearly, men and women are not aware of these physiological and molecular hagglings; they respond to the urge for sex with a combination of brain and body and mind and experience and hormones and mood. They can and will seek intercourse whether they are fertile or not, able to have children or not. But in some significant ways, we have evolved physiological stratagems that help confer on men and women maximum advantage in securing the best genes.

If a woman, for instance, gets pregnant by a lover instead of her spouse, she may *say* she doesn't want that to happen, but all of the biology that has been conserved to drive her to infidelity—to whatever degree that is the case—sends signals to her body and brain to prepare her vagina and cervix through masturbation, nocturnal orgasms, and sperm retention or flow-back to give the edge to her lover's sperm. By the time a woman begins foreplay, her body has mucus, a cervical filter, and a certain level of sperm—from none to millions—in her cervical canals, womb, and vagina, depending on her hormones and the phase of the menstrual cycle. During rape, when a woman rarely reports orgasm, the sperm she would retain is decided by the cervical filter she already had before the intercourse began. By avoiding an orgasm, her body is telling her not to change the nature of what is already in place, making failure to climax a successful strategy.

For all of the uncertainty that still remains in "orgasmology," the science so far seems to affirm the notion that orgasm has properly become the focus of so much of human sexual activity because, like courtship negotiation, orgasm is a *natural strategy*. In their quest for mutual pleasure through sex, each gender gets something—well beyond 0.8-second interval contractions—that it can't get without the other. He gets a somewhat more secure fix on the paternity of his children; she gets the best genes and resources for her egg; and both, if they stay together, will perhaps get children who are better able to triumph over other children in the next generation's race for fitness and resources. It's not that such sexual cooperation is altruistic or good, only that biologically it is a great winning strategy. The pleasure we experience in arousal, copulation, and orgasm coevolved so that no one gender overall has an edge, even though individual members of a gender will be better sexually fit than others.

We are not driven solely, as are most animals, by immutable rules of biology. Men and women literally do learn by experience to be better lovers, both physically and physiologically. Hormones secreted by and influenced by the pituitary, the adrenal glands, the testicles, and ovaries are primed by experience to come when they are called. The more you call them, the more easily they come, traveling the neural and hormonal circuits that make their hosts sing with pleasure.

The better men and women are in this enterprise, the more they will be sought out by the opposite sex. That we have Lothario and Venus,

Don Juan, and the nymphs as part of our legacy bears witness to the fact that those with reputations as great lovers succeeded in producing more offspring than their less apt sisters and brothers. Coital positions and all of the tricks of the Kama Sutra may do very little to alter the chances of conception, but they do a great deal to encourage us to keep the act of intercourse and the experience of orgasm as the best strategies for winning the chance to try.

Every time we use the phrase "sleeping together" or carnal "knowledge," or describe people as being "intimate," we acknowledge more aptly than we might suspect nature's playbook in the sport of sex. Intercourse and orgasm are not only for the purpose of conception, but also for attaching us to each other long enough to provide stable resources for our offspring to mature and reproduce as well. And also for unattaching us when it is in our strategic interest to do so. In the very throes of orgasm we still are in the contest by which men and women negotiate commitment, resource sharing, and parenting and by which they *stay* or keep falling in love with their mate. Or with someone else. "Infidelity" may be a cultural construction, but monogamy and promiscuity are a matter of biology, too.

10

NOW AND FOREVER

"The good news is that human beings are designed to fall in love. The bad news is that they aren't designed to stay there."
—ROBERT WRIGHT, *THE MORAL ANIMAL*

Sexual statistics seem never to add up. Consider, for example, that the vast majority of men and women in every society make passionate declarations of never-ending commitment. They marry, they have convenient sex whenever they like, and they have children. Then, as the demands of their couplehood multiply like rabbits, the bonds of trust, warmth, mutual dependency, and companionship become niggled by jealousy, sexual dissatisfaction, ennui, and the irritation of dirty socks. All of the urgent compelling reasons our brains and bodies and hormones recruit to get us together with that one single object of our lust and love seem to dissipate. At various points throughout our lives, our strongest commitments waver, bonds will weaken, the libido will go limp. In survey after survey, while pair-bonding and mating for life emerge as very human traits, so too does adultery.

In one study, for instance, an estimated 90 percent of undergraduates say they have lied to their partners about their sexual histories or fidelity, suggesting that very early in our sex lives, we learn the power of deception to protect our primary relationships. But we also learn to tolerate sexual lies in others. We say we want honesty in our mates, especially the ones we love and want to love "till death to us part," and yet we almost all lie—and accept lies—at some time in our lives. Even accounting for the widespread lies people tell pollsters and social scientists about their sex lives, best guesstimates are that roughly half of all men and women have extramarital affairs, some lasting years or decades, as with the recent revelation about CBS correspondent Charles Kurault, whose thirty-year romance in a Montana hideaway came to light after his death. Cheating hearts are so widespread, so universal, that only those prepared to argue that nature went strangely awry in designing the most important and dramatic aspects of human sex—love and commitment to offspring, along with infidelity—could believe that these traits were mere inventions by self-centered, immoral twentieth-century humans. The vast majority of human societies chronicled in history, permit polygyny.* In the most advanced nations and the most primitive ones (Ache of Paraguay, !Kung of southern Africa, Eskimos), women copulate with more than one man frequently. They do so not least of all to make sure that some of the guys invest in their offspring, which they will do as long as they think the kids are theirs or even may be.

In the words of Mae West, the only hard part of marriage is the period after the wedding. Hard enough that in the United States, fully half of all marriages end in divorce. And an impressive percentage of the survivors get married all over again. Nature has clearly not evolved us to rigidly pair-bond forever. Deep attachment may be accompanied by deep resentments. They will think about and probably seek other partners, sometimes for sex, sometimes just for close relationships. They will love their children and despair of them. If they are lucky and persistent, they will achieve an enviable intimacy that can last for a lifetime. But the issue of "now" or "forever" can arise anew each day because of changing circumstances and the ebb and flow of their individual makeup—genetic, biochemical, psychological.

Being attracted to or even lusting for a wide variety of the opposite sex is an essential prelude to genetic immortality. So nature understandably devotes a substantial part of our biology and psychology to *getting* sex. In a similar vein, but less appreciated, perhaps, are the biology and psychology devoted to making sure that sex pays off. Nature isn't likely to have organized a system that stops working at the point of orgasm or

---

*Polygyny means literally "many females." A man with one woman is more accurately mono*gynous* than mono*gamous*—from *mono*, or one, and the Greek for female, *gynos*; or, he may have several women at once, which is more accurately poly*gynous*. Women—heterosexuals—are more properly mon*androus* or poly*androus*. Monogamy simply means "one spouse," and polygamy, "many spouses."

even conception. It needs to get a pair to stick together and stick around, at least long enough to rear and protect whatever young are born of their liaison, to give their children whatever edge they can in the quest for the best DNA of their generation. So embedded in those statistics is an apparent paradox. If we're biologically and psychologically organized for commitment and fidelity, we also—by dint of the almost universal existence of *infidelity* among us and most of our animal ancestors—seem constructed for straying.

A few popular recent books have argued seductively that we—both men and women—are by nature born to love and leave. A few hardy optimists suggest that with a little help from social rules and pressures, we can overcome and modify whatever biology we inherited. These point to experiments with promiscuous-by-nature fruit flies forced in the lab to pass through fifty generations of sex with just one partner, which found that the offspring in such fly societies are healthier and sexual conflicts reduced. That could theoretically work on people, too, but it's hard to imagine how it could be accomplished without armies of sex police. And who would police them? The reasonable conclusion is that humans are naturally equipped for and capable of both behaviors, and as time and circumstances have changed over millions of years, our wires have overlapped if not tangled. Consequently, while a single person's quest for sex may be measured in moments of orgasm, a *couple's* sexual success is found somewhere between the pleasures of now and the bonds of forever. Our sexual operating systems, therefore, must have evolved to cover not just physical sex but also *emotional commitment*; in fact, researchers have identified biological and physiological foundations not just for the former but for the latter as well.

Anthropologist Helen Fisher has said that we humans and our closest mammalian relatives have evolved three distinct but overlapping "neurophysiological emotion systems" for mating. The first two have been discussed already: the libido or sex drive that makes us crave sexual gratification, facilitated by a variety of sex hormones; and attraction, marked by "increased energy and unfocussed attention," by "elation, exhilaration, and intrusive thinking [in which] the object of your attraction keeps popping into your brain," by emotional dependence, and by help from dopamine, norepinephrine, and serotonin.

The third part of the mating process is attachment, characterized in mammals by close body contact and separation anxiety and in humans by reported feelings of calm, security, social comfort, and emotional union, all linked to such peptides as oxytocin and vasopressin. The sex drive motivates us to chase and enjoy sexual gratification with *anyone*. Attachment is there "to tolerate one another and to sustain . . . connection long enough to complete species-specific parental duties."

It doesn't take a rocket scientist, or even an evolutionary biologist, to see the contradiction and conflict inherent in this system. Even among gibbons, a male can groom and woo a particular female and at the same time show attraction to another female. Among people, a person can feel deep attachment to a spouse and still be deeply attracted to others— sometimes lustfully, sometimes platonically, but often with consequences either way. So widespread are these feelings that pop psychology likes to draw distinctions between "romantic" love and "mature" love, but the symptoms are the same from a physiological point of view. We pine, fantasize, and lose our appetite for food. We idealize the object of our affections and feel the rush of high energy and pleasure created by dopamine and endorphins.

The contradiction and conflict make sense only in the context of that overarching need for flexibility, and there is indeed evidence that while Fisher's mind-brain emotion systems overlap, they also operate independently of each other for maximum versatility. For instance, experiments have been conducted in which women, injected with testosterone, show a predictable rise in *sex drive*, but in keeping with Fisher's hypothesis were *not* more likely to fall in love or feel strong attachment to anyone they had "better sex" with. Call the scientific search to resolve these issues the Columbo Effect. Like the legendary small-screen detective with his scuzzy raincoat and bumbling manner, those asking the hard questions about sexual behaviors are fully aware that you can rarely tease out the truth, or reconstruct the mind and behavior of people, directly. Instead, you build your evidence from what is more easily accessible, from the small, subtle, deductible clues that emerge from the *variability* of behavior, the patterns of behavior, and the deviations from both the expected and the unexpected.

Which brings us to a central theme emerging in the search for chemicals and genes that facilitate our sexual and emotional attachments: wherever they exist, they are fellow travelers with chemicals and genes for detachment, divorce, and alienation. Where there is the capacity to fall in love, there is also the capacity to fall out of love. Where there is fidelity and monogamy, there also are promiscuity and jealousy.

Although candidates for molecules and genes of fidelity have been discovered in a tiny rodent, such a biology of "forever" exists in sharp contrast to equally compelling evidence for the evolution of polygamy in both its forms and in adultery. In fact, the sexual themes played out by Lady Chatterley and Count Vronsky, by Bill Clinton and Monica Lewinsky, are classic not because they are "shocking" but because they are all too familiar. And the prairie vole that appears to have a monogamy gene has a species first cousin that has conserved a "promiscuity gene"— something that better fits the conditions it faced in its early habitat. In humans, however, it appears that sexual strategies developed to assure a continuous desire for sex with anyone *and* for just someone. As Charles Scriver has noted, we are less aptly called *Homo sapiens* and more aptly called *Homo modificans*. We are nothing if not flexible.

It is likely that the contradictions inherent in our sexual IQs are in part the result of an evolutionary "arms race" in which men and women evolved to both cheat and detect cheating, manipulate and be manipulated. Why else might we have a reproductive system that hides both female ovulation and the certainty of paternity? Before the days of paternity testing, a woman who cheated and got pregnant by her lover might still calculate correctly that her husband would support the child he wasn't sure was *not* his own. In the animal world, the story is pretty much the same. Male langurs and other primates will sometimes kill a rival male's offspring to assure the supremacy of his own, but will rarely do so if he's not sure whose baby it is.

Recent studies on monkeys and apes have resulted in some novel ideas about the natural predilections of human mating behavior, notably that the cheating and lies so prevalent in us may have roots in sperm and sperm delivery systems.

Consider the story of gorillas and chimps, two species very close to us biologically, yet very different from each other in their social organization. Gorilla societies are essentially organized with an alpha male and a harem, while chimpanzees live in brotherhoods and sisterhoods. Scientists have long observed that these two species' sexual behaviors and sexual bodies are in stark contrast to one another, as well. Female chimps, in their fertile periods, are allowed to—and do—have sex with any and every unrelated male in the male troop that lives with her and the other females. One has been known to mate with seven males eighty-four times over eight days. For a male chimp's sperm to stand a chance of impregnating any given female under these conditions, he needs to deposit large amounts of sperm to either block or squeeze out competing sperm. And nature seems to have accommodated him. Chimp testicles are exceptionally large for their body size and they are capable of ejaculating immense volumes of potent sperm many times a day, compared to gorillas, which ejaculate small amounts of relatively less potent sperm. Gorillas have testicles smaller than a man's; in proportion to their body size, this makes them substantially smaller than humans. Presumably, the head of the gorilla harem doesn't need his sperm to compete with any other gorilla's to assure paternity. By this reasoning, the human male's testicle size, sperm volume, and potency would seem to suggest that he's somewhere north of gorilla but well south of chimp. Says Alison Jolly, a Princeton University primatologist and author of the 1999 book *Lucy's Legacy,* men evolved "to defend at the sperm level against their mates' having another lover in the same fertile period—not many others, but one or two." Assessing another well-known trait of polygynous males—their larger size compared to females—Jolly concludes that men, too, may be naturally inclined to have more than one sex partner. In monogamous primates— and there are a few—males and females are about the same size. On the other hand, male chimps are a third larger than females and male gorillas twice as large as females. Men and women are clearly not 1:1 in size, but somewhere around 1.15:1, leaving Jolly to conclude that "human males 'expect' perhaps two or three mates." Aligned with this view is the work of Chung-I Wu (see chapter 6) and his team of geneticists who reported that several genes involved in sperm function have evolved much faster in humans and chimps than in gorillas, evidence that there has been pressure on these sperm-related genes to adapt and compete more.

Further support for sperm competition's role in human sexual "flexibility" might be found in the fact of concealed ovulation. Chimps, for example, have visible swellings of their reproductive parts that signal their periodic ovulations, while women's big breasts and behinds serve as permanent—if "deceitful"—signs of fertility, according to some evolutionary psychologists. Robin Baker, author of *Sperm Wars,* suggests that absence from a partner may not make a man's heart grow fonder but may clearly increase his sperm count, perhaps as a way to overwhelm any other man's sperm that found its way into his mate's body during his absence (see chapter 6). Or, of course, absence may simply give a man a better shot at impregnation if he has been celibate for some period and thus unlikely to have inseminated anyone. Baker also believes that the abnormal sperm found in most human ejaculate may serve to sabotage competing sperm, and that concealed ovulation gives a woman the ability to promote sperm competition in men. He has collected vaginal fluids, semen, and secretions from volunteer couples to support the idea that such sexual behaviors as monogamy and polygyny change our biochemistries.

Like Baker, University of Michigan psychologist David Buss notes that men's relatively large testes are a "solid piece of evidence" that women in evolutionary history were promiscuous, that they had sex with more than one man in the space of a few days. When the sperm from two or more males are in the same reproductive tract of the female at the same time because she has mated with them, he says, such competition exerts a "selection pressure" on males who can make the largest ejaculates with the most sperm. "The attribution made in many cultures that a man has 'big balls' may be a metaphorical expression that has a literal reference." Buss recruited thirty-five couples who agreed to give him ejaculate resulting from sexual intercourse, either from condoms or from the mass of seminal fluid that is spontaneously ejected by a woman at various points after intercourse. "Men's sperm count increased dramatically with the increasing amount of time the couple had been apart. The more time spent apart, the more sperm the husbands inseminated in their wives when they finally had sex. When the couples spent 100 percent of their time together, men inseminated only 389 million sperm per ejaculate. But when that time was reduced to 5 percent, men inseminated 712 million." The implication is that sperm increases when there is the biological risk that some other man's sperm might be inside the wife's body, made possible by a wife's affair in his absence. "This increase in sperm is precisely what would be expected if humans had an ancestral history of some casual sex and marital infidelity."

Critics note justly that such interpretations of sperm counts are still controversial and assume a universality across cultures that has yet to be tested. But if Buss and Baker are right that men evolved a biological and behavioral strategy to protect them from women's infidelity, then it would also appear to be true that, as Baker suggests, the overall best evolutionary strategy for women is to use sperm competition and concealed ovulation to hedge their bets: they "marry" or stay committed to any one of the limited numbers of good providers who are willing to stay with them, but rely on promiscuity to get the best genes for their offspring.

Whether or not sperm competition truly exists—or exists any longer— in humans remains an open question. By some estimates, only 4 to 10 percent of children in Western societies have what is called "discrepant paternity," a biological father different from the man who thinks he is the father. That's not very strong evidence for sperm competition, but nor does it rule out a less-than-monogamous behavioral legacy.

If it's clear we're not as promiscuous as chimpanzees, neither are our biologies semper fi. Indeed, lifelong monogamy appears to be as rare in us as in the animal world, at least among the so-called alpha or most powerful males and females. What Bill Clinton, John Kennedy, China's Chou dynasty emperors, Augustus Caesar, Aztec king Nezahualpilli, and Egypt's Ramses II all have in common is the urge to philander and the *power* to do so. That urge evolved among our distant ancestors as a way to assure survival of the species, long before it became possible to prevent pregnancy with contraceptives or have abortions. Their behavior may, in today's terms, be "immoral," but as *The Moral Animal* author Robert Wright has put it, it's "hard if not impossible from an evolutionary and biological standpoint for such men to resist the fawning adoration of women. . . . Though few men share an alpha male's opportunities for sexual addiction, any smoker who has kicked the habit rather than die young, only to fall off the wagon, knows the mighty logic that can make a presidency self-destruct."

Dennis Hasselquist and Staffan Bensch of Sweden, studying the great reed warbler, a migrating bird, found DNA evidence of this same balanced-cost accounting. The female warbler enjoys indulging in "extra pair fertilizations," which is to say she is willing to step out with males who clearly "covet thy neighbor's wife" for a lot of one-night stands. The gigolo beats wing out of there while the cuckolded mate even raises the young of the cheater, contributing food and protection. Of course the cuckold is no saint; he covets his neighbor's wife—and pursues her—with the same élan possessed by the Don Juan who dropped into his nest. To get their way with such willing tarts, the males sing fantastically. Hasselquist and Bensch not only analyzed the DNA of adults and nestlings, but also got to know which males sang the most complex songs and which parent belonged to what offspring. They found that the babies sired by males with the largest storehouse of songs were twice as likely to survive their long journey to sub-Saharan Africa, where they winter over, and to return healthy to Sweden the next spring to mate themselves. "It is not the number of fledglings that matters, but the number that come back and breed" that is the key. Neither the unfaithful females nor the offspring born "out of wedlock" get any direct benefit in terms of their size, for instance. Nor did the better singers give more food to the offspring or the mother. Nor did the females give more nurturing to the babies fathered by the outsiders. Instead, it seems that the babies fathered by the best or better singers seemed to have better genes, that the females are using the song repertoire as a "marker" of genetic fitness. The deal isn't for herself but for the chance that her offspring will live to sing another day.

Other evidence for a natural tendency to *infidelity* emerges from how easily and simply our behavior and our biochemistry can be subverted to the game. One investigator, William Tooke, has analyzed the widespread course of deception—most of it committed by men, and much of it transparent—that goes on between the sexes. Tooke's compelling conclusion is that such lies would persist only in an evolutionary context, and that they only make evolutionary sense if they are viewed as part of men's evolved need to believe or at least accept their own sexual lies as a way to avoid conflict in a committed relationship while hedging their bets with other women.

Again, in keeping with the evolutionary truism that highly conserved sexual traits get that way because they are mutually beneficial to men *and* women, Tooke believes that the *reason* such cover-up capabilities evolved was to meet and match the increasing sophistication of women's "deception detector" capabilities during courtship. "Among our primitive ancestors, males who were *themselves* able to believe their sexual lies were believable increased selection pressure on females to increase deception detection, which in turn selected for even more self-deception as a means of avoiding guilt and conflict." That is, males may have evolved to get rid of guilt and avoid aversive feelings about their greater vulnerability to arousal. They evolved self-deception to help them lose their sense of guilt over telling the lies, at the same time that females evolved to detect the lies.

This would all be a Mexican standoff without the underlying differences in male and female mate-choosing strategies (see chapter 6), says Tooke. Men more than women may be, for evolutionary and cultural reasons, prone and willing to cheat sexually on a partner, but tell a woman you love her, signal some *commitment*, and she becomes more willing than she might be otherwise. Guilt and shame always seem to be part and parcel of sexual cheating, but it's also true that in most cases, she knows that he knows that she knows. And in exchange for commitment, she's a willing partner to it. Notes Tooke, only a few societies in all of recorded history have ever legally allowed women to have more than one husband, even though a random survey of two thousand Americans in 1990 found that 31 percent admitted to having an affair.

Yet evidence for the prevalence and reward of promiscuity in females is considerable. One study of Eastern bluebirds found that 20 percent of females consistently have offspring by other than the primary mate. Cornell University anthropologist Meredith Small in her study of female Barbary macaques has found the females remarkably promiscuous and at first blush "indiscriminate" in their choice of mates, seemingly defying the "male chases, female chooses" hypothesis of many evolutionary biologists who want to explain reproductive and mating strategies, as well as monogamy and polygamous behaviors. When the macaques are in heat, they run after males, swinging their behinds into male faces. "The day I watched a female copulate with three different males in the span of six minutes, I knew it was time to reevaluate the current concept of female choice" and pickiness, she says. She concludes in her study of these monkeys and of twenty-five other species that female choosiness may operate only with some animals, like birds, or maybe rarely for mammals as well, especially primates like humans.

Sara Blaffer Hrdy, author of *Mother Nature* and other feminist biological books, was surprised by what she found in her study of primates: that females aren't necessarily choosy at all, although there is so much complexity to the macaques' mating habits that it could be these seemingly indiscriminate ladies are actually selecting multiple males for very specific purposes.

Female savanna baboons often select males who are socially bonded to them, but other apes and monkeys seem to swing from high-status males to low-status, familiar males or strangers. It's possible that female choice isn't so important, but it's equally possible that these girls are choosing things the experimenters haven't yet discovered. The females may be so clever that they are literally "hiding" their choices. Hrdy suggests that females who are promiscuous might be trying to make all males "think that the result of any pregnancy is theirs alone, entrapping males to provide resources while keeping them uncertain about paternity." It's the macaque version of "Mama's baby, Papa's maybe," and all the putative papas chip in.

Similarly, the evidence for such behavior in human females as well as males is in fact easy to come by provided someone wants to look for it. As Helen Fisher notes, philandering by women has brought the wrath of

authority down on them in ways cruel and unusual, from public whip-
pings to genital mutilation; chopping of noses and ears; branding; divorce;
and even death by stoning, burning, and drowning. But the biologically
telling point is that women continued to do it in huge numbers. One
reason, Fisher says, is that monogamy and fidelity are *not* the same thing
when it comes to biology. Nor does monogamy imply or require sexual
fidelity. In modern Western society, the institution of marriage stabilizes
pair-bondings for rearing children, for example, or providing economic
stability. Fidelity, or promiscuity, on the other hand, are sexual concepts.
And in humans and most animals, adultery and infidelity—what Fisher
calls "nature's Peyton Place"—are widespread, common, tolerated, and
in fact reinforced by our biology. Only if promiscuity really maximizes a
woman's reproductive edge is it worth both the risk and her having
evolved those subtle deceits, such as hidden ovulation and the capacity
to hide or fake orgasm.

Our ancestral mothers may not have had the same opportunities to
have extramarital flings without risking jealous mates and desertion, but
they were biologically well-adapted for playing around. For one thing, in
times of war, famine, or other stress, promiscuity might have meant survival for women, getting them extra resources, gifts if you will, and a
hedge against a killed or ill mate. Second, given the kinship and nomadic
patterns of Stone Age man and woman, her status and ability to gain
mating with the "alpha" men in her group might not, for one reason or
another, have been very good, so she would be motivated to try and
upgrade her offspring's genes by making herself available to males outside
the group, sneaking off to get another man's genes without costing that
man very much at all. A woman who did this kind of sneaky sex with
skill could have had many offspring by many men, kept her mate around
to provide resources for them all, and given *her* genetic line a better shot
at producing at least one child who could adapt to an unpredictable
future. When a man in her society was powerful and resourceful, a woman
would have competed with many other women to get him. She wanted
powerful (read: promiscuous) sons to give her grandchildren.

Like behavior, biochemistry has been recruited to the cause of promiscuity and infidelity. Scientists at Indiana University, Ohio State University, and James Madison University were able to compel the highly
monogamous males of a species of bird called the dark-eyed junco to risk
their primary relationships and even the lives of their offspring simply by
implanting small bits of testosterone under their skin. In this species, the

male's testosterone normally rises in spring in order to enhance his appearance and attract a mate, and then falls when the eggs hatch and the male is expected to do his share around the nest. The implants simply maintained the male birds' spring levels of testosterone, creating males that went astray and abandoned their family responsibilities. While the babies produced by their primary mates died more often due to lack of paternal support, the extramarital flings yielded enough other offspring to compensate. It's not hard to imagine changes in habitat that would favor a shift of a species or a flock within a species to adopt a promiscuous lifestyle; in fact, there are many birds whose testosterone levels remain high naturally. What we have here are clear instances of what the investigators called "physiologically mediated trade-off between parental care and mating effort."

In an intriguing and somewhat parallel finding, scientists at Johns Hopkins have found that a monogamous male prairie vole's sometimes violent defense and protection of his pups—an oddity in male rodents, who usually leave that to the female—may be due to a brain chemical already linked to aggressive behavior in mice, and in particular to nursing female mice. When the monogamous male voles and the nursing mouse mothers are approached by a stranger, both experience increased production of citrulline in the brain. Citrulline is a by-product of the production of nitric oxide. The same Hopkins researchers, led by psychologist and neuroscientist Randy Nelson, had previously discovered that male mice specially bred to be lacking NO in their brains were unusually and relentlessly aggressive to other males, and sexually aggressive to the point of rape even among unresponsive females. In female mice, however, loss of NO led nursing mothers to ignore their pups, leaving them vulnerable to attacks. The scientists switched to the study of voles because, more like humans, male voles are relatively monogamous and help take care of their pups. And also, unlike mice, their generally lacking aggression only comes dramatically alive when they have pups to protect. Among the mated male voles and female mice with pups, they found consistently higher levels of citrulline in the paraventricular nucleus of the hypothalamus, a part of the brain where environmental stimuli are evaluated and integrated with instinctive behaviors. In some way, it seems, monogamous behavior, wherever it exists, is due to an unusual brain chemistry that probably evolved to accommodate a very specific set of environmental and mating circumstances.

In such ways and for such biological reasons, humans and other animals can be monogamous, polygamous, polygynous, and polyandrous alternatively and even simultaneously throughout their sexual lives and even beyond their reproductive lives. Not an obvious formula for golden wedding anniversaries.

If we are going to make decisions about short- or long-term commitments, *strategic* thinking is obviously critical. And, in fact, studies of animal and human behavior have found that the he and she who are flexible, who have a kind of "situational ethics" with respect to mating, probably do best.

Birds within a single species, for instance, can alter their fidelity, depending on the ecological and social circumstances. In the pied flycatcher, to name one, some individuals mate monogamously while others engage in polygyny. For females, polygyny generally requires them to share a single male, with one being essentially the "first wife" and the other(s) fighting for the remnants of provisions for her offspring. Over time, this kind of disadvantage of being the "second wife" could reduce her reproductive success by half, compared to her sisters who were in monogamous relations or were primary females, getting all of the provisions brought to the home nest.

If the birds in this story are typical, it's quite likely there is natural selection pressure on individual animals to ease such disadvantages by not just sitting there but doing something. Some studies do show that females in such a bind will frequently set out to "trap" males, clearly a strategic move.

One scientist tested this hypothesis by removing the males and essentially "widowing" the twenty females in the pied flycatcher group just described. Ivan Roskaft and his coworkers in this way produced a generation of "second wives," to see what they would do. Seventeen were visited by neighboring males and six solicited copulations from the new males, three of which were successful. What is especially interesting here is that none of the females were fertile at the time, having already laid their eggs. This strategy is never seen, the researchers say, in monogamous or primary females; they see the second wives' club behavior as a strategic effort to make the males think they might be fathering some birds at the girls' nests and that they should bring some goodies home to those nests, too. "Males should be most easily deceived when the deception is most plausible," says Roskaft. "Thus females widowed immediately after egg laying should be more successful in deception than females widowed when they have nestlings." And that turns out to be the case.

There's abundant further support for the idea of a "natural" infidelity among both sexes. Men are fertile for decades longer than women, who all undergo menopause. Not exactly conducive to keeping him at hearth and home. The "family values" crowd shakes its head at the "unnatural" high divorce rate that forces women to rear children alone. Yet the fact of our biology is that women can and do raise offspring successfully without a full-time, committed mate.

It's highly likely, for instance, that women evolved to sustain a pregnancy on the skimpiest of resources or with substitute mates, given the dangers confronting ancestral males on long hunts. Wherever scientists look, adultery, mate stealing, and jealousy are hallmarks of animal society, and it's possible that in the absence of social restrictions, promiscuity is an equal-opportunity behavior for both sexes that is not always "situational."

Such behavior makes the most sense, in fact, in light of what evolutionary biologists call "hypergamy" and psychologists call "marrying up" or the "sexy son" strategy, pursued by a woman whose chances for a committed mate are marginal for any reason, whether biological or environmental. Think of the never-married, aging single woman. If she has sex with a man who has traits she admires—including a successful marriage and healthy children—she could hope that male offspring from such a union might carry such admirable traits.

The concept of strict monogamy as "moral" or "natural" is a relatively recent, "socioreligious" construction in the history of mankind, and no more or less "natural" than any other mating system. *Homo modificans* use what can work in a given exigency.

Among the Bari people of Venezuela, for instance, the male who biologically fathers a child is considered the "primary" dad, while any extramarital lover mom takes during her pregnancy becomes a secondary dad, who also has a socially mandated obligation to help feed the child. It turns out that the Bari are protein-poor, eating mostly plant foods, so that secondary fathers who provide extra fish and meat are highly valued by all. From an economic standpoint, humans were evolved to deal with *scarcity*, not abundance; with abundance, the imperative we evolved to stay together and the urge to ignore economic opportunities that liberate individuals disappears.

Indeed, studies show that over the past few hundred years of Western

civilization, monogamy and polygyny ebb and flow with wealth and poverty, markets and economics. Says anthropologist and reproductive physiologist Laura Betzig, after a lengthy survey of European and other
civilizations, all political ambitions and all forms of economic competition
can be viewed as sexual and reproductive competition—powerful men
using that power to win at all three. She specifically poses the idea that
preferences for or against polygamy have evolved to favor "maximization
of the reproductive value."

In her survey of the sexual practices of Western civilization, Betzig
analyzed brides' dowries relative to the value of their husbands' estates,
and female succession in the absence of a male heir in the British aristocracy in the late medieval and early modern periods. She believes that
the consistency of her findings—that the most powerful and "noble"
males were the most likely to dominate the economic and sexual lives of
society—attests to their biological underpinnings and drivers.

If she's right, her analysis would help explain why those alpha males
throughout history with the most sexual power were often the most brutal—they had the most to lose economically. And, says Betzig, it was only
when overseas trade and colonial economic engines made it possible for
the first time in history for *lower*-status males to aggregate capital and
attract females with some wealth that the old tyrannical, baronial system
of "protection," primogeniture, and sequestered wives fell apart. Women
almost instantly went after the relative freedom that marriage to a busi-
nessman could give them. Call it bourgeois, but it was clearly a lot more
fun than sitting home in the castle in a loveless marriage. And it certainly
prevailed.

What parts of our biologies work against commitment are sometimes
surprising and even counterintuitive. Never more so, perhaps, than in
the case of orgasm. The "logical" idea might be that orgasm is so
rewarding when it occurs mutually between a couple that it should
cement relationships, enhance commitment, and encourage fidelity. And
so it does (see chapter 9). But it also works *against* commitment, in yet
another example of nature's conservative tendency to use what it has to
do a variety of jobs.

During orgasm, a man's penis has blood vessels that release blood
back into his body, disengorging the penis to a limp stage and subjecting
him to a relatively long "refractory" period in which erection, ejaculation,
and orgasm are relatively unlikely. But after a woman climaxes, her genitals do no such thing. With some practice, and if she knows how, she

can have orgasm after orgasm in quick sequence. The fact that female chimpanzees and human women seek and enjoy a lot of nonreproductive coitus suggests that a woman's high sex drive, coupled with hidden ovulation, sneaky sex, and orgasmic capabilities are a highly conserved and continuously rewarding biological tactic to win additional paternal investment, insurance against a father's infanticide of her offspring (out of fear the offspring are not his). Every sexual anthropologist from Alfred Kinsey to Helen Fisher has concluded that given a level social playing field and an absence of severe punishment, women, not men, are the promiscuous sex, seeking variety. "Even in those cultures which most rigorously attempt to control the female's extramarital coitus, it is perfectly clear that such activity does occur. . . . with considerable regularity," Kinsey wrote.

Similarly, the physiology of a woman's orgasm is a reflection of her evolutionary blueprint to seek extra-"marital" sex. Says Buss, in *The Evolution of Desire,* "Once it was thought that a woman's orgasm functions to make her sleepy and keep her reclined, thereby decreasing the likelihood that sperm will flow out and increasing the likelihood of conceiving. But if the function of orgasm were to keep the woman reclined so as to delay flowback, then more sperm would be retained when flowback is delayed. That is not the case. . . . there is no link between the timing of the flowback and the number of sperm retained."

Although it's clear that nature's sexual plan provides strategies and biological support for *unattaching* us when it is in our strategic interest to do so, as noted earlier, it is also true—and intensely more significant— that there would be little point to the whole exercise unless we also had the brain and chemical machinery to successfully negotiate intense commitment, resource sharing, and parenting responsibilities. In other words, our bodies and minds must know how to *stay* in love, or keep falling in love, with our mates.

The evidence for a biology of family commitment—parents to each other and each parent to the children—is strong and comes from studies of genes, biochemicals, and widespread observations of kinship and attachments in hundreds of animal species, including our own.

Biologist Diane Bianci, for example, has shown that during pregnancy, the placenta produces substances, including endorphins, that spill into the maternal blood and may stay in the body for up to a decade. Other studies show that what also sticks around, if women give birth to or even have miscarried pregnancies with sons, are pieces of the male Y chromosome. According to evolutionary biologist David

Haig, studies showing the persistent presence of the Y chromosomal material in such mothers—in one case tested for twenty-eight years—may be far more than what scientists dub a "fascinoma," a mere purposeless curiosity. It may be a way that males influence maternal feelings from the womb. Whether or not this turns out to be so, other research, from Swedish investigators, shows that small proteins found in breast milk are chemically related to sedatives, perhaps guaranteeing babies will sleep and let their mothers sleep after childbirth, enhancing physical intimacy and contact. When babies cry and activate the biochemical "letdown" of milk to the mother's nipples, it may be a way of telling the mother's brain, "If you don't feed me, you won't sleep, we won't feel content together, and I won't survive."

More direct evidence for a biology of commitment has emerged in studies of pheromones and oxytocin, the "cuddle" hormone, in male underarm sweat and the probable existence (see chapter 4) of a vomeronasal organ in women that can detect them. If indeed men and women coevolved a sort of sixth sense to help them subconsciously bond, nature may have been motivated to produce more and more females with the ability to emit pheromones that promote paternal behavior in their mates; and since good parenting of his children are in the male's interests, males would be selected to be seduced by such female pheromones and to pair-bond with the mother of any children he fathered.

Some corroboration comes from the fact and nature of human orgasmic experience and our willingness and desire to "practice," to learn how to sexually satisfy each other consistently over time.

Consider what's in it for a man to help his partner achieve orgasm when she wants one, and why men consistently find it sexually exciting to them to bring a partner to orgasm. From a man's standpoint, keeping his partner happy and sexually satisfied, keeping committed, makes reproductive sense. It's logical, says Robin Baker, that "[a] man is most prepared to cooperate over foreplay . . . when he has the least to lose, that is when the woman either has no sperm in her or when the inseminate he is about to produce has little chance of becoming involved in sperm warfare." So it isn't surprising that men are most likely to be cooperative over foreplay orgasms when they have spent most of their time with their partner; and he can consider her unlikely to be straying from the marriage bed when she is most insistent about such orgasms.

"As we have seen time and time again," Baker adds, "most of the strategies shown by men and women in relation to ejaculation and orgasm are subconscious. But the conscious element has a place nonetheless as women and men learn by trial and error how best to satisfy their feelings."

"Men," he goes on, "have to learn many things, from the basics of penetration to the subtleties of the female orgasm. Women have to learn how to climax, how to encourage men to help them to do so and when and how to fake it. Both sexes have to learn the strategic subtleties of infidelity and prevention of infidelity."

This may well be why practice with the same partner is adaptive, and why biologically, the purpose of intercourse and orgasm is not only to fertilize an egg but also to encourage pair-bonding, closeness, intimacy, and mutual learning and trust.

If science has established that we have a biology for both "now" and "forever," it hasn't yet come to any certain conclusions about what "now" means and how long "forever" is for any given species, couple or individual. The moment sex joins two gametes, the sexual and social mathematics become far more challenging to master than simple addition or multiplication.

If we look to other animals and to the primitive experience, the only consistent clues we find is that our biology intends us to make long-term commitments at *some* times and short-term commitments at *others*. It intends us under some circumstances to protect our children and our spouses, while under other, relatively rare and extreme conditions, such as threats of torture, war, starvation, and illness, to abandon them. *Sophie's Choice*, the novel in which a woman is forced to send one of her children to the gas chambers in order to save the other during the Holocaust, is a case in point. Our biology and psychology may recruit us to use tactics that favor our grandchildren over our children or to favor our mates over our children.

When making love makes a baby, men and women face complicated and consequential choices. Our biological as well as our social legacy reflect how our ancestors worked out the answers to such questions as whether a particular mate is right for a day, a year, a decade, or life. But the road to commitment is rocky, full of jealousies, negotiations, infidelities, decisions, pushes, and pulls. It may not satisfy those who demand consistency in sexual mores, but in Western culture, at least, perhaps the best that evolutionary biology can tell us, in the words of David Buss, is that we seem to do best with "one wife and hardly any mistresses."

This is no apologia or biologically determinist rationale for serial monogamy. Keep in mind that the human capacity for making socially responsible, and what we now call moral, rules evolved right along with our capacity to leap strategically into promiscuity. If it is true that we may have built-in tendencies to a seven-year itch, it's also true that we can alter our sexual strategies at any phase of our lives in response to culture, environment, and opportunity. The DNA of those who failed along the way to make not only flexible choices, but choices that the wider community supported, probably didn't survive.

Serial monogamy, or the commitment to a primary mate for some extended period of time, but not forever, may be simply an emotional "deal" between men and women that ultimately permits polygyny. In this view, divorce American style may be not only understandable and politically "correct" but biologically sensible as well. How? By providing a way for nature to sort out for women who gets access to her, perhaps reducing her risk of getting bad genetic material out of her mate. Consider that the average woman can bear children for only about twenty-five years. If some men get more than twenty-five years' worth by either marrying several women over time or tying up a young woman without impregnating her for some years, or by having much younger wives, it means some man is doing with fewer than his allotted twenty-five years of fertility in a woman. This means the most successful men get the extra sex resources and the less-endowed men go without as many. What's good for society is not necessarily good for the individual man or woman.

Humans mate in every conceivable kind of arrangement, but serial monogamy at least appears to have the evolutionary edge in most studies of successful societies. Nonmonogamous strategies have declined with modernization and the leveling of the economic playing field for—and between—both sexes, and wherever polygamy has been tried in advanced civilizations, it has pretty much failed. Think Mormons. And Hutterites, where patriarchal polygyny is more an economic than a biologically strategic commitment for a man. He has multiple wives who have no real choice but to accept the life of a harem, join a religious order if they can, or kill themselves. Even if she occasionally gets in some sneaky sex, penalties are severe. In this scenario, rewards are best for him if he has more wives than other males, while the rewards for females parallel the strength of any punishment she might get for not adhering to his whims.

Some looser patriarchal systems may have arisen as a means for poor and unattractive men to ensure they got access to wives, but for sheer variety and flexibility monogamous members of societies win the lottery. Some are relatively libertine and others more or less puritan. Under the looser model, the community church and state have generally not approved adultery, but have not enforced this norm of fidelity for men. Often there is also an implicit prestige for the sexually promiscuous man, the old double standard. But in this model, internalized rewards (of macho prestige) are muted by official prohibitions. Men can openly support only one wife; if they have mistresses, they need to sneak resources to her. In most cases they can't really use them openly to strut their stuff, have their children, or help do the housework, even though they may cost him as much as his wife. Compared to polygyny, this is not as great a deal (for men) as it would seem. Then there's the strict puritan mode in which infidelities are reported and punished severely, but more severely for women than men. As Laura Betzig might argue, since the Enlightenment and industrial revolution, the growing economic equality of women has contributed to the downfall of such strict codes, which originally arose perhaps to give less attractive or less fit men access to long-term mates; the men enforced this by mate guarding. In the women, it's enforced by the uncertainty of paternity and hidden ovulation. There's an equilibrium. Under this scenario, the cost of infidelity is the highest for men, in which she takes him for all he's got in divorce court.

Finally, there are subsets of each of these: societies in which there is a monogamous majority and libertine minority; societies where spouse swappers are prevalent; and those in which both sexes are equally deviant but in different ways.

There are punishments and payoffs—genetically and socially—in each of these strategies. And all may have built-in mechanisms that, depending on what's happening with resources, bring a society into reproductive *balance*. A study at the University of Chicago, for example, examined numerous current and historical societies that ranged from "free love" groups to more puritanical ones. Struck perhaps by the fact that even the most conventional societies (Victorian England, for example) tolerated and even celebrated a certain share of libertines in their midst, they sought to determine if such tolerance had some adaptive purpose. Apparently it does. They found that if a society punishes polygynous or promiscuous people *severely* for doing something they like (having sex their way) but lets at least *some* libertines get away with violating the rules, members of the monogamous majority are less likely to cheat or cheat too often, but the libertine minority is *more* likely to, betting, logically, that if they get away with it once without getting punished, they will again. Over time, even those with strong anti-promiscuous codes will be content to be tolerant of the libertines, because nonlibertines will have freed *themselves* from internal and external costs of such deviance, while lessening the likelihood that if the libertines should want to risk the occasional affair, the ones who approached *them* for sex will be left alone.

Eventually, however, as game theorists have shown, if every member of each group—the non-promiscuous and the libertines—pursues the best situation for themselves, then all the libertine arrangements eventually collapse and the libertines are forced to revert to non-promiscuous behavior.

To escape from this dilemma, the libertines would have to organize and have a very strict, authoritarian type of organization to create a non-monogamous community big enough to stay so loose. But the same libertarian desires that drive such libertines to pursue "free love" turn them from such authoritarian strictures. In the fabled commune of Oneida founded by John Humphrey Noyes in upstate new York in 1837, "free love" was rejected in favor of strict rules. If a man or woman wanted sex with another member of the commune, he or she had to petition the community elder to carry a message to the party desired; if the other consented, the couple could meet a few times, but they could not continue seeing each other if any feelings of attachment occurred. If they violated the rules, they were expelled. Men were forbidden to ejaculate in order to contracept and help female sexual pleasure, and there were no pregnancies. Not surprisingly, the commune lasted only thirty years.

As a consequence of such dilemmas, argues Hughes, which almost surely reflect evolutionary pressures to maximize fertility and mating in a balanced way, "we now have many anecdotes describing free love veterans of the Sixties backsliding to a revisionist advocacy of monogamy. Public disapproval of adultery among Americans has risen from 80 percent to 85 percent since the Seventies." But what Americans say they favor and what they do are two different things. What seems to be the case is that human societies do best when they live and let live, up to a point, in order to keep our social responsibilities and our biological drives in some balance.

Compared to what we know about individuals looking for a relationship, studies about sex behavior among couples are emerging at record speed. Their results are destined to inform our future social debates about human sexuality, marriage, infidelity, family life, and morality. Among the more intriguing are those that have tracked the biology of attachment, gender cooperation, and family making.

A family is defined as a social condition in which parents and offspring have an ongoing relationship with each other and their cousins, grandparents, uncles, and aunts—all of whom give up some self-interest sexwise for the good of the family. Kinships form through mutually beneficial attachments. The bond is tough, but not so tough it can't be broken. Less Super Glue than rubber cement. Families are a mix of couple love and group need. And they are almost universally considered the foundation of civilization and the best venue for rearing young and protecting couples and their offspring.

From a purely biological standpoint, however, there's a cost to staying around the homestead and limiting your reproductive choices. And as any biologist worth her field trip can attest, where there are "costs" to a behavior, there are likely to be rewards as well. It is at this intersection of costs and benefits that biological "choices" become social and psychological choices as well.

A variety of animals, including people, expend huge amounts of energy forming and reforming families, a telltale sign that the activity has its roots in sexual biology and not just culture. Today's politicians and social commentators like to say that the family has fallen on hard times; cynics claim it is dead, at least as humans have known it for the past five thousand years or so. But in his research, Stephen Emlen of Cornell University and his colleagues have found that while families are in many ways fragile things, and unstable, the fact that they persist as an ideal even in the face of our promiscuous underpinnings suggests they are adaptive and biologically rooted. The world of Beaver Cleaver (sort of) lives, even among beavers.

In a landmark article in the *Proceedings of the National Academy of Sciences* in 1996, Emlen argued that families are a natural offshoot of sex and parenting, combining the biological and emotional aspects of attachment and bonding. According to Emlen, the animal that wants to stick around the parents, the momma's boy or girl, isn't likely to go off and reproduce its DNA, neither fulfilling its own nor *its parents'* destinies. In fact, families, extended ones, may form in times of relative scarcity, when food or nesting areas are limited and offspring born to the grown-up kids would have a slim chance of surviving. As soon as the economy gets a little better, they're moving out. On the other hand, if there is an abundance of resources, the equivalent of a noble, wealthy family, the kids may stick around for the inheritance. Some acorn woodpeckers build such dynasties and leave big extended families that would put English kings to shame.

John Bowlby, Konrad Lorenz, and Harry Harlow, who spent their careers watching human, bird, and other animal babies bond to their mothers, set the stage for our modern notions that when we are very young, we first form attachments to our parents and siblings, and that forever after we forge a fear of abandonment so strong that we will go to enormous lengths to avoid it, grieving as piteously as an orphaned baby when we lose it. A corollary of this kind of "instinctive" love is that it is biologically a part of our survival armamentarium, linked inevitably to the emotional "maps" we use to guide our way to subsequent loves, including our mates. When we form a strong attachment to someone other than a parent—to a lover, a mentor, a friend, or especially to a mate—we are said to have truly matured and indeed we have, since none but the latter can help us beget the next generation. And when conflicts arise between a couple marching toward, or in, a committed relationship—as surely they do—we behave just as we might have as infants and children: with jealousy, irrational rage, violence, petulance, anxiety, depression, insecurity. In what Diane Ackerman calls a "behavior the body knows by heart," we go into denial, fury, sadness, and finally acceptance when we lose love, so strong is the capacity of humans and other animals to attach to some significant other person.

Over time, we get to be happy and to love again, but the strong reactions we have to the *loss* of kinship and mating relationships affirm our fundamental and natural capacity for making and keeping those relationships. It is not insignificant to remember that we define the mentally ill often as people who cannot bond successfully with others, who are "sociopathic" or "withdrawn." Their brains and minds are deeply affected, damaged chemically in many cases, and the result is lack of "attachment," the inability to maintain relationships, to "fall and stay in love."

It's become a cliché and the foundation of the pop psychology industry to say that people who had bad childhoods, bad parents, will have marital problems in abundance and will not treat their children well. Cliché or not, it is to some extent true, because attachment and love have a chemistry that we cannot unload as easily as a wedding ring.

Attachment theories hold—and may also account for—the fact that unrelated women may be relatively less likely to develop certain social bonds with each other, in part because they are competitors for the best males and in part because the strongest affiliations are most likely to occur in all animals between those who are related, who share genes and interests. Many women no doubt feel that they bond *better* with other women than men do with other men. But it's the emotional content and goals of the relationship that may differ. Thelma and Louise were not Butch and Sundance. College sororities are rife with cliques. Fraternities are less so. Male-male bonds are more often displayed in group activities and teamwork, female-female bonds in emotional sharing and protection.

Amy Parrish's studies of bonobos, whose sexual behaviors are close to our own, suggest that the fabled tensions and battles between mothers-in-law and daughters-in-law are a direct outgrowth of their unrelatedness genetically. On the other hand, human females, it may be argued, have other ways of building relationships with both relatives and unrelated women. Among bonobos, or pygmy chimps, females associate with each other, groom together, and even have sex together by rubbing their genitals vigorously in each other's embrace. Many scientists believe this female-female sex behavior has evolved in great part to reduce conflict and tension, and in that way bonobos may be more evolved than we. Reducing conflict among each other gives females more edge with males who apparently don't want to come live and travel with them if there is conflict.

On the other hand, it may be that female-female sex behavior evolved as a logical extension of what females found they *could* do to help themselves. Parrish looked into relationships among female bonobos and their control of food and mating between 1990 and 1992 at the San Diego Wild Animal Park, the Frankfurt Zoo, and the Stuttgart Zoo. "It is obvious from the studies . . . that female-female bonds are possible among unrelated females on a routine basis in pursuit of common goals that confer a reproductive benefit to all female group members. Where female aggregations are possible and coupled with an incentive to cooperate for common benefit, bonding can occur even without close genetic relationships among the participants. The extent of routine bonding expressed among bonobo females probably exceeds that found in most human groups and demonstrates that [such] bonds are not unique. . . . Relaxation of constraints on bonding may have preceded modern-day humans and their relatives in the hominid line."

Within the "extended" family, of course, lies the nuclear family—

mother, father, child—that does the heavy lifting in the reproductive relationship. And here again, the two sexes, male and female, behave according to their individually evolved biologies.

Mothers first bond with their unborn infants in physiological ways
impossible for fathers, sharing endorphins and the cuddling hormone oxytocin, nutrients, and their very blood. Fathers love and commit to offspring only when paternity is certain or in cases when it's to Father's
advantage to suspend such certainty, such as gaining access to a child's
mother for sex and emotional reasons. Fathers feel protective, duty-
bound, and their hormones drive much of this behavior.

The brain and body differences between men and women, particularly
those within the limbic system, may explain what makes a woman of
almost any age, but particularly one who has experienced childbirth, irresistibly attached to even photographs of babies, human and otherwise.
Studies show that even women in old age and in pain from illness will
show dilated pupils and other evidence of emotional reaction to such
pictures, while men will not.

Because all our behaviors and reactions are facilitated by our biochemistry, it's no surprise that scientists have found a few molecules playing key roles in attachment. Of these, it's no surprise that oxytocin should
play a key role. We've already seen how it helps mediate attachment
between couples in chapter 4, but its first influence in building attachments comes soon after birth. While the infant is milking Mom's oxytocin
stores, the hormone contracts her uterus, shutting down bleeders that
sprung during the separation of the placenta, and shrinking mom's figure
back to the waist-to-hip ratio her mate loved in the first place, reinforcing
the links a woman already has between the physical and emotional sides
of love and sex. Moreover, when fathers bond with their infants, their
levels go up, too. In short, mothers and fathers *chemically* fall in love with
their babies. According to Thomas R. Insel and Lawrence Shapiro at the
National Institute of Mental Health, male and female montane voles deficient in neuroreceptors for oxytocin abandon their babies and each other
at a high rate.

As we've seen, the mere fact that families have persisted throughout
human evolution provides ample testimony to the value of forging long-
term relationships between men and women. According to social scientists
like Stephen Emlen, such relationships are a means of maintaining peace
between breeding couples at least long enough to get the children out on

their own. Emlen points to the obvious Darwinian argument that those behaviors that enhance survival and reproduction—either of the individual or its close relatives—should be favored. Thus, those who practice such behaviors will leave more children, and whatever neurological mechanisms predispose the individual to express these behaviors (at the right time and place) are mechanisms that have survived.

"If there've been tens of thousands of years to select for the predispositions we have," Emlen adds, "one way to find out just what those predispositions look like in our brains and chemistry is to look at animal species that have similar types of social organizations to those of humans. The premise is that species living in similarly structured societies will be those most likely to have evolved similar sets of algorithms and decision rules. And it turns out that there are many such species." He studied many of these species and put together a synthesis of expected behavior patterns, hypothesizing that if he found universal behaviors or at least convergences across many different kinds of birds and mammals that live in family organizations, those patterns would be good candidates for predispositions that humans might have, too.

For one of his major studies, Emlen extended observations and comparisons of biologically intact families and "reconstituted" or stepfamilies. He began with the premise that there are good biological reasons to expect extended intact families to be more cooperative and more harmonious than any other kind of social organization we find in nature. That's because there is first of all genetic kinship, a common descent. "Individuals basically form families when offspring stick around and continue to interact with parents, yielding continued interaction that gives rise to very close kinship levels, and highly cooperative, altruistic behaviors through reciprocity." Among the animal species he studied, more than 90 percent of all those that live in genetically related multigenerational family structures cooperate in the care of their offspring. In contrast, such mutual sharing of helping the young is virtually unknown in any society that is not based on genetic family organization.

He also hypothesized that a second characteristic of extended genetic families is that sexual competition is reduced, and again, Emlen found that across the animal kingdom natural selection has in fact favored "inbreeding avoidance situations," or incest, such that close genetic relatives generally display little or no sexual interest in one another.

"This is the good news," Emlen says, and if scientists are right about families being an adaptive and hardwired strategy for encouraging long relationships, cooperation, and family building, then that should all change in stepfamilies, where genetic kinships are weaker. That in fact appears to be the case, both in animal and human societies.

"Families once broken cannot easily be reconstituted," he notes. "Step parents are unrelated to the current offspring already in the family and if that couple now reproduces on its own, the new offspring will not be as closely related to the former offspring. If in fact in this new partnership both partners bring children or offspring from a previous mating and come together in the equivalent of a blended family they will be step offspring with no genetic relatedness whatsoever. And when we compare these kinds of family situations in animals, we have been able to predict some of the very same social dynamics that we see in human families: increased conflicts of interest between different members of the step family situation."

Emlen found that replacement mates, stepparents if you will, always invest less in step offspring than parents will in their own biological offspring. This is overwhelmingly supported by literature on birds and mammals, and if one extends the definition of families to involve kin groupings of multiple generations, it is also true within social insects. Scientists have investigated in birds, for instance, the relative amount of nutrients given to nestlings and demonstrated with genetic tests that even though birds are willing to "adopt" orphaned or abandoned offspring of others, the primary parents' own biological babies are better fed. Not only do stepparents in animal families not invest as much, they may also hurt their nongenetic offspring. Infanticide occurs among birds, rodents, carnivores, and some primates. And, Emlen has found, replacement mates in a stepfamily are also more likely to compete for sexual attention with their nongenetic offspring. The incest taboo only seems to be strong with genetic offspring. "When a replacement mate comes in there will be no inbreeding costs of reproducing with an offspring and often there can be very great fitness benefits in doing so. For a male it could become polygynous by rearing offspring both with the new mate and the daughter and in essence have two partners." In the animal world, even from the offspring's point of view, there can be advantages if an unrelated male comes into the household as long as there is little chance of inheriting the home spot.

Finally, if, as Emlen believes, family life is so biologically valuable that our sexual tool kits are always maintaining or trying to fix them, we can predict that stepfamilies would be inherently less stable than intact ones.

Emlen also found that to be the case in two key ways in the animal kingdom. First, he found that offspring in stepfamilies leave home earlier, and second, that such offspring have a harder time maintaining successful mating relationships. The offspring of stepfamilies have probably gotten less nutrition and other investment from stepparents and thus had less motivation in terms of a fitness advantage to help indirectly raise more young *within* the family. There's more advantage to going off on your own.

Emlen believes that the wide expression of these behaviors in animal species makes them applicable to humans if we accept that nature conserves "what works." As it turns out, human stepparents *do* invest less in offspring from a previous marriage than in their own biological offspring in a later marriage. Moreover, a growing number of sociological studies find that stepchildren are at greater risk for physical harm or of being ignored by each parent's biological offspring. Perhaps the most alarming behavior, he says, is that stepchildren are at greater risk for sexual abuse by stepparents. In one study, stepdaughters were found to be at eight times greater risk of sexual abuse than biological daughters. And stepchildren do leave home earlier.

Despite the evidence supporting biological foundations for stepfamily difficulties, such evidence can never be applied directly to the human condition, notes University of Arizona scientist Patricia Gowaty. "People and animals can and do learn to be terrific step parents, to avoid step sibling rivalries and abuse by discussing feelings and negotiating with the natural parent ways to treat children evenhandedly." The biological foundation of the discrimination may remain, and ancient experiences probably altered brain circuitry to hardwire certain responses to certain environments—such as blended families—but the brain is ever able to learn.

Despite all that has been made of their power up to now, evolution and genetics are never the whole story in human relationships. "Before we get too discouraged," Emlen cautions, "it's important to remember that we're talking about behavioral predisposition and we can consciously suppress those that impact negatively on other individuals. The question is how best to improve our ability to do that. Evolutionary models give us the tools to predict what might be, and thus arm us with information that can help us avoid the downside."

"Right now," Emlen says, "we're seeing large increases in numbers of step families and in single-parent families as the divorce rate goes very high in the United States. For many thousands of generations we and our ancestors have lived in societies where the extended family was part of the essential core. In an extended family there are a large number of helpers at the nest. Women ancestrally were not selected to be sole child rearers at all. There's a great deal of discussion about the disappearing family as we go to more single-parent families but in reality the disappearance of family occurred with the breakup of the extended family. The bulk of that built-in workforce disappeared, and as we go from the nuclear family to a single-parent family I think we've created almost a mission impossible for the single parent whether it be male or female. From an evolutionary standpoint, the nuclear family without any interaction with other genetic relatives is very new, with no emotional, social or biological antecedents. Whatever our heritable predispositions and decision might be they were never adapted for single-parent child rearing. We've created a culturally new social situation for which our inherited predispositions have poorly prepared us, and there are two take-home messages in this. One is that we are creating an impossible situation for single parents; the other is that we are trying to socially construct alternatives, surrogates for that work force that is now missing."

One major alternative, he predicts, will be the strengthening of the grandparent-grandchild bond.

In a study at the University of Michigan of "relational uncertainty," or "RU," investigators looked at the number of times in the line of descent between two relatives that the genetic relationship between them could be severed by cuckoldry. They found that for fathers, the RU was 1— they were "pretty sure" they were the fathers. For mothers, it was 0. They absolutely knew they were the mother and who the father was. For grandparents, RU can range from 0 to 2. When the investigators looked at various families and how involved the grandparents were in investing in their grandchildren, they found that RU predicted grandparental investment and the emotional closeness between grandparents and grandchildren. The more certain grandparents are about who the parents of a child are (their children), the more they care. The findings were so consistent, they reported, that it's hard to argue against some biological foundation for them.

The need for kinship and attachment to those most genetically like ourselves is a big part, no doubt, of why long-term mating relationships persist. From an evolutionary standpoint, it "takes a village" indeed to successfully rear offspring. The irony, of course, is that such a biological strategy runs counter to a fundamental characteristic of sexual reproduction, which is to mix genes in new ways, to protect the species from common disease genes or other deleterious DNA. The counterbalance to kinship, then, may be incest avoidance, and the whole structure of dominance and submissiveness between the sexes and *within* each sex, nature's modus operandi for making us *too* attached to our relatives.

One consequence of this is that humans and other animals have biological as well as cognitive means of identifying kin and close relatives, of bonding to our children, of jealously guarding our mates, of investing in the long-term survival of our offspring, and of using social glue to keep our DNA intact in future generations.

Marriage, for example, makes it possible for children to potentially identify hundreds of additional kin on both their father's side and mother's side, conferring so many supportive benefits that marriage is a core family "value" in about 95 percent of all societies, and long-term pair-bonding is widespread in other animals as well.

If large extended families get in the way of young love at first, that, too, is both a mark and a "benefit" of kinship. April Gorry, a social scientist at the University of California at Santa Barbara, has investigated families and concluded that chaperones and family busybodies who try to manipulate their younger relatives' relationships may in fact be playing a biological urge to manipulate reproductive interests that will serve their DNA. In her cross-cultural study of modesty codes and chaperoning, she concludes that these are ways older women have of doing what a will does for a wealthy man: enhancing the reproductive success of her offspring long after she may be gone by manipulating the signs of sexual availability and material and social resources that attract men.

David Pfennig and Paul Sherman, experts on the genetics and evolution of "kin recognition," note that in a variety of species, the animals have genes and cultures that teach them to sort out who is related, who is not, and all the degrees of relatedness. From sea squirts to gorillas, they can identify their relatives using a variety of chemical signals. Some cannibalistic species will even taste other members of their species and avoid eating their brothers and sisters, presumably to optimize their own families' reproductive fitness. Studies on the primitive sea squirts, who reproduce asexually in interconnected colonies, have even identified particular genes in their immune system that recognize kin and non-kin and keep them from fusing (having sex with, essentially) individuals too closely related. Similarly, Wayne Potts and Edward Wakeland have shown that house mice tend to mate with males that smell different, even if they later go back to their home nests to be with their sisters for help in rearing pups.

It seems that animals have such built-in "stranger" detectors to help them strategize about forming bonds that enhance their fertility. For example, a study at the Philadelphia Zoo showed that when, after four infertile years, Jessica, a rare lowland gorilla, was moved to the San Diego Zoo, she became pregnant right away and kin discrimination may explain why. Jessica, like unrelated boys and girls who room together in college dorms or Israeli kibbutzim, would not mate until she was introduced to males other than those she had lived around since birth. In nature, such individuals would usually be relatives and Jessica may have viewed her companions as such.

The lengths to which our biological and emotional legacies have taken us to succeed at sex has also been made clear by hypotheses linking menopause and the grandparent bond. Although controversial, the ideas, most notably advanced by evolutionary biologist Jared Diamond, offer an alternative to the facile explanation that menopause is a mere fallout effect from recent increases in the human life span. Diamond reasons that menopause only makes sense if, as a woman gets older, "she can do more to increase the number of people bearing her genes by devoting herself to her existing children, her potential grandchildren and her other relatives than by producing yet another child." As evidence, he points to what any exhausted mother will tell you: Raising a child to be self-sufficient can take decades, before which time they are dependent economically and emotionally on parents. Ancestrally, as a mother aged, each successive birth disadvantaged all of her children in the competition for scarce resources. Among one of the last remaining hunter-gatherer tribes in the world, the Hadza of Tanzania, anthropologists have watched the huge amount of time women must devote to getting roots, honey, and fruit and seen that most who did this were postmenopausal, spending long workdays bringing home food for their children and grandchildren. After menopause, these women needed far less food than they did when they were childbearing, so for the same amount of effort they were able to feed many more of their children and grandchildren. Grandmothers (i.e., older, postmenopausal women) could also teach the younger members of society everything from how to predict bad weather to where the best sources of food were. They could help arm the younger generation against uncertainty by passing on valuable information about seasonal cycles, sources of food, and danger. Concludes Diamond: "The old men were not at risk from childbirth or exhausting responsibilities of lactation and child care so they did not evolve protection by menopause. But old women who did not undergo menopause tended to be eliminated from the human gene pool because they remained exposed to the risk of childbirth and the burden of care."

If families and genetic attachments are sometimes cause for avoidance in sexual liaisons, they nevertheless may also directly influence fertility rates. Karen Kessler and James Boone of the Human Evolutionary Ecology Program at the University of New Mexico were among the first to explore the observation that humans, especially wealthier humans, often appear to have fewer offspring than would be expected by the theory of "fitness maximization," and the observation that lowered fertility can actually lead to higher rates of fitness in the long run in some families.

Wondering why this was the case, Kessler and Boone noted that it's easy enough to speculate about why poorer people would have fewer children—lower fertility triggered by starvation, few resources, and fewer relatives to help out; and why in such cases a higher number of offspring might sometimes occur. If you're poor, and weak, you might well adopt a strategy that says "have as many as you can drop out of you as fast as possible and hope a few will make it through the crises and the genetic bottlenecks."

Not so easy to understand is why the wealthy would not always "maximize" fertility, especially when there was abundant food and relative peace and calm. After all, the central goal of life and sex is to continue life and sex.

The answer seems to be that there is much to be gained by *families*, if not individuals, in *limiting* fertility as well as enhancing it (a strategy that, not incidentally, may also have helped set the biological stage for infidelity and interest in nonreproductive sex among some family members—or at least did nothing to disable them).

In marmosets, for instance, the dominant female of these new-world monkeys is the only one who gives birth, while other group members, male and female, help care for the babies she has. This matriarch doesn't fight overtly to keep other females from bearing young; she makes her subordinates infertile physiologically—by influencing ovarian suppression, which is widespread among subordinates. Their ovaries literally dry up and so does the egg supply for fertilization.

Biologist David Abbott has discovered that a combination of phero-mones, visual cues, and bullying are at the matriarch's disposal in her mission of suppression, and that similar physiological ovarian sabotage is prevalent in other monkeys, naked mole rats, and a variety of primates. Similarly, it is likely that humans who best survived periodic catastro-phes were those whose bodies and brains cooperated and figured out that reducing fertility in the group would give some individuals a better chance of having at least a few of their offspring get through the scar-city bottleneck.

Human patterns of reproduction fit this model, too, in at least three ways: Human history is marked by local periods of growth punctuated by crashes such as climate calamities and famines; it's a strategy that can increase the probability of survival through a bottleneck but requires diverting resources away from making more children; and the benefits of increased survivorship through a crisis must outweigh or equal benefits that would come from putting the same resources into having more off-spring during periods of growth.

But it turns out that over the long haul, increases in survivorship can outweigh the benefits of higher fertility in individuals, even if crises are neither frequent nor harsh. To understand why this is so, Abbott and others have analyzed information from primitive societies still in existence. Because humans are so bent on keeping growth low and steady, it must have been the case that our ancestors—and their bodies—learned from the experience of repeated, often unpredictable calamities, that they needed to alter their reproductive strategies. They did so via their hor-mones and physiology, to make sure the species survived such resource bottlenecks and came out the other end—one by one if necessary, but alive. In short, those that could afford to do so hedged their biological bets, giving up the short-term gains for the long term. In fact, according to one calculation, if societies that experienced frequent scarcity had not suppressed fertility as a reproductive strategy, they would have grown from, say, a starting population of ten thousand to the present population of the entire earth in just 1,849 years.

A reduced fertility strategy would play itself out by having smaller numbers of eggs available for fertilization (ovarian suppression), so there would not be resources (of food or energy) wasted on trying to keep a larger clutch of eggs (in the case of birds) alive. It might also be why human females only rarely ovulate multiple eggs.

This strategy also would play itself out in the context of families. In times of shortage, lower-ranking families will be the first to be adversely affected; higher-ranking families, last. It's not that greater wealth and status per se confer more fitness on offspring, but that wealthier, higher-status individuals tend to have fewer offspring and therefore invest, proportionally, more in each, or guarantee the offspring a likelier chance of surviving on their own through a crash or a bottleneck. Otherwise, we would not see so many high-ranking societies that invest so much time and energy purely in status reinforcement, resources that could go into making more babies. Why? Because such reinforcement of status maintains the families' place in the front of the line to get more resources during bad times. This is the genetic equivalent of a baron with a castle's and a year's worth of food in storage when the siege or the plague arrives; and there is evidence that such "urges" are widespread and conserved in our psyches. Some American Indians of the Pacific Northwest engaged in what may be the ultimate form of conspicuous consumption (in case you thought that consumers of the twenty-first-century bridal industry invented it). During ceremonies, the elders competed rigorously to see who could give away the most of their possessions and wealth, destroying their own property in public, even to the point of impoverishing themselves for a while, until the next such ceremonial in which other elders would then be compelled for status's sake to outdo the previous potlatcher.

MOLECULES OF MONOGAMY

Perhaps the most fascinating biochemical story of commitment and love in recent years focuses on a species that should now be exceedingly familiar: voles. To summarize what we've already noted, voles come in relatively monogamous and promiscuous types and have evolved means of reading their social circumstances to regulate their sex hormones and reproductive biology. Unlike hamsters, for example, who have estrus cycles that make them fertile for four days, regardless of what's going on with food, famine, predators, or drought, the monogamous prairie voles have biochemicals that make them "maternal" or "paternal" not only four days a cycle but throughout their lives. Male montane voles run from female to female while male prairie voles pair off with a single female, probably for life. Here, then, is a species whose monogamy, says Carter, had to be grounded in processes that allowed their brains and bodies to respond to others around them on an *ongoing* basis.

We've also seen that many of the differences in the two species so far appear to be found in molecules that influence the brains of these creatures, notably oxytocin, but also vasopressin, a less well known hormone but one also believed to trigger maternal behavior, even in male prairie voles. A chemical that stimulates absorption of water by the kidneys and decreases urine flow to regulate fluids, vasopressin is currently the subject of intense research by neuroscientists.

Thomas Insel, for example, has found that compared to the brains of polygamous voles, the monogamous prairie vole brains have different distributions of receptors for oxytocin, and that the receptor proteins that sit on the surface of nerve cells clasp, like locks, both vasopressin and oxytocin. In a groundbreaking report published in the journal *Nature*, Insel and Sue Carter set out to nail down whether either of these hormones could directly induce or affect pair-bonding in the prairie vole.

Monogamous voles spend most of their time with one mate, and the males not only protect their nests ferociously but also help in child rearing, attacking any strange male that comes close to new offspring. Before offspring arrive, however, the voles are pretty wussy, so something must be transforming the timid guys into protective Rambos.

To pinpoint that "something," Insel and Carter chemically blocked either vasopressin or oxytocin activity in the brain, then let the voles mate. Like untreated voles, males that got an oxytocin blocker attacked other males after successfully mating, suggesting that oxytocin isn't involved in the aggression end of the story. The voles given the vasopressin blocker stayed timid and didn't take the part of the fierce protector of home and hearth, suggesting that vasopressin is what triggers this aggression. In fact, they found that blocking vasopressin made the monogamous males more like their promiscuous montane cousins who don't guard their females so jealously but are more interested in impregnating more females. Further studies showed that the more exposure a prairie vole has to vasopressin, the more he becomes a jealous and attentive husband.

Turning to what hormone or neurotransmitter might be involved in the pair-bonding and parenting action of the prairie voles, James Winslow set out to see if vasopressin, so important to these animals for the mate-guarding and nest-guarding function, was also the neurotransmitter involved in monogamy. They gave male prairie voles infusions of either vasopressin, oxytocin, or a hormone-free fluid as the control and housed each male with a female vole whose ovaries were removed. The males could not successfully reproduce, of course, but the pair cuddled and groomed during the treatments. After treatment, the researchers put each pair in a cage designed to give the males a partner choice. His female partner during the infusion was tied in one area and a different female was tied to another, the lab experimenters' version of temptation. Males were allowed to freely approach either female and the team measured how much time each spent with each female. Males that got the oxytocin or the hormone-free fluid split their time fairly evenly between the new female and the "wife." Without vasopressin, the males failed to form a strong attachment to their previous female mate. Although they sometimes had a fling, the males who got the vasopressin spent 75 percent of their time with the wife, suggesting that this is indeed a monogamy chemical in males.

Some scholars believe that vasopressin plays a role in how the brain lays down memories. What the males may be doing, then, is "remembering" familiarity, which in this case breeds not contempt but pair-bonding. Sue Carter's latest studies have focused on interactions of vasopressin and oxytocin within the same receptors in the brain, and she has been able to manipulate these molecules to create strict mate choices among female prairie voles and aggression to strangers in the males. Human brain chemistry, of course, is not a direct analog of rodents, but given their 85 percent homology in the genome, it's possible, Carter says, that some neurotransmitters or the same ones are playing out their roles in bonding and bonding disorders.

Gregor Mendel's contribution to science was to show that there are strict patterns of inheritance, and that, by and large, a gene is a gene and will carry the same code whether it comes from mother or father. But Mendel was only partly right. There are genes, known as "imprinted" genes, that are expressed *differently,* depending on whether they came from the maternal genome or the paternal genome.

In recent years, scientists have struggled to figure out why such genes, which exist only in mammals, exist at all and how they work. So far, they've found that these genes appear to regulate other genes, and that they may regulate some growth genes. Most recently, Princeton's Shirley Tilghman, Harvard's David Haig, and others have discovered evidence that they may play a role in helping evolution level the biological playing field between sperm and eggs when each may be vying for supremacy during conception and fetal growth. And Tilghman has postulated that mammals' use of imprinting to regulate their development in some way may also explain why mammals are only rarely monogamous.

According to Tilghman, one of the deepest ideas in evolutionary biology is that the evolutionary advantage of an organism depends upon its reproductive fitness. "Sending as many of your genes into the next generation as you possibly can—that's the goal. Sex evolved, remember, to enhance the opportunities for reproductive fitness by putting egg and sperm in competitive positions, to let them vie, as it were, for the best assortment of complementary genes, by having the bodies and brains that house them attract and negotiate with the opposite sex's bodies and brains."

When sex evolved, one consequence was that the "meiotic drive"—the need to sort and mix genes—of males and females became disconnected, and Tilghman believes they became disconnected in part because nature did not want them to mate for life. Mating for life, she reasons, would be incompatible to some degree at least with the purpose of sexual conjugation, because in such a state, competition would be seriously reduced, along with incentives to seek out "new" genes that might confer reproductive advantages to subsequent offspring.

A second big idea embedded in evolution is that the disconnection between the sexes, the fact of sexual reproduction's evolutionary survival, must confer some advantage to the parents as well as the offspring, but must do so in a way that maintains some balance in the war between the sexes. Otherwise, if one sex came to predominate—genetically speaking—the competitive conditions would fade, eventually destroying the value of sexual reproduction and possibly the species.

In Haig's admittedly controversial view, what we got was what he dubs "fetal attraction" facilitated by imprinted genes, and in Tilghman's view, that means monogamy loses.

Here's how. According to Haig, the sexual strategy of the paternal genome is to have its genes dominate any pregnancy, consuming as many nutrients from the maternal eggs and resources as possible during gestation. Dad wants his offspring to grow as big as possible so that they can send more of his genes into the future. So, his genes are sending signals into the embryo to grow, even at the expense of the mother. That's just dandy for him, because he will produce an almost limitless store of sperm and can find, well into his old age, younger females with lots of young eggs to impregnate. He can pursue a bimbo strategy and not worry whether all of the moms survive to nurture someone else's offspring, or even more of his own.

Mom's genome, meanwhile, has an understandably different idea of the situation. Unrestricted growth would so consume her she might never go on to have other children and send more of *her* DNA into the future. She needs to compensate somehow for Dad's strategy. According to Haig, imprinting is the mechanism by which this war between the sexes is played out in our genes.

In fetal attraction, the fetal genomes—formed my meiosis, or mixing of paternal and maternal genes—will have genes from Mom and from Dad, each selected to increase the transfer of nutrients to the fetus (Dad's) and to limit transfers in excess of some maternal optimum (Mom's). If, in fact, fetal actions are opposed by maternal countermeasures, imprinting, since it occurs only in mammals and largely in genes concerned with growth, may indeed be the means to preventing a lethal escalation of the genomic battle of the sexes.

In this way, genomic imprinting reflects a conflict within fetal cells between those genes of the fetus expressed from Mom-derived genes and those derived from Dad. "During implantation," Haig writes, "fetally derived cells invade the maternal endometrium and remodel the endometrial . . . arteries into low-resistance vessels that are unable to constrict. This invasion has three consequences. First, the fetus gains direct access to its mother's arterial blood," so Mom can't reduce the nutrient content of blood reaching the placenta without starving herself. "Second, the volume of blood reaching the placenta becomes largely independent of control. Third, the placenta is able to release hormones and other substances directly into the maternal circulation."

Placental hormones, he says, manipulate maternal physiology for fetal benefit, and if Mom can't deliver, gestational diabetes is the consequence for them both. Preeclampsia or toxemia of pregnancy is an effort by a poorly nourished fetus to get more nutrients by increasing the resistance of its mom's blood circulation.

According to Tilghman, the genetic evidence for Haig's hypothesis is that genes that have been found to be imprinted are, for the most part, involved in fetal growth and that most of those that promote growth are *paternally* expressed. They're genes that say "grow big, really big." The maternally expressed genes that are imprinted are for the most part those that *retard* growth; they're going to be saying "slow down."

Corroborating evidence emerges from the fact that imprinted genes are not observed in animals that lay eggs, because once that egg is laid there is no control over the nutrients. The nutrients have all been provided before the father had anything to do with it. He can't control that.

Tilghman argues that imprinted genes are likely to be absent, at least relatively, in monogamous species, because if there are mammals that mate for life there is no need for this war between the sexes. This war is driven only by the fact that there is polygamy. And, because imprinted genes have indeed been found in humans, they make a pointed case for a polygamous aspect to our nature.

To show this, Tilghman went on a hunt for a truly monogamous species in order to document the *absence* of imprinted genes. "This was really hard. We even looked into wolves, which mammologists insisted were monogamous but in fact turned out to be just as promiscuous as other species. They just sneak around better than most."

With persistence and luck, she and her colleagues found a wild mouse called a deer mouse that exists in two populations in North America. One of the populations is very widely distributed across the North American continent, in Canada, even up into the Yukon, and through most of the Midwest and Southwest United States. The other species survives only in a small area in Florida and Georgia.

The southern deer mouse is considered genuinely monogamous, mating for life. The others are like any other wild mouse, polygamous.

Tilghman recounts that before she began her gene hunt, the conventional wisdom held that mice were what is called polygynous. That is, if you went into a barn that had mice in it and looked in one corner, there would be a bunch of females and one male. The one male would father all of the children in that part of the barn. If you then went to another part of the barn, you would find the same thing: a different group of females and one male who would father all of the children in that part of the barn. And this was, in fact, the way it was believed mice organized their society.

"The DNA revolution came and it was actually possible, like it is with human beings, to go in and establish paternity. And what was discovered in these barns would make a brothel look like a house of good repute. These mice at night were not hanging around their little game in the corner of the barn. These mice were all over the place. And not only were the females scurrying around or the other males scurrying around and mating outside their den, but even in the lab they would have multiple fathers in a single litter."

What that means with respect to supporting Haig's hypothesis is that not only is the male gene competing with the female to get resources for his progeny, but now suddenly in the wild there are going to be two sets of male genes within a single litter, competing not just with the female but with each other as well, in order to get the maternal resources for their progeny. So mice became a perfect setup for the kind of in utero competition that is reflected in imprinting.

If Haig is right, moreover, there should be *no* evidence of imprinting of growth genes in the monogamous mouse species, and there should be definite imprinting of growth genes in the promiscuous species.

In fact, that is what Tilghman found. So far so good. But to test the hypothesis further, Tilghman crossbred a monogamous mother and a polygamous father. Although the resulting embryos had growth promoter and growth retardation in balance, they could only have inherited the growth *promoter* gene from the father; so if imprinting *had* created any imbalance, this embryo should have become really big because it's got way too much growth factor, which was not balanced in any way. By the same reasoning, if they were to breed a polygamous mother and monogamous dad, they should get really tiny mice.

Again, that's exactly what happened. The mice embryos conceived with a monogamous dad were predictably very small. "These were some of the smallest mice I've ever seen, really tiny, with an average size of 12 grams. (Normal is 18 grams.) These are runts." On the other hand, the average size of the embryos from the monogamous moms was up between 24 and 32 grams. "These were rat sized, not mouse sized. These embryos are so big that the mothers can't deliver them." For Tilghman, these studies support the idea that the monogamous species has in fact relaxed its imprinting, while the polygamous one has retained its imprinting and the two now are incompatible with each other. And she believes the mice's biochemistry is simply following necessity. "The polygamous species is using energy to do this imprinting and if it doesn't need to do it will stop doing it."

It should be noted that not all experts in imprinting and growth factor genes buy into Tilghman's conclusions. For one thing, not all growth factor genes seem to obey the imprinting rules she has identified. For another, as Tilghman is quick to point out, the behavior of the two mouse species she worked with is not what geneticists call "robust." It's not easy

to breed them successfully. It could be that the monogamous and polygamous behaviors evolved *coincidentally* in these species, or randomly, and
much more work is needed to rule out such alternatives. As in most of
biology, no observation is without its exceptions, and coincidences can
indeed be just that. Nevertheless, there is substantial agreement within
the scientific community that evidence in support of her ideas bears further investigation.

One of the hallmarks of sexual behavior is that much of what goes
on under the bedsheets is hidden from all but the couple involved, creating serious disconnects between what people *think* goes on, or what
they *say* goes on, and what is *really* going on. Game theorists have developed mathematical models that shed some light on the truth.

Take the case of strict patriarchal polygyny, in which the women in
the harem are sequestered and guarded vigorously. If the sheikh goes away
for a long time and members of the harem risk infidelity (and studies
show they do), one might think that the wives would all stick together
and run around having affairs, while protecting each other from being
found out. But in fact studies show that infidelity is not more common
among polygynous households, because if one of the more reproductively
ambitious wives should break protective ranks and report another
woman's infidelities to the sheikh, there would be major hell to pay for
everyone in the form of increased vigilance and repression. In other
words, game theory suggests that "cooperation" for infidelity is risky for
all. It may sound counterintuitive, but only if the wives are mutually sus-
picious, rather than cooperative, might infidelity be more common among
the individual wives.

Under a looser and less patriarchal system, men will in general choose
to cheat to the extent they can afford the extra costs. Women will not
cheat to the degree they believe they'll be caught and punished. Under a
libertarian mating system of swingers and group sex, each partner can
choose—or not choose—to have an affair with a single person, but only
in the beginning; that is, as long as there are single people. The supply
of single people is a critical variable, but as long as there are plenty of
them, the nonmonogamous swingers do better than if they remain monogamous. They get more for their resource expenditure.

Now, if everyone has a good chance of finding single others who are
willing to stay monogamous with each other, then everyone's best strategy
is to break off relations with involved others and seek out single others.
But if everyone does *that,* then the environment changes again. When no

one will maintain a relationship with someone who's involved with someone else, the greatest number of sustainable partners is one.

The bottom line is that being cooperative will get you less stress and more undivided attention from one partner, but overall slightly less matings and at lower cost. It makes economic sense. And, when cooperation and commitment are successfully negotiated, and become a mutual strategy for couples to conduct their sexual and parental lives, the war between the sexes takes on a chronic rather than episodic mode. And it mutates into a more subtle struggle for *ongoing* power and dominance.

"Marrying up," a man or woman surrenders some power to the higher-status spouse, but relationships over time create dramatic shifts in the power struggle. In lore, legend, and experience, the real power is determined not by wealth or by the highest born but by other kinds of power, from a rapist's brute strength to a woman's wiles and her ability to conceal fertility and paternity.

Is there much evidence for these strategies in the real world? Laura Betzig has found that throughout two thousand years of Western history, a strong case can be made that polygyny, promiscuity, and harems went hand in hand with despotism, nasty behavior, torture chambers, and the rise of medieval canon law that focused almost obsessively on punishing promiscuity and demonizing sex. What such religious obsessions really reflected, Betzig says, were the economic imperatives of our sexual natures. "Canon law was really about regulating marriage and inheritance and property rights, not about sex, which to contemporaries was much taken for granted. Bastards were produced left and right, even by the clergy, but only one wife was permitted because of the powerful ruling classes' obsession with controlling who might inherit."

In just a few years of seventeenth-century British history, she argues, the power most associated in popular lore with "repression" of women's sexuality changed dramatically. "Where it all fell apart," she says, "is with the Enlightenment; when Charles I loses his head, the reformation takes off throughout Europe and the whole nature of man is reevaluated in the context of revolutionary changes in politics and economics."

The key change, she says, is the fall of kings and rise of parliaments, marked by the pivotal moments between 1642 and 1649 when Cromwell rose and fell and the monarchy was restored, but with a much weakened role. "Before that you had Henry VIII and all the wives who lost their heads at his whim. Today, we have Charles and Diana and no one takes them seriously."

The changes that took place were real enough for noblemen and commoners alike. Historical records show that household size among the nobility shrunk along the same time line as power. "The less power you had, the fewer adolescent females were dusting your tables, conceiving your bastards and running off with stable boys."

"What caused this cataclysmic change? Why in the whole of history do we have endless accounts of powerful men grabbing girls and chopping off heads and suddenly it stops? It was so entrenched. This is a constant in history. When you read any chronicle or source, the same male historians who dare to report that another male is chopping too many heads off is really saying the head chopper is sleeping with too many women for the chronicler's taste, and keeping the common man from getting his share. The first novels carried this theme as well. They described the noble aristocrat and the common girl who gets him, a theme in which the master demands sex of the servant maid and she says she would rather die a thousand deaths, but eventually dreams of marriage and gets him."

For Betzig, the answer is "blindingly simple" if you know your evolutionary biology and anthropology, in animals and in humans. "Animal societies and all societies are put together to regulate social behavior—sex, parenting, group living—and make it more *fair*. And what makes animal societies fair across all taxonomies from insects and birds to fish, mammals and humans, is mobility. If you've got big strong guys and little weak ones who have nowhere to go if the habitat is saturated, then the big ones beat them up and do all the reproducing, getting all the females, and using the little guys to help them keep and feed what they have. But when there's lots of space to move around, and lots of resources available, the little guys say go to hell and go over the hill and you get little democracies instead of wretched despots."

That, in essence, is what she believes happened in seventeenth-century England when exploration of the New World opened up and more and more "little" powerless guys were needed to man the big ships. They were the ones who told the kings and barons to go to hell and went over the ocean, instead of the hill, to New England.

Even more important than geography, however, was their mobility, because with the new mobility came opportunities for trade. There were not only new lands but new money from new markets, to be shared with the little guys who stayed in England. The industrial age burgeoned with sweatshops, artisans, textile makers, and manufacturing of all kinds. "They were amassing gobs of money," says Betzig. "Until then, trade had accounted for only 2 to 5 percent of the occupations of people on land. Now most people had access to trade and capital. Charles I lost his head when his revenues dropped and they dropped like a stone because the advance of trade meant there was more money to be had from customs fees than from feudal fiefs. But the tradesmen said to hell with the king, we're keeping our money and we're free to tell the noblemen to bug off."

Concludes Betzig, "Wherever there is exploitation of power there is exploitation of sex. It's Darwinian and automatic. When power drops, men lose women." At the same time, the female has not been passive in all this tumult, Betzig argues, but instead "coevolved" her own set of survival strategies to manipulate what power she could get to, namely his.

"Knee-jerk liberals would say the decline of polygyny was good for women, because the best of all options is for her to have sole access to the most powerful man and no competition. That doesn't happen, however, because when a man is #1, *she* has to compete with hundreds of women for him."

Thus, polygyny in this scenario might be good even for women, as long as they get enough of a powerful man's resources through succession for their children, especially their sons. That's because her biological imperatives, her selfish genes, want powerful, sexy sons to give her many grandchildren. In this view, wives didn't do badly under the old system and one can argue that women might have fared much worse.

In a piece of research that supports Betzig's historical view, James Boone of the University of New Mexico and his colleagues have investigated elite families to see if parental investment is at the root of much evolutionary biology.

In a bid to test whether elite families invest differentially in sons versus daughters, he and his colleagues analyzed the sex roles among the fifteenth- and sixteenth-century Portuguese nobility. He has looked at the idea that patriarchal family structures among elites in highly stratified societies originated as a form of parental investment favoring male children. Scouring genealogies, he found that among the higher nobilities, males outreproduced females, whereas among the lower nobility, females outreproduced males, and that the tendency to concentrate investment in male offspring correspondingly increases with status. Thus, primogeniture has the social effect of generating intense competition among males for available titles, which results in increased warfare, higher death rates among men, and, indirectly, increased cloistering and guarding of women.

Keep in mind that in those times, males, especially the firstborn, inherited all of the land and real property; daughters, if they inherited anything, got only portable assets, jewelry, and cash, and usually only once—in the form of a dowry.

Evolutionary pressures—first documented by Robert Trivers and Daniel Willard in nonhuman mammals—lead parents to invest the most resources in whichever sex of offspring are most likely to produce the most children. In patchy environments, where there are clumps of super-rich and clumps of poor, the effect would also predict different parental investments: among the rich, more in males; in the poor, more in females, because the overarching effect of the economic condition is on the male's ability to find a suitable mate.

So, under this effect, if males are really wealthy, the wealth determines the number of potential mates, and poor parents will therefore invest more in daughters who can marry up.

Taking all of this into consideration, Boone looked at four levels of Portuguese nobility, and that model tested out pretty well. Dukes, counts, viscounts: among them parents invested more in sons than daughters. In lower levels of nobility, such as unlanded knights descended from titled families but marginalized, parents invested more in daughters. (Those historical romance novels play on this theme, too. A recent one told the story of an impoverished medieval girl and her brother, who were descended from a nobleman who lost his land and castle to an evil knight. The girl's father had actually invested a great deal in her to make her educated and resourceful, as well as attractive.)

Boone, who began his career as an archaeologist working on a settlement in Morocco near the Strait of Gibraltar, had to go to Lisbon to look at some documents and got interested in what was then known as sociobiology. "I had a very good genealogical data set on Portuguese nobility and had read some literature that among animals such as marsupials, females who are well nourished will breast feed more of her male offspring than female and if she is not, vice versa."

Boone concludes that the best bet for most women in some of these situations is what might be called the Romance Novel or "Tess" Effect. More scientifically, it's called hypergamy, the practice named for marrying at least at the level of your own social status, if not higher, originally referring to Hindu women who married into their own or a higher caste. Tess of the d'Urbervilles, you'll recall, self-destructively schemes to get the rich guy (although ultimately, in a memorable plot twist, rejects him). She is willing to sacrifice almost irrationally to secure her future.

If it is helpful to study religious and cultural trends of the past, relying on the explanations offered by these same religious and cultural institutions can be misleading. For instance, the finger of the family values crowd wags at premarital intercourse, and for centuries, it seems, the theory of "sex conservation" commonly taught to boys argued that if they started intercourse too young, their semen and sperm would be diluted and wasted. But the Stone Age biology we are heir to disputes that and so, fortunately, does modern science. The peak of male and female sex drive is in the late teens, not the marriage ages of twenties and thirties common in the West today. Boys who have early puberty and begin sex activity earlier have the *highest* rate of sex activity and continue to do so to an older age. So do women, by the way, giving the lie to those who would proclaim it "unnatural" for very young adolescents to want or have sex. Whether we like it or not, it's been natural for at least forty thousand years. Douglas Kenrick and his coworkers, in a study in which they analyzed five samples of men and women to determine their "sexual peak," acknowledge the common perception that men and women peak sexually at different ages, men around twenty and women a decade later. But their analysis suggests that what men and women mean by "sexual peak" is very different, with male peaks more likely defined by men as a "desire for sex," and women's as "satisfaction with sex." So it seems that for men, the frequency of intercourse is the goal in their youth (which fits with their age-old reproductive strategies), while for women, sex is best in a committed relationship (likely to come when they are older), which fits with their programmed strategy.

Conventional scientific wisdom has held that promiscuity, particularly polygynous matings (one male, many females), is caused by the sex imbalance of the species, far too many females to males, for instance. But in fact, in all polygynous species so far studied, the ratio of males to females at birth is still fifty-fifty and this ratio persists until parenting is completed. It is probably the case, therefore, that the polygynous relationship causes the imbalance, not the other way around.

From evolution's standpoint, it makes sense that sexual selection is the dominant factor in a polygamous and promiscuous species, resulting in a greater mortality of males than females. One male can impregnate many females, thus lowering his selective value and making her more "valuable." So, susceptibility to sexual selection—in which animals display traits and functions not apparently tied directly to survival—always favors a polygynous and promiscuous system unless it puts females at a fatal disadvantage. In most mammals, including humans, polygyny, or serial monogamy, is the dominant mating system because the male, in a relative sense, is less essential for direct parental care on a day-to-day basis.

In the primate world of families, status and prestige clearly count and the mothers in these groups are using all the stratagems they can muster to give their offspring the edge. Frans de Waal, the great Dutch prima-tologist and author of *Chimpanzee Politics*, who has studied bonobos for much of his career, believes that "bonobo society evolved to reduce infan-ticide," which is frequent among chimps and generally carried out by males who are not the fathers of the murdered infants. The females defend the babies, and in order to do this they in part obscure paternity through frequent, promiscuous sex and prolonged female sexual receptivity. It's the bonobos' reverse version of Aristophanes' *Lysistrata*, in which all Greek women refused sex to their men until they stopped making war. Written in 411 B.C., the play depicts the seizure of the Acropolis and the treasury of Athens by the women who, at the eponymous heroine's insti-gation, declared a sex strike.

So where does this leave us at the start of the twenty-first century? If promiscuity is so universal among mammals, if it's so natural (and it is), why do guilt and shame so often accompany it? The same human beings participating in all that adultery are the very same ones who developed, manufactured, designed, or agreed to adhere to religious and secular moral systems and social contracts, giving support to the notion that shame, guilt, and concepts of sexual morality *evolved* just as surely as our tendency to stray. Indeed, de Waal has identified behavior traits in chimps that he calls a "sense of justice" and "animal morality." "These animals are sensitive to social rules, help each other, share food and resolve con-flict to mutual satisfaction. Because of the quid pro quo that regulates their behavior, they can even be said to have a crude sense of . . . fairness: the right of the strongest has given way to mutual obligation." In his book *Good Natured: The Origins of Right and Wrong in Humans and Other Animals*, de Waal reports that dogs can look guilty and gorillas will put themselves at risk for a wounded mate or to save a human child, as one did several years ago when a youngster fell into a zoo gorilla's habitat. As with so much else, de Waal says, human morality could not have developed as some unique aberration of our neocortex, but evolved along with everything else. "Natural selection is indeed a harsh and competitive mechanism, yet it has produced successful species that rely on cooperation and mutual assistance for survival."

Essentially, it's a you-scratch-my-back-and-I'll-scratch-yours biology for humans and other animals. The reasons may not always be obvious to us, but there are compelling examples of self-sacrifice, and of helping the next guy or the next generation at the expense of oneself, even when it comes to sex.

In the extreme are such examples of self-sacrifice as redback spiders, whose males don't just risk death to have sex, but do acrobatic feats to make sure death occurs! The males first dance around the web for hours, then spend some time climbing on and off the much larger females' egg-swollen abdomen. After nearly eight hours of such courtship, the male uncoils one of a pair of globular structures from his head and inserts this palp into one of two openings in the female's belly. Then he pivots himself around the palp to offer his abdomen to the female's fangs. While she literally chews and digests her erstwhile mate, he transfers sperm for up to twenty-five minutes in a display of sexual suicide by voluntary cannibalism. Conventional wisdom has held that this might be a case of a male's offer of his body as a "nuptial gift" to nutritionally supplement the female that will carry his progeny. But in the case of the redback, the female is so huge in comparison to the male that, as one writer put it, "he hardly even counts as a square meal." A singular act of altruism this is not. It turns out that it's another case of overcoming sperm competition. The male who gets to father more eggs is the one who can prolong sperm transfer. His willingness to be devoured is his way of keeping Madame Redback distracted and happy with his palp instead of someone else's. Not all males die during mating, but the ones that survive parent far fewer eggs. And in one study, twenty-two of twenty-three females whose male mates survived sex went on to mate with another male fairly soon. It also turns out that after one mating, the tip of the palp breaks off, making the male functionally sterile in that organ and giving him not much motivation, biologically speaking, to try sex again or survive. What's best, it seems, for such males, is to give his life to his offspring.

Such traits are found in some fashion in many animals and even in a colony of bacteria known as *Bacillus subtilis*. According to Eshel Ben Jacob, each *B. subtilis* colony's requirements are satisfied by a genome that ensures that each member of the colony has the right shape and communication system so that the entire colony can, when needed, switch directions to find food and compete for it. When experimentally scientists stopped feeding the colony, the stress impels ten thousand groups of cells to uniformly start foraging for food, not in an "every bacillus for itself" way, but with built-in search patterns that make sure each group explores a different piece of the landscape and shares what it finds so that the whole colony survives. Like bacterial colonies, according to some biologists, humans and many animals are, in addition to being individuals out to maximize their own self-interests via sex, "complex adaptive systems" that can and will, when it serves their needs, cooperate and even sacrifice for each other.

Whether or not promiscuity is "economically" sensible, it is certainly natural and constrained mostly by social rules, which more or less repress our natural sides. April Gorry of the University of California at Santa Barbara once conducted a study of "sex tourism" among white tourist women vacationing in tropical locations such as the Caribbean, East Africa, Indonesia, and Greece. After field research (we can only guess) of three months in the Caribbean, she found that tourist women tend to engage in more promiscuous behavior than they would at home, taking one or more lovers in the span of a few days, choosing men of lower status than themselves, providing payment for "romantic services," and prioritizing male appearance and reputed lovemaking ability over the mate-selection criteria. They show a strong preference for dark-skinned Rastafarian men who they describe as "manly," exotic, primitive, and taboo. The women attribute their behavior to the effects of a novel environment where the opportunity for romantic and sexual relationships is plentiful and their actions will only temporarily affect the way they are viewed by others.

In the face of guilt and consequences, why would women be promiscuous? Part of it clearly is that the rules are nonexistent, or if broken won't have negative consequences, at least in the short term. It's also possible, or even likely, that women, like men, seek novelty in their sexual relationships and use such relationships to "test" and "retest" their own judgments about the opposite sex, their own "sexiness," their fertility, and their future or past choice of permanent mates to father their offspring. Clearly, they may also be sowing oats they forswore in order to make the necessary trade-offs to produce good offspring who are well-provisioned by men with resources and status. It's also possible that this is atavistic behavior, a throwback to natures that evolved when resources or competition for the best males might have been such that it was in the female's best interests to have sneaky sex with someone not her husband, either to achieve an optimal pregnancy or to thwart the prospect of offspring with men who were less than optimal DNA-wise, but who had the women in their grips nonetheless.

Pascal Gagneaux of the University of California at San Diego, along with David Woodruff and Christopher Boesch of the Basel Zoological Institute in Switzerland, have discovered that among chimpanzees, fully half of all babies are conceived by such "sneaky fuckers" when females sneak off for risky trysts with males outside their own social groups. "When they can get away with it, they sneak off and try to expand the pool of possible fathers," said Gagneaux of their studies in which they painstakingly worked out the genetic family tree of a group of chimps in the Tai Forest of the Ivory Coast of West Africa.

Between 1991 and 1995, he and his colleagues collected DNA samples from all fifty-two members of the troop. Paternity tests yielded the surprising results. Because of the animals' ferocious territoriality, such "extragroup" couplings might have a practical benefit. A female that has such a tryst creates the possibility that her offspring may be related to a neighboring male. That might lead the neighboring male to show mercy in a future encounter with her or her children. Clearly, says Gagneaux, "females are not some sort of resources just waiting around like fruit to be picked. They have their own agenda." Indeed they do.

What do we make of "swinging"—a modern version of promiscuity that rarely if ever results in pregnancy? Have swingers truly put aside the evolutionary pressures that attract men to fertile women, and women to powerful wealthy males who are ready to invest in their joint offspring, as the price of sex? Leanna Wolfe of Antioch College says that in fact swinging and polyamory give people "ethical" ways to access multiple sex partners. Male swingers can be like alpha bull seals with their harems of willing ladies, while female swingers can be like queen bees with stables of partners. Thus, swinging may achieve the sensation of an effective Darwinian reproductive strategy without the outcome. And since swinging doesn't tend to attract people who are already rich, powerful, and successful from nature's reproductive standpoint, those who do participate get to play out their "royalty" fantasies without the financial and social costs. "With the absence of both prestige and progeny, swinging cultures fail as true reproductive strategies," says Wolfe. These people haven't been willing to invest in true courtship and mating strategies and are in truth "pretenders" to the throne of reproductive dominance.

Joanna E. Scheib has suggested that women trade off their choices of mates using criteria in the context of both long-term and promiscuous relationships. In long-term mateships, women may benefit by pairing with men who provide high-quality gametes, expressed in children as increased survivorship or superior beauty or "sexy sons," and who give benefits to the mateship such as material resources. If adulterous pairings were a consistent part of women's histories, then mate choices could be expected to offer sufficient benefits to females and outweigh potential costs or penalties imposed by the "husband," such as physical injury or paternal investment. She argues, from her studies using photographs and detailed descriptions of handsome and not so handsome men, that if adultery for women functions as a beneficial way to attract quality gametes, then adulterous women would be more likely to choose for such liaisons really good-looking men, because such pairings would have to be sneaky, fast, and she'd have to rapidly assess his "fitness" in the same way that males mostly assess a female's "fitness"—by looks.

One could argue that Scheib's study was unnecessary, but it confirmed clearly what the tabloids tell us every day about movie and other pop stars. If they fool around, it tends to be with even better-looking men and women than they are married to.

One reason may be that a person's willingness to engage in adultery or swinging depends in some part on what the group around the individual is doing, suggesting that in the game of life, it doesn't pay to stay behind. The train is moving on and if you want to get to your destination, you might have to ride with people—or behaviors—you don't always like. Bram Buunk and his colleagues in the Netherlands say evolved gender differences in sexuality suggest that among men, more than among women, the perceived prevalence of extra-pair copulations would be strongly related to one's own willingness to engage in these activities. His surveys show that among men, but not women, exposure to a message that 47 percent of the population had engaged in this behavior led to a higher inclination to do it, as compared with exposure to a message that only 3.8 percent had done so. Again, there may be very important survival genes involved in such "going along to get along" behavior.

It might sound oxymoronic, redundant, or Clintonesque, but there are prices to pay for inappropriate promiscuity, not necessarily promiscuity in general.

As we've seen, the physiology of human reproduction is marked by a numbingly complex series of biological and chemical events organized by various molecules in the brain and body. Today, we exploit knowledge of these mechanisms with chemical contraceptives that alter those molecular signals and with fertility treatments—for men as well as women—that mimic them with respect to ovulation, sperm concentration, sperm motility, uterine competence, and so on.

In terms of commitments to parenting, long-term relationships, and at least serial monogamy, fertility is a key for many couples. We've long since given up the practice of lopping off the heads of consorts who can't produce an heir, but fetal loss and differences over number and spacing of children are issues that determine whether relationships are transient or long term. Genes that cause birth defects or genetic fragility are the targets of gene hunters, presumably to enable "better" choices for those considering long-term mates and childbearing. Scientists are learning the relationships between the age of sperm and the age of eggs in producing healthy versus less healthy fetuses. Increasing emphasis is being placed not only on the maternal factors that can induce fetal loss and birth defects (older mothers, smoking, alcohol) but on the paternal ones as well, including smoking, alcohol, occupational or environmental hazards, and stress.

In recent years, scientists have found evidence that male primates alter their mating relationships, their reproductive "strategies," becoming more or less conservative or more or less daring, depending on whether females' reproductive rates are slow or fast or whether there is enough food or mates to go around.

Others studying humans have found that stresses within families can alter hormonal functions to the degree that male hormones become more sensitive to the presence or absence of a father than to female hormones. Boys who grow up without a father have low cortisol levels during infancy and high or defective cortisol profiles during teen years; they also have high cortisol and low testosterone levels during adulthood compared with those of males raised with Dad at home. This may possibly be "adaptive," according to Jane Lancaster, in that reduced testosterone levels lower energy costs of muscle maintenance in subsistence populations where males are often gone or die young. "The family," notes Mark Flinn and coworkers who conducted the testosterone–family stress studies, "is the most critical source of physical and social resources for children," and the response to stress may be a way for children to allocate energy to a variety of body functions, including sex.

If our physiology informs our family commitments and parenting strategies, it is likely that it also underpins some of the emotional glue that keeps us negotiating our status within the context of close relationships. Jealousy comes to mind. David Buss, as well as Todd Shackelford of the University of Michigan, have looked into the biology behind the high divorce rate in the United States—about half of all marriages. Specifically, they focused on behaviors that appear to lead up to divorce, or separation of a bonded pair. In scientific language, they looked at "mate retention," "mate guarding," vigilance, and jealousy as markers for these behaviors, finding that patterns in married couples that also are found in animals as diverse as spiders and birds are powerful clues to the fact of ancestral nonmonogamous mating.

In one study, for example, they found that men's but not women's mate-retention efforts are a function of the partner's youth and beauty, regardless of how long the two have been married. This fits with the biological propensities men have to become attracted to youth. Women's mate retention in their sample of 214 married individuals, although less predictable than men's, was, as evolutionary biology would predict, linked more with the effort allocated by their husbands to the problem of their status and success as it related to others. Overall, men reported using more resource display, vigilance, and threats against their wives to retain them. Women reported using more cosmetics, plastic surgery, emotional manipulation (guilt), jealousy induction, and derogation of competitors.

Others have investigated the phenomenon of "mate stealing" and its evolutionary and biological implications. Several surveys of married and dating couples, for instance, have found that among men, the choice to fool around or engage in what scientists would call "extra-pair copulations" is more strongly related to a *female's* willingness to do so than is a male's willingness among women. In other words, males looking for promiscuous sex are more likely to approach a female who is promiscuous herself, thereby increasing their chances of having sex. Unfortunately, the secretary out to get your spouse has an edge. Social scientists at the University of Groningen in the Netherlands, for example, in studies with eighty-four men and women, showed that males were not only more motivated to have sex with physically attractive females, but also they perceived such females to be more promiscuous than unattractive ones.

This was true even though they also thought it riskier with such women because of the risks of AIDS and other sexually transmitted diseases.

As described in earlier chapters, the emotional and motivational mechanisms that mediate sexual arousal and attraction differ between men and women because men's and women's reproductive strategies must be different. Recent studies suggest that even when modern, independent, contracepting women choose short-term sexual mating strategies, in order to test, for instance, the long-term interests of any given guy, their innate preferences for a husband as opposed to a casual lover will win out. In one survey of 243 male and 298 female college studies that included interviews with twenty-eight highly sexually active female college students, the results showed that even when females voluntarily had casual sex, they felt anxious about their partners' willingness to invest in any future offspring. And the more partners, the worse these women felt.

Toronto's Nicholas Pound has developed a taxonomy of mate retention and tactics, evolutionary models that predict such things as harems and genital mutilation. "Males," he says, "have been very creative in coercively controlling females to achieve mate retention. But women have tactics, too." Among those used by both sexes are vigilance, concealment, violence, emotional manipulation, derogation of competition, spending money, seduction, submission, apologies, promises to change (mostly by males), derogation of mate to competitors (what a "bitch" she is; what a "jerk" he is), threats, stares, and of course physical abuse.

Studying married couples and comparing what both subjects and their spouses separately reported, along with interviews of the couples together, Pound attempted to link the degree to which one suspects a mate of infidelity and the use of mate-retention tactics. The results, in the case of a one-night stand of infidelity, were more or less predictable: Males who suspect their wives of infidelity gave a lot more effort to mate retention; but women whose mates indulged did nothing. In fact, the only time women consistently increased their mate-retention tactics was when they suspected their mates were devoting a lot of time—with another woman or not—to achieving higher social and economic status than to the wife. Evolutionary biologists argue that jealousy is adaptive, that its almost universal appearance in higher-order animals is proof that it promotes mate retention and therefore relationship stability. Social psychologists, on the other hand, argue that jealousy is maladaptive, that it pulls people apart instead of together. Rather than refer to benign little green-eyed monsters, they reference Tolstoy's murderously jealous man in the Kreutzer Sonata, who is aroused to homicidal rage by the sight of his wife accompanying a young violinist at the piano. Bram Buunk and Ralph Hupka of the Netherlands gave questionnaires to 877 men and 1,194 women studies twenty to twenty-one years old in Hungary, Ireland, Mexico, Netherlands, the United States, USSR, and Yugoslavia to look at how common jealousy is and therefore how adaptive it might be from evolution's standpoint. The items covered jealousy related to flirting, kissing, dancing, hugging, sex, and sex fantasies. "Results indicate that for nearly all subjects, kissing, flirting and sexual involvement evoke a jealous response, while dancing, hugging and sexual fantasies on the part of the sex partners do not."

As every movie treatment of the subject probably attests, men are more jealous over their partners' physical infidelity, while women are more jealous over their men's emotional attachments to other women. Shackelford and his coworkers at the University of Texas tested 214 married people to assess the usage of nineteen mate-retention tactics, ranging from vigilance to violence. Men but not women's mate retention was linked to the partner's youth and physical beauty; women's but not men's mate retention was linked with the partner's income and status striving. Men were more jealous if they perceived a partner's infidelity and men more than women used gifts, submission, debasement, and threats to retain their mates—while women used "appearance enhancement" and verbal signals of possession ("You're mine!"). Heather Claypool and Virgil Sheets of Indiana State University surveyed college students in a romantic relationship regarding jealousy when they started a semester and resurveyed them at the end of the school year to assess the status of the relationship they had in the fall. They found that the more jealous someone was, the more unstable the person expected the relationship to be, but in fact the jealous behavior tended to elicit reassurances and different behavior in the partner, thus stabilizing the relationship—a finding more consistent with the idea that jealousy is indeed programmed into us as a means of continuing the negotiations through marriage to keep a mate, when original infatuation and lust have long flown the coop.

Because men relate a woman's beauty to her fertility, while women link a male's dominance to his worth as a mate, and because jealousy is triggered by real or imagined characteristics of a rival that we perceive is important, it follows that if jealousy in males is biologically wired, it should be influenced by the rival's dominance, while jealousy in females would be influenced by a rival's physical appeal.

To test this, Bram Buunk and his coworkers conducted an experiment in which participants were presented with a scenario in which their current real or imagined partner was flirting with an opposite sex person. They then gave them one of four profiles of the flirt with photos: low attractiveness with low dominance, low attractiveness with high dominance, high attractiveness with low dominance, high attractiveness with high dominance. And in keeping with jealousy as hardwired, "females exposed to physically attractive rivals reported more jealousy than females exposed to average looking rivals. Males exposed to handsome rivals did not report more jealousy. But males exposed to powerful, dominant males reported more jealousy than to rivals who were failures, while females exposed to rivals high in dominance did not report more jealousy." It's probably safe to bet that whether we're talking about birds, insects, or people, jealousy, like every other sex-related behavior, has its uses and its costs, depending on its intensity and the situation at the time. The willingness to expend a lot of energy on jealousy is finite, among desert sheikhs with harems or Wall Street bankers with mistresses. But the willingness nevertheless is abundant.

Is married sex always a formula for fading passion? Do old (or older) married couples fit the humorous stereotypes of hair nets and hairy armpits in place of honeymoon negligees and taut bellies, and roving eyes instead of commitments?

There are hints that in the course of sexual relationships, personalities will have an impact on the persistence of romance and sex, or its disappearance. In marriages, there are novelty seekers and sensation seekers, and there are less thrill-conscious sexual partners. And the "best" (i.e., longest-lasting) marriages tend to be made up of partners who are well matched, whose temperaments are more similar than dissimilar, and presumably whose genes and hormones played a role in their getting together in the first place.

In one study of sensation seeking in female sexuality, scientists compared samples of high and low sensation seekers on frequency of sexual intercourse, marital satisfaction, sexual desire, arousal, assertiveness attitudes, and satisfaction. Although the high sensation seekers showed a lot more sexual desire, were more easily aroused, and liked sex better, there were no differences between the groups in frequency of sexual intercourse or sexual assertiveness.

Recent studies suggest that only a minority of animals actually "divorce" because so few form lifetime pair-bonds in the first place—about 5 percent. But in those that do form monogamous relationships, the divorce rates range from a few percent to more than half. The key, among lower animals, is predictably related to resources and habitat. Where the pickings are slim, divorce rates soar, each member of the pair driven to forage for himself or herself. Where there is abundance, the split rates are low.

But even among animals, it's fair to ask what possible benefit a divorce would bring. Wouldn't one or both at least be better off "economically" if they stayed together? To study what benefits there may be for birds in a divorce, Andre Dhondt of Cornell University counted the number of offspring of divorced male and female blue tits and found that after separation, females improve their breeding success, but males don't. Female chickadees divorce to get males with higher status, according to other scientists, and kittiwake gulls divorce when a particular pair can't hatch at least one egg. Some birds after they divorce find it so daunting to find a new mate that they stay "single" for years.

The parallels to human species are eerie, but far less is known about why people divorce these days at such a high rate.

In her work documenting not a seven- but a four-year itch, Helen Fisher proposes the existence of a biochemical foundation to adultery, taking her cues from the patterns of divorce and philandering she has investigated in animals and humans in dozens of cultures worldwide. Data on divorce between 1947 and 1989 in the UN yearbooks of sixty-two peoples show similar patterns regardless of the actual divorce rate: among these, divorce risk peaks in age categories twenty to twenty-four and twenty-five to twenty-nine, and remarriage peaks from twenty-five to twenty-nine and thirty to thirty-four. "This pattern of cyclic pairbonding during reproductive years is governed by neurophysiologic mechanisms in the limbic system of the brain that generate the emotions for attachment and detachment and these neural systems evolved one by one to predispose ancestral hominids to bear and rear young with serial mates."

Fisher says marriages that stay strong through the first four or five years have a much greater chance of staying stable than if they wobble in the first years, when the couple is having kids and spending a lot of time taking care of children instead of bonding with each other. While humans seem programmed to be monogamous, it is more of a serial monogamy, in which we stay together long enough at least to beget a baby and get it to the point of school.

Roger Sullivan and John S. Allen, New Zealand anthropologists at the University of Auckland, are happier with six-year itches, rather than

Fisher's four or the fabled seven. Studying supreme court statistics of divorce in their native land, they analyzed statistics over a sixty-year period to test the idea that men will apply for and be granted divorce on grounds of adultery at a higher frequency than women. Significantly more men than women are found to have gotten a divorce on grounds of adultery; however, it is believed that the cause is not a greater preponderance of adultery in New Zealand women, rather a gender-specific sensitivity to cuckoldry.

They also analyzed statistics over a twenty-four-year period to test the idea that women will apply for and get divorces more than men and that many more women than men are found to have been granted divorce by court order and protection from domestic violence. Their findings with regard to the duration of marriages over a twenty-five-year period support the idea that six, not four years is about the point of natural and cultural exhaustion for a marriage.

One explanation for the consistent limits on a couple's staying power—whether four, six, or seven years—may be in the nature of biochemical baths and our body's use of them. If every time a biochemical hormone or neurotransmitter was recruited—say for fight or flight, for cuddle or battle, for attachment or fear—it went on without brakes, we'd soon be literally poisoned by chemicals and rendered paralyzed and soon dead in the doorway. Instead, our bodies and brains have evolved receptors for these chemicals that literally "fill up," and when they are filled to capacity, they flash a "tilt" sign up- or downstream, using chemistry or simple mechanical overload signals to feed back to the source and say "Stop! I've had enough." This is essentially what happens when we are full of food; we feel "sated" and not hungry anymore. At least for a while.

The same, so the argument goes, may be true of love and attachment. After the arousal and mating chemicals that saturate our bodies and brains stop firing, our libido and our limbic systems are dampened for love. We are tired of responding. At this point, we hope endorphins, the attachment chemicals, will carry us over the tide of ennui, along with the cultural reinforcers of Dad with a shotgun or a cute baby's eyes. But not everyone's chemistry is in it for the long haul; not everyone can wait it out and let small absences from each other's sexuality and empathy trigger deep needs for each other again. Some people have brain chemistries and genes that lead them to seek novelty more often and more intensely than others.

And so we go to adultery, sneaky sex, classical "tune-outs" of our spouses and lovers, and sometimes divorces. Even though the process is usually painful to both parties, we seem compelled to engage in it, whether the end result is a formal legal split or a "loveless" marriage that lasts decades. We pull away from the individual who once represented our romantic and biological destinies. We go from passion and past contentment to displeasure and dissatisfaction, accompanied frequently by the urge to seek greener pastures.

One clue to such urges emerges, ironically, from studies of happiness, a subjective sense of well-being that, according to molecular biologist and behavioral geneticist Dean Hamer, nevertheless can be measured on psychological scales of optimism, pessimism, anger, and contentment. "Obviously, at least one component of happiness is the capacity to see the future as a positive thing, something that will bring rewards rather than punishments," he says. "Most people think they're happy on countless surveys even though it's clear to others that they are deceiving themselves about it. Something in us wants us to be happy. Evolution has conserved this desire. From there it's a short leap to suggest that from a biological perspective, we evolved some physiology to keep us *wanting* happiness and pleasure in spite of all evidence that it's elusive, even during times of famine and war, and that we *seek happiness* in others. It's quite likely that just *pretending* we're happy makes us more appealing to others, including potential mates. Depressed people don't attract as many mates, perhaps, although as it turns out anxious, novelty-seeking people may attract more."

From a biological standpoint, we are still prehistoric, trying to stay alive as either hunters or gatherers. If a man had to be a hunter to survive, he would certainly have evolved certain biochemical assistants (hormonal, brain chemical, etc.) to make sure that he was a problem solver, that he was not only willing but on the search for novelty and adventure; that he was more Vasco da Gama than Grandma Moses; that he was not content to sit by the fire and starve but eager to slay the dragon.

On the other hand, if he were a gatherer, he would soon have some ideas about agriculture and cultivation; he would, if he were the most innovative of the group, begin to sow, plant, and reap, building ever larger families because he could feed more of them more efficiently. He would be a problem solver of another kind, but nevertheless bold in his conviction that he didn't need to spend his days in the treacherous pursuit of a wild animal to feed his growing clan. From here it's a short intellectual hop to understanding that innovative people, adventurous people, survived more often and in greater numbers than more passive, unimaginative people—or at least that the former were the leaders, the alpha males and females. And from there, it's not hard to imagine that *Homo sapiens, Homo modificans,* applied whatever physiological mechanisms made them the major world players on the human evolution scene to that most central and essential of all human traits—sex.

Along with the committed, it seems, the novelty seeker, the adventurer, the *promiscuous* were fated to survive in our genomes, our biochemistry, our neurobiology, our hormones, and our minds.

11

BAD SEX

*"Adaptation is a special and onerous concept that should be used only
when it is really necessary."*

—EVOLUTIONARY BIOLOGIST GEORGE WILLIAMS

As befits a species that puts rape-across-the-continent "romance" novels
on best-seller lists, makes a heroine of a hooker, finds apache dancing
sexy, enjoys John Wayne spanking Maureen O'Hara as wide-screen Tech-
nicolor howl, and declares marital rape a judicial oxymoron and infidelity
a cause of justifiable homicide, humans are endlessly conflicted about and
aroused by the links between erotic sex, aggression, coercion, and pain.
We chemically castrate pedophiles and jail fathers who have sex with their
daughters, but wink at old men who marry teenagers, market sexy bikinis
to preteen girls, and even hold beauty pageants for six-year-olds.

How has something so fundamentally life-sustaining as sex, marriage,
and parenting also given us rape, epidemic domestic violence, incest, child
abuse, prostitution, and perversion? The better question might be, "how
could it not?"

As the story of sex in these pages recounts, male and female have
different and often conflicting biologies and reproductive "strategies" that
evolved to protect and promote their hosts and their interests during
courtship, mating, and child rearing. Women get pregnant and men don't.
Women have wombs that contract to permit parturition. Men don't. Men
produce, on average, levels of testosterone ten to twenty times higher than
most women to fuel their strength, sex drive, impulsivity, and daring. Men

have to solve the problem of uncertain paternity to know where to invest their resources. Those who didn't evolve the mechanisms to do so aren't with us anymore, and neither are their offspring.

Darwin's second evolutionary theory, that of sexual selection, predicted that differences in sexual behavior between males and females will always mirror the strategies each sex finds most useful in the mating game. Modern science has begun to fill in Darwin's theories with evidence that the emotional, cognitive, hormonal, genetic, and physical traits of each sex are different precisely in those ways that represent adaptations to the different problems each faced in producing the next generation of their DNA. And most recently, scientists have amassed evidence that these same "natural" biases may trigger not only "normal" sex but also bad sex. Indeed, because sex *is* so essential to survival, men and women are likely to be in peaceful, easy balance only in those biological, mental, physical, and cultural domains that confer no particular reproductive advantage or disadvantage (both have sweat glands and tastes for sugar, for instance), and are less likely to be so in those domains that are principally part of sex. "To an evolutionary psychologist," David Buss writes in *Sex, Power, and Conflict*, "the likelihood that the sexes are psychologically identical in [those areas where they have faced different evolutionary conditions] is essentially zero."

"We are all descendants," Buss goes on to say, "of a long line of ancestral men whose adaptations led them to behave in ways that increased their likelihood of paternity and decreased the odds of investing in children who were putatively theirs but whose genetic fathers were other men. . . . We are all descendants of a long and unbroken line of women who successfully solved the challenge (of securing a reliable or replenishable supply of resources to nourish them through pregnancy during scarcity), by preferring mates who showed the ability to accrue resources and the willingness to channel them to particular women." Or, as Buss's colleagues Martin Daly and Margo Wilson have stated the case, what survives over generations "are successful *attributes*, not individuals. If humans and their animal relatives display certain attributes, the one absolutely positive thing you can say is that they were 'designed' by historical circumstances and genetic variation to contribute to our ancestors' ability to 'outreproduce' other members of their societies in certain environmental circumstances."

From this foundation, it's not hard to imagine how sexual strategies, under the right kinds of stress, could convert the "cold war" between the sexes into a hot one that gave the perpetrators and collaborators some reproductive edge, while incidentally causing casualties—among adults and between adults and children. In such a context, conflict is not only natural but inevitable; and there is growing evidence that marital conflicts and abuse, incest, sexual harassment, rape, prostitution, and pedophilia are in some important ways biological and physiological echoes of selection.

In fact, just about every creature studied by scientists, and every human society, demonstrate sexual aggression and coercion. The adaptive reproductive fitness strategies men and women "use"—subconsciously and physiologically—boil down in many ways to a case of reluctant women and eager men, of warlike men and milder but manipulative women. If that is the field on which the battle of the sexes is waged, then inevitably this leads to differences of opinion about the timing, frequency, and conditions of intercourse and conception, and differences of opinion about every aspect of sex.

Without question, however, human beings' large brains also evolved to create minds able to modify or overrule many instinctive behaviors. Perhaps the premier accomplishment of evolution, as UCLA psychiatrist and anthropologist Walter Goldschmidt has noted, "has been to render us capable of suppressing . . . negative [sexual] impulses in favor of an equally important inherited quality called love. It has long been known that social mammals require the licking and petting that mothers give to their infants and that without such expressions of affection . . . the nerve cells [to the brain and in the brain] will not mature. It is this need for affection . . . that has helped to establish the comparatively high degree of social interaction regularly expected of human beings and that has laid the foundations for the uniquely human trait we call culture. And it is culture, finally, that prompts people everywhere to behave in ways designed to counterbalance the beastly behavior of the human animal."

But inherent in Goldschmidt's scientific logic is the fact that the "beastly" remains in the first place, that it is there to require culture's counterbalance. Thus, many sex-related actions must be governed at least in part by inherited penchants and preferences, inborn temperaments, and persistent physiological or hormonal biases in how we assess conditions and make choices.

Several recent popular books—notably Randy Thornhill's and Craig

Palmer's *A Natural History of Rape*—have captured imaginations and attention by emphasizing the biological and evolved psychological foundations of rape, incest, and other forms of bad sex. Some feminist tracts have argued that evolution had nothing much to do with it at all. But drilling deeper into the evolution of sexual behaviors, scientists who carefully examine *all* possible interpretations of the evidence consider the polarization of views unfortunate and unnecessary. The truth will likely reflect a complex range of perspectives on human behaviors that hurt or offend so many, corresponding to a complex range of behaviors themselves. Stephen Emlen of Cornell University feels "there has been a paradigm shift" in our thinking about the impact of nature and nurture on behavior. "We realize now that what really happens is most organisms have in their repertoire the ability to behave altruistically, selfishly, and spitefully, in a variety of ways in different contexts. And what gets 'selected' for in evolution are the *decisionary* rules, the neurological mechanisms that *influence* or *predispose* the individual to express certain behaviors and not others in certain social contexts with certain other participants. That's what gives us what we see of social behavior with all of its incredible plasticity and flexibility."

What this view translates to in the real world is that rapists, for example, appear to have a very different set of responses than non-rapists to everyday sexual signals and cues. Tests suggest these men perceive signals that for them alone have sexual significance. But that neither makes rapists "typical" nor, simplistically, victims of their biology. To make either of these cases, it would be necessary to "explain away" the fact that a lot of common "bad" behaviors aren't very adaptive, such as cigarette smoking. And it would be necessary to explain away the fact that the vast majority of men don't rape or engage in incestuous relationships with their stepdaughters.

What the more polarized investigators would consider the ultimate payoff—the discovery, say, of a discrete number of genes or chemicals substantially responsible for rape or violence—remains elusive and likely does not exist at all. (Some of the search has been hamstrung by outbreaks of political squeamishness, notably the cancellation in 1992 of the University of Maryland's planned conference on "genetic factors in crime," including rape. Then National Institutes of Health director Bernadine Healy, a conservative, knuckled under in part to feminists and other social critics angered by any suggestion that violence against women may be "natural" and yanked the conference's government funding.)

But with the new paradigm in mind among the most widely respected scientists, the search for the biological roots of sexual violence, incest, rape, and other forms of "bad" sex has been invigorated in recent years by advances in molecular and behavioral genetics, and in biochemistry. And there are discoveries to report.

One that helped scotch the University of Maryland conference came from Gerald L. Brown, a psychiatrist and clinical director of the National Institute on Alcohol Abuse and Alcoholism, who has studied the neurotransmitter serotonin. Among other things, serotonin regulates sleep, sexual behavior, appetite, and impulsivity, putting a "brake" on behaviors that swing too far from the basal level. In 1979, Brown and his colleagues had made the first link between low levels of serotonin and aggression in American soldiers. The relative absence of serotonin "disinhibited" behavior of all kinds, including violent and sexual behavior. Overall, he found serotonin is 20 to 30 percent less *available* in the brains of men than in women. Its level is higher in newborns, lower in teens, and rises again as we age.

Another advance in this area was made in 1993, when a mutation that disturbs serotonin levels was found in a large cohort of men of a single Dutch family with a history of violence, although this family also had overall low IQ. And a few years later, the Dutch family story took on new significance because of a group of mutant mice bred at the Pasteur Institute in Paris known as Tg8. The mice were first bred in 1994 by Olivier Cases and his coworkers, who were trying to develop a novel gene therapy by injecting a one-celled embryo of a special lab strain of blind mice with a foreign gene. They hoped to get a mouse pup strain with strong immune systems, but instead they got a strain of males with really nasty dispositions. The mice first bit the researchers, and when put in the same cages, the male transgenics bit and clawed each other to shreds. When males and females mated, the males more or less "raped" the females and the females screeched in protest at the male brutality. The French researchers recalled the Dutch family story in which four generations of males were either suspected or convicted of rape, deadly assault, and arson. These males were missing an enzyme called monoamine oxidase A, or MAO A, which breaks down neurotransmitters, such as serotonin. Lacking the product that can break down serotonin, these men had higher levels of the chemical available for uptake by receptors in their brains. Although the Dutch mutation is a rare one in humans and in no way explains the violent sexual or other behavior in most people, it cleared the path for further study of how imbalances in brain chemistry might be involved in bad sex. The French scientists recruited Jean Chen Shih of the University of Southern California, a renowned expert on MAO enzymes. She tested the Tg8 mice and discovered that indeed they lacked the gene for the MAO just as did the Dutch men.

Randy Nelson, a scientist and behavioral psychologist at Ohio State University, called the finding one of the first to document a biological root for aggression in animals, including sexual aggression, and further studies published in 1996 showed that the serotonin defect actually affects the structure of mice's brains beginning in fetal life and extending afterward. In an article in the *Los Angeles Times* describing the Tg8 mice, Nelson points out that "anybody who knows the difference between a pit bull and a Labrador retriever knows that aggressive behavior has a genetic basis."

Along similar lines, studies by C. F. Ferris and his colleagues at the University of Massachusetts Medical School in Worcester revealed a role in sexual and other violence for other brain chemicals. They injected dominant male hamsters with a drug that blocks arginine vasopressin (AVP) receptors in the back of the animals' hypothalamus. AVP is associated with the ability of dominant males to "flank mark," that is, to bite or nip flanks of subordinate males, letting them know who is boss, particularly in competition for females. When AVP was blocked, the scientists could provoke an immediate bout of flank marking by usually subordinate animals and could turn on or off such dominant or subordinate behavior by alternately blocking and releasing AVP.

Because AVP receptors in the hypothalamus are testosterone-dependent, studies of castrated hamsters suggested that the reduction in aggression could be due as much to a loss of AVP as of testosterone. The investigators also found that serotonin antagonizes AVP activity in the central nervous system; giving the hamster the equivalent of Prozac inhibited the AVP-induced, unprovoked aggression as well.

Another neurotransmitter, nitric oxide, implicated in sexual violence, along with the gene that codes for it, was identified at Johns Hopkins University in 1995, when scientists first bred male mice in which they biochemically "knocked out" this gene. The mice are not only vicious to other males, even murdering them, but also relentlessly sexually aggressive to females. The males repeatedly mount the females even when the females are not fertile and even when the females scream in protest and run frantically to get away from the males. "When you look at them," said Lasker Award–winning neuroscientist Solomon Snyder, one of the investigators who produced the mouse, "you can't help drawing an analogy between what they do to female cage mates and what we would call in humans, repeated brutal rape."

Snyder and Randy Nelson, who, while at Hopkins, conducted some behavioral tests on the nitric oxide (NO) knockout mice, speculate that nitric oxide, which in some forms, as we've seen, paradoxically plays an important role in male erection and orgasm, may serve as an inhibitor or motivational brake on impulsive or out-of-control behaviors of many kinds. Thus, a *lack* of the chemical leads to wild and aggressive acts. (It's a matter of some molecular irony, perhaps, that it's a gush of NO from a sperm penetrating an egg that leads the egg to dispatch calcium ions, which then free sufficient calcium stored in the egg to facilitate division and development of the ovum after fertilization!)

Snyder, Nelson and their coworkers were drawn into this story of behavioral genetics by accident when they set out to investigate strokes. During a stroke, the brain literally explodes with nitric oxide, and Snyder speculated that by blocking NO synthase, an enzyme that catalyzes NO production, stroke damage can be reduced. Early experiments in animals showed the damage could be cut by half. Pursuant to their studies, they genetically manipulated some garden-variety lab mice to delete the gene responsible for NO synthase, hoping to make a mouse colony that would tell them more about brains with and without NO.

To their chagrin, at least at first, the males looked perfectly normal. But Snyder recounted in a long interview that his lab technicians began to notice that on many mornings, when they checked the cages of the male knockouts housed together, a couple of the mice were bloody, wounded, or dead. Intrigued and concerned, they began to watch more carefully and discovered the males were fighting each other. That led them to deliberate experiments with the mice under different "social" conditions, says Nelson, who found that unlike normal mice with intact NO production, the knockouts *never* conceded inferior status to one another by rolling over and exposing their throats, for instance. They just kept fighting. When put in cages with females, the knockout males weren't so much violent— as they were toward other males—as they were terminally aroused.

Even more intriguing, it seems that knocking out NO has no effect of this kind, or any kind so far as they can tell, on the females. If in fact NO works by combating hyperaggression and hypersexuality, this makes sense, because females *normally* are less aggressive and less sexually interested in each other or males. They never, for example, initiate sex with males. More recent studies looking at the brains of these mice suggest that NO works selectively in the limbic system of the brain, where emotional and sex drives are controlled.

In another set of experiments, Snyder and his group tried to account for the difference in behavior between the knockouts and other mice, and between males and females, by seeing if the process caused damage to other known triggers of hypersexuality and aggression. One of these is the animals' smell system, which they use to detect females in heat and genetic kin. But when the investigators hid fragrant cookies (Oreos are the mice's favorite), the knockouts found them just as easily as did other normal mice, suggesting their sense of smell was intact.

The cookie test, a convenient and standard one in behavior genetics circles, was also used with another knockout that appears to be linked to sexual violence. This strain was produced at the Jackson Laboratory, where systematic breeding of purebred strains of mice are contributing to behavioral genetics as well as research on genetic disease.

Elizabeth Simpson, the researcher formerly at Jackson who focuses on sex determination genes and behavior, used a genetic engineering tool to knock out a gene on chromosome 10 in mice. In males, the protein made by the gene is involved in sperm production, suggesting that it may be associated with or influenced by other genes and hormones that act differently in males than in females. A very similar gene is believed to exist on human sex chromosomes as well. Without this protein, mice, both males and females, in Simpson's lab engaged in a lot of bad behavior. They too were overly aggressive, never gave up in a fight. The males, like those in Snyder's cages, attacked females, killing many of them. Female knockouts stopped caring for their pups and abandoned their inborn maternal instincts to nurture and nurse.

Simpson planned to send her mice for "professional" physical and behavioral testing, conducted her own preliminary "cookie" test, and ruled out smell-based brain damage that could account for the aggression. "We hid an autoclaved chocolate chip cookie under four centimeters of wood chip bedding and gave the mice ten minutes to find it." To their bemusement, in the first round no mouse found the cookie, not even normal mice, a situation they concluded must be due to the fact the mice weren't hungry enough. (Maybe Oreos are as preferred in Maine as in Maryland.) Simpson and her technicians then withheld food for twenty-four hours to see if that would motivate the search. This time, the knock-outs almost died, leading to speculation that the animals may be diabetic. But on a second round, all the Jackson mice, including the knockouts, found the cookies. Their smell brains appeared to be intact, and measures of testosterone and other male hormones were normal as well. The males were not "hypermasculinized" and the females were not masculinized at all as far as Simpson could tell.

Simpson and her colleagues conducted another behavior test, known as the intruder challenge. With normal mice, introduction of a foreigner or intruder to a home cage seems to confer a home cage advantage during the inevitable competition that follows, with the intruder always losing. But if Simpson used a knockout as the intruder, introducing it into some other home cage, the knockout would always win, even if it was a female knockout put into an all-male home cage. "The females will attack the male and though they don't manage to kill the males, they cow them. They step on the males' backs, too, what some consider an attempt to sexually 'mount' the male." She and her colleagues again measured tes-tosterone levels to see if she inadvertently bred a population of masculin-ized females, driven to testosterone excesses. The mice were hormonally normal.

A brain-centered approach to examining the physiology of bad sex was revealed in a conference at the annual meeting of the American Asso-ciation for the Advancement of Science in 1996, where experts gathered to examine the limbic system's role in normal and abnormal emotion and behavior. In particular, they presented evidence for a neurobiological underpinning in all kinds of destructive and violent behaviors in the con-text of evolved "mental mechanisms." Some of their data came from stud-ies of patients with partial limbic seizures and some from animal studies that identified imbalances in the limbic system involving the amygdala and its links to the temporal and frontal lobes.

One of the presenters, Paul D. MacLean of the National Institute of Mental Health, explained that the forebrain of humans and other advanced mammals has evolved a three-part structure that reflects ana-tomic and chemical ancestral commonalities with reptiles, and both early and late mammals, which, like birds, evolved from reptiles.

Animals' movement up the evolutionary tree from reptiles to mam-mals, he said, is marked by the development of integrated but distinc-tive layers of the brain that parallel integrated but distinctive kinds of behavior. The limbic brain represents an intermediate level, establishing communication between such things as infant nursing, maternal care, and a "family way of life," along with connections to Broca's area (for language) and the brain stem's more instinctive drives. Quite possibly, he said, we might track bad sex to limbic-generated emotions that guide us in fight, flight, and sexual arousal, specifically to limbic brain patterns influenced by "the painful nature" of the separation of infants from their mothers after weaning. His hypothesis not only fits the differences known to exist in the organization of the male and female human brain (see chapter 4) but also might explain the sharper female sense of "empathy and altruism," the different appetites for violence, and lack of male sympathy for their female victims of sexual abuse.

George Washington University neuroscientist Frederick Goodwin has contributed compelling evidence to the argument that aggression and violence are highly conserved human behaviors. Moreover, they are subject to heritable concentration—that is, they will be highly reinforced—under certain conditions. He has found, for example, that 6 to 8 percent of children of all social classes and races show aggression and impulsive conduct at an early age and these children are at high risk of becoming repeatedly violent as teens, especially if they grow up in bad environments.

In reporting on this data in some casual remarks at a meeting of the NIH Mental Health Advisory Council, Goodwin drew understandable anger when he compared the behavior of male rhesus monkeys (half of whom die of violence at the hands of other males by adulthood) and violent young men in inner-city neighborhoods. "That is the natural way of it for males, to knock each other off, and there are some interesting evolutionary implications of that because the same hyperaggressive monkeys that kill each other are also hypersexual so they copulate more." Although Goodwin did not make specific reference to black males or to the high rate of unwed motherhood in inner-city black communities, his remarks were enough to achieve his resignation. Ultimately, these comments were to tarnish the advice he gave to the Department of Health and Human Services, which had financed the University of Maryland Conference on violence and led to the political firestorm that canceled the event. But his attempts to draw lessons from evolution, and to draw links between sexual strategies and violence, albeit clumsily articulated, are validated by the good science linking genetic mutations and biochemicals to the sexual and general violence that runs in families of both humans and animals.

Consider, for example, studies of the behavioral impact of testosterone. Humans who use testosterone injections to combat everything from HIV-induced fatigue and weakness to competitive disadvantages in sports uniformly report not only greater strength but also more self-confidence, feelings of physical superiority, arrogant self-esteem, impulsivity, lust, willingness to risk all kinds of dangers—including social opprobrium—and a chemical rush of irritation, anger, and readiness to fight. Female rat pups injected with testosterone not only developed penises from the same tissues that made up their infantile clitorises, they also mounted other females, while male rats deprived of the hormone shortly after birth became sexually passive.

Studies have found, too, that the more "butch" partner in a lesbian relationship has higher levels of testosterone than the more "femme" woman; actors have more than ministers; single men more than married men; armed robbers more than burglars; litigators more than civil lawyers; and fans of winning soccer teams more than fans of losing teams. And in both women and men prison inmates, those with the highest testosterone levels tended on the whole to have a history of more violent crimes.

But it's important to emphasize that testosterone has never been shown to be the sole or even the primary *determinant* of behavior in any particular individual. Indeed, *no* study has ever found a direct cause-and-effect link between levels of testosterone and violence, sexual or otherwise.

It may be true that, as told in John Colapinto's book *As Nature Made Him*, David Reimer was never able to feel or behave like a female after a botched circumcision led his parents to remove his penis, give him female hormones, and rear him as a girl. Testosterone presumably fashioned his brain and body forever male. But it's impossible to know just how much of David's male psyche was due to nature or nurture, to subtle cues given or not given when he was being reared female. Neither does the obverse appear to be true, that giving large amounts of testosterone will predictably and universally make a man a he-man; and this throws a monkey wrench into the theories of sex-hormone determinists. Female-to-male transsexuals have reported feeling like he-men after testosterone shots, but what happened to the presumably equally strong fetal effects of relatively lower testosterone and higher estrogen levels on their bodies and behaviors? Moreover, whereas those inmate studies did show relatively more testosterone among violent male and female prison inmates than among those incarcerated for, say, theft, these studies have never identified any overall high testosterone levels among inmates when compared to non-inmate populations. And, finally, if nature had intended the most testosterone to make the most manly men, why is it that studies have universally found that men with the highest levels of testosterone, expressed in highly muscled physiques, heavy brows, and jutting jaws, are the *least* attractive to women. Women are far more attracted to men who have somewhat softer features. Similar studies show that both men and women prefer mates who are not overly testosteroned or estrogenized. It's as least as likely that both sexes prefer mates with variable hormone levels: men who are both risk takers when they need to be and nurturing when they need to be, and women who are strong when they need to be and risk averse when they need to be. While some evolutionary psychologists would have you believe otherwise, the most likely to succeed women may not be those who are willing to wed the sensitive milquetoast while getting the good genes in the bed of the daring warrior–bungee jumper. It's just as likely, perhaps more so, that today's men who have high enough but variable testosterone levels are the descendents of similar men who could not only bring home the bacon but cook it, too. In studies of birds by John Wingfield, it's the males with the constantly high levels of male hormone that are the worst fathers.

In short, if testosterone were the be-all and end-all of the story of sexual violence, it's unlikely that we would be seeing such strong evidence, at least in Western society, of widespread interest in more peaceful existence, and harsher punishments of and intolerance for rape and violence. Society *is* becoming more egalitarian, more cerebral than physical, more "feminine" if you will. It's hard to imagine this could be the case if such nonhostile and cooperative tendencies and traits were not also selected for and had not become part of our evolutionary heritage. At the very least, we have within us as many seeds of nurturance and of anti-testosterone behavior as of testosterone-determined behavior.

Whatever biological fuels are stored in our behavioral gas tanks, the engines that both arouse and push action must be cranked by situation. M. P. Kafka of Harvard is among those who have examined the category of neurotransmitters known as monoamines (which include serotonin, dopamine, and norepinephrine) to develop a "monoamine hypothesis" for what is happening in the physiology of men with paraphilias, or sexual perversions, another form of sex widely condemned by most societies. The hypothesis is based on the fact that these chemicals are all involved in the sexual arousal of male lab animals and on the observation that alteration of these neurotransmitters modulates everything from impulsivity, anxiety, and depression to compulsivity and other behaviors. In humans these might be considered part of the distortions reflected in a paraphiliac's interest in the more bizarre aspects of kinky or dangerous sex. Indeed, a few studies show that raising serotonin levels with Prozac ameliorates paraphiliac sexual arousal and behavior, lending credence to the monoamine hypothesis.

Scientific experiments and observations have also contributed to evidence that biochemistry is beneath some tendencies to sexual predation and addiction, the tabloid behaviors we sometimes refer to as nymphomania and Don Juanism. Like the alcoholic, junkie, or compulsive gambler, such individuals are said to have cravings, no matter what it does to their lives and despite repeated bouts of self-loathing, guilt, and suicidal depression. The behavior can play out as not only sexual intercourse but exhibitionism, voyeurism, masturbation with pornography, child molestation, or crippling fantasy. Kafka and his Harvard colleagues conducted a comparative study of men and concluded that there are several types of sexual addictions and that paraphilias may or may not be involved. But where they are involved, they tend to become complete substitutes for sexual intercourse, rather than "enhancements" of what we consider normal sex.

The concept of sexual addiction is controversial. The American Psychiatric Association, for example, does not recognize "sex addiction," largely because there are too many complexities and differences among those who claim its label. But some objective evidence comes from psychiatrist Fred Berlin and his colleague James Frost of Johns Hopkins who have analyzed PET scans of the brains of men (including sexual deviants) during sexual arousal and documented the release by cells in their brains of opiate-like substances.

Integrating evolutionary biology and psychology with genetics, endocrinology, anthropology, and animal studies, sex researchers are creating something of a unified theory of what we might call "bad" sex. At the very least, we are evolving a fairly universal consensus about what is considered "bad" from the standpoint of biological pragmatism.

In their chapter in *Sex, Power, and Conflict*, Daly and Wilson describe how the very traits all humans abhor came to exist in sufficient abundance for us to coevolve the tendency and cultural institutions to deplore and punish them. Suppose, they say, that hundreds of thousands of years ago, there was a group of peaceful hominids, and suddenly a new mutant variety of male appeared who was combative and aggressive and able to fertilize more females than the pacifist variety on average, despite the fact that his combativeness encouraged his pacifist cohorts to join together to kill off him and his like at a fairly young age—though not before sexual maturity. In such a case, the combative mutation—whatever the confluence of genes and hormones that conferred the combativeness—would spread through the population over generations. One big consequence would be the decline of male life span. Does this mean that men are power-mad beasts today because their genes make them so? Of course not. For one thing, the imaginary population could just as well have been a combative one in which a pacifist mutation arose that conferred more fitness on pacifists who learned how to cooperate and band together for the common good of the group and thus produced more abundant off-spring. And one big consequence of the combative mutant is shorter male life span (shorter than females, anyway), which might not confer much long-term fitness in the end in cases where the females, say, banded together to protect themselves from the more combative males, or secured the help of less combative males to protect them in groups. Females and males, after all, *co*evolve. If any one sex's attributes or traits become completely out of synch with the other's, a dioecious (two-sex) species would die out.

Eventually, traits such as these may balance out in the battle between the sexes, but the scenario illuminates some of the possible ways that mild polygyny, serial monogamy, and other negotiated solutions, along with the very "need" for the conflict, were established. Remember too that humans—and most other animals—produce fused gametes that undergo meiosis to sort and mix their genes and yield embryos and offspring that are *by nature* genetically different from the mother and the father. This, according to much current thinking among evolutionary biologists and psychologists, makes parent-offspring conflict literally "natural" or at least endemic.

"Insofar as males are specialized (physically) and psychologically for violent competition with other males, and insofar as male fitness is largely determined by the frequency and exclusivity of mating access, it is hardly surprising that males also commonly apply their aptitude in violent conflict to the task of direct aggressive control of the females themselves." In this rather blunt statement, Wilson and Daly are unflinching in their assessment of the origins of rape: Given nature's sexual mandate for males, it is hardly surprising that we have it. Yet like most in their field, they are quick to point out that there is enormous variability in such willingness or capacity for violence; and wherever there are pair-bonds in which males and females care for their young cooperatively, as in foxes, some monkeys, humans, and other mammals, such differences and conflicts are weakened or even absent.

Not surprisingly, in the face of such ambiguities, the nature of rape is perhaps the most hotly debated in the realm of bad sex, with some factions—political and scientific—arguing that rape is solely or principally about abuse and use of power, a form of violence, while other factions— also political and scientific—insist that it is solely or at least also about sex. The evidence today is weighted toward a Solomonic answer—it's about both.

In the research for his book with Craig Palmer outlining the case for rape as an adaptive male trait, Randy Thornhill, his wife, Nancy Thornhill, and their team conducted laboratory studies to test the idea that men who rape do so because they have evolved a specialized sexual biology and psychology to do so. The Thornhills speculated that if their idea was true, if in fact rapists were naturally programmed to rape, then a number of other things would also have to be true.

First, nature would have had to evolve men to be highly sexually motivated, aroused, and able to perform equally well in both rape and non-rape situations. Second, it follows from this that getting physical control over an unwilling partner by force would be sexually arousing for men. Third, that a man's age would affect his willingness to use force on the premise that reproductive strategies change with age. Fourth, that men's willingness to force sex would decrease with increased social status (because if he were an alpha male, the females would go for him preferentially and he'd have no need to rape). Fifth, a man's willingness to use sexual force would be very sensitive to the probability of detection and to the likelihood that his mate would be punished if she were suspected or found to be unfaithful.

Their results at least partly support all of these predictions. For example, in a series of studies and surveys, they found that men are as sexually aroused in laboratory settings using depictions of rape and force as they are by other forms of noncoercive sexual images; that men are aroused by physical control of an unwilling mate; and that men far more than women dominate and generate moral religious codes that support the so-called double standard in sexual conduct. This is because it is ultimately in men's best interests to put a moral appearance on their use of force to "cook the social rules" and avoid punishment *by other males* or society in general for forcing sex. There is also a stunning amount of social research suggesting that men who rape their wives or other women say or believe they do so because of suspected infidelity, or to punish "all women" for their "deceptive" ways. In fact, a Maryland judge, Thomas Bollinger, twice let off offenders for killing an unfaithful spouse, affirming the male unity in such an approach. And he stayed on the bench.

It's important to note that even Nancy Thornhill does not accept such data without large shakers of salt. In a published review of *A Natural History of Rape*, Frans de Waal points out that if, as the anthropological evidence suggests, our ancestors lived in small communities, it is likely that rape was harshly punished so that a male's future reproduction would be reduced, not enhanced, and evolution would have favored men that don't have a tendency to rape. He further noted that most men don't rape, and that Thornhill ignores "the complex mental life" that also evolved in us to make us act in all sorts of ways depending on the environment. "To focus on just one [sexual] behavior, (rape), is like seeking to understand why the kangaroo has such tiny front legs while ignoring what happened to its hind legs and tail." De Waal argues that the real connection isn't between violence and sex, but between power and sex. "This age old connection may explain how power and sex get mixed up in the minds of men and occasionally spin out of control together, not because men are born to have coercive sex but because power in general is a male aphrodisiac." Other studies show that when women fail to become aroused by coercion, which is frequently, men often fail to stay aroused. In addition to victim response as a factor in rape, laboratory depictions of rape are never as violent as the real thing and border more on fantasy and erotic pornography, whose arousing images rarely if ever produce real-life rape and never show the victim in pain or resisting. Thus, the romance novel in which the dashing hero "forces" the virgin heroine to "submit" to his seduction usually ends the way most fantasy porn in the lab ends—with orgasm for both parties—and also with the heroine's eternal gratitude and love for the hero.

Nevertheless, the widespread acceptance by both men *and* women that women "like" powerful, dominant men and, not infrequently, some roughhousing or "pretend" coercion in the bedroom, suggests some bio-logically grounded preferences for sexual coercion, at least up to a point. Says Nancy Thornhill, "the adaptations are solutions to environmental problems that have consistently impinged on human endeavors to survive and reproduce.

"The suggestion," she goes on, "that male sexual psychology is a coer-cive one [i.e., envelops rape adaptations] and that women's minds include adaptations for response to rape assumes that rape has been a consistent and recurrent issue in human evolutionary time. It suggests that rape, like finding a mate, protecting offspring and finding food and shelter, has been an integral part of the human adaptive environment; that the general mat-ing strategy of men sometimes includes sexual coercion and sexual coer-cion of one form or another may be involved in many matings in humans. It might be that obtaining sexual intercourse by coercion is as much a part of men's sexual behavior as men's use of noncoercive sexual approaches. . . . Current knowledge of men's sexual behavior doesn't pro-vide evidence of psychological adaptation to rape itself. It seems likely that the occurrence of coercive sexual behavior by men could as easily be the combined effect of species-typical adaptation to coerce desired rewards and sex-specific adaptation for sexual behavior."

Keep in mind also that the picture of the evil, brutal testosterone-sloshing male perpetrator of rape and abuse is largely mythical. For one thing, most men do *not* rape or beat their wives. For another, women under certain—but rarely sexual—circumstances are equally or more vio-lent toward spouses than are men. This is not a particularly "tidy" idea, as clinical psychologist Judith Sherven and her coworkers like to say. But the facts are that half of all spousal murders in the United States are committed by wives and not all of them are in response to physical abuse by men. The 1985 National Family Violence Survey supported by the National Institute of Mental Health shows that women and men are phys-ically abusing each other in roughly equal numbers and that women often use weapons to make up for their relative physical disadvantages, as dra-mas about penis-slicing attest. A study on family violence published in the *Journal of Marriage and Family* in 1986 found that 1.8 million women suffered assaults that year from husbands or boyfriends, but 2 million men were assaulted by a wife or girlfriend. The study also found that mothers abuse their children twice as often as do fathers in some surveys, and that one reason this seems so shocking is because men who are taught to be "manly" are terrified to admit they have been abused by a woman, so their numbers are vastly underreported.

Every high-profile case of male-against-female violence brings with it a predictable round of public comment and explanation with a numbingly similar theme. Just days after the death of Nicole Brown Simpson, for instance, an editorial cartoon in *The Baltimore Sun* was typical of the genre. It showed a cowering woman and shadowy, club-wielding man with the legend, "Who knows what evil lurks in the hearts of men?" A more chilling example came from a prominent defense attorney who suggested that Simpson's lawyers might overturn a jury's anger over 911 tapes recording his violent threats against his ex-wife by asking each juror, "Who hasn't yelled at his wife or screamed 'I could kill you!'?" After all, the attorney noted, "it's human nature."

The death of Simpson, like the rape of a young woman by boxer Mike Tyson, perpetuate the almost universal specter of the brutal, strong-armed male out of control in his rage against a weaker, peaceable woman who disappoints, disobeys, or fails to do his "reasonable" bidding. Abusers' public pronouncements are scripts right out of Screenwriting 101: "I snapped." "She pushed my buttons." "She knew I was tired." Celebrity incidents, especially, strongly reinforce the common belief that while society may—and should—severely punish such behavior, the men who hurt women are also victims, of their human natures, their sexual biology.

Less celebrated statistics support the view that male-against-female violence—and probably vice versa—is both biologically *and* culturally entrenched among *Homo sapiens*. (As with any other facet of human behavior, the answer to the question "Is rape nature or is it nurture?" is a definite "yes.") Worldwide, experts say, between 20 and 50 percent of women are victims of abuse. *Scientific American* reported in 1994 that although underreporting of violence by women makes accurate estimates hard, regional studies suggest horrifying epidemics of domestic abuse. A survey in Papua, New Guinea, found, for instance, that 56 percent of urban wives were battered and 18 percent beaten badly enough to require hospitalization. Their rural counterparts fared even worse, with 67 percent beaten. In Bombay, a fourth of all deaths among women fifteen to twenty-four are from burnings carried out by their husbands or families as a way to secure bigger or more dowries through subsequent marriages. Lori Heise of the Pacific Institute for Women's Health in Washington, D.C., recounts the 1991 incident in Kenya when young boys in a boarding school attacked a girls' dormitory, raping seventy-one and killing nineteen.

Afterward, the school principal "explained" that the boys meant no "harm" and only "wanted to rape."

Lest we tend to think Western societies are somehow immune, according to a U.S. Senate Judiciary Committee staff study published in 1990, 3 to 4 million women a year in the United States are beaten; of 36,100 murders of women eighteen to thirty-four between 1976 and 1992, a third were committed by boyfriends, husbands, or ex–sex partners, half if you exclude unsolved cases. A 1990 study by M. P. Koss and colleagues at the University of Maryland estimated that if current trends continue, one in five adult American women will be raped in her lifetime; one in 3.5 will be attacked by a rapist; one in seven women now in college has been raped; and more than half of college rape victims are attacked by their dates. Women, the Senate Judiciary study concluded, are ten times more likely to be raped than die in a car crash.

However much we may wish that the nature of bad sex is simple enough to be addressed by tougher laws or medical treatments, the evidence strongly suggests that it reflects a complicated gemish of brain chemistry and self-will, biology and culture, sexuality and violence, feelings and actions. And that neither simple punishment nor tolerance is the answer. In his book *Listening to Prozac*, psychiatrist Peter Kramer describes the professional man who takes the mood-altering drug Prozac for depression and finds not only that he feels better, but also that he has lost his taste for the hard-core pornography that informed his erotic life for decades. The serotonin reuptake inhibition of the drug may indeed have had an impact on his brain.

One implication is that if the medication alters natural brain chemistry in ways that alter sexual personality, there may in fact be natural brain chemicals and structures, perhaps even interactions among a collection of single genes, that actually code for increased predisposition to such mind-brain-body compulsions as the urge to dominate women, wife battering, impulsivity, helpless rage, shyness, or thrill seeking. Indeed, research studies have suggested that some people with an extra-sensitive amygdala are more likely to be shy. Studies by behavioral geneticist Dean Hamer, author of *Living with Your Genes*, seem to demonstrate that single genes or gene complexes are involved in sexuality, shame, guilt, happiness, shyness, and homosexuality. A report in *Science* by Antonio and Hanna Damasio identifies potential sites in the prefrontal cortex of the brain that house molecules of "morality," tissues that track our ability to weigh right and wrong. In such ways, sex crimes, including rape and pedophilia, may be, as sexologist John Money believes, disorders of the mind caused by problems in the brain, and that even such things as our moral sense are physically vulnerable to disordered chemistry, injury, and emotional trauma. Looking forward in time, Hopkins's Solomon Snyder suggests that if alterations or distortions of the products of single genes, such as nitric oxide synthase, can produce in rodents such complicated and sexually aggressive behaviors as rape-like behavior, then drug treatments *can* be designed to treat psychopathic sexual violence and abuse now considered wholly volitional and criminal, rather than medical, in nature.

Rape or at least rape-like acts clearly exist in many species, giving additional weight to both rape's "natural" roots and its "value" in our biological and psychological legacy. Geoffrey Parker of the University of Liverpool in England and his colleagues, for instance, have reported that insects, birds, and nonhuman primates sexually "harass" and threaten females of the species no less than politicians and business executives do women, and that males will do whatever they must or can to get access to female sexuality. They might try seduction or they might try force. Animals may rape immediately or slap, bite, and get the females to "give in." Says Parker, "It's all very squalid and depressing, but there we are."

There we are indeed. Barbara Smuts and her coworkers have found similarly that among many kinds of monkeys and apes, females in heat are a target of male aggression and sexual coercion; they are especially attacked by "alpha" males when these females associate with lower-ranking males. Among some baboons, males try to control females *all* the time, whether the females are fertile (in estrus) or not, herding the females jealously away from any bachelors around and threatening them by staring, gesturing, and biting them in the neck.

Among wild chimpanzees, too, "rape" is a mainline tactic for getting sexual access. Jane Goodall in studies of her Gombe chimp troops explains that male chimps remain in the troop of their birth so that all the males are related to one another, but females *transfer out* of their natal home at sexual maturity. When a female becomes fertile, which is for only a few months every five years or so, the males compete and the winner, or alpha male, usually keeps all other males from mating with her. Lower-ranking males try to get females into the forest for sneaky sex and sometimes the females are willing. But if they resist, the males commit what Goodall calls a "fair amount of brutality" to coerce sex, including outright attacks.

The picture is more brutal among orangutans, who don't even bother to try peaceable means of obtaining sex before they attack. Nearly all copulations by adolescent males and half by adults happen only through extreme aggressive behavior from the male. Male coercion among non-human primates can also involve infanticide, a common practice in which a male will kill the offspring of other males. Since sexual cycling of the female is curtailed by nursing an infant, killing the infant has a dual benefit to the killer: it removes another male's offspring and brings the female out of lactation and into heat.

An almost inescapable conclusion from such evidence is that males, faced with the constant "problem" of competing for sex, getting sex, and keeping *other* males from getting sex, evolved means of "solving" it in a variety of ways, including a psychobiology that lets them overwhelm their own tender proclivities and "rape." In keeping with the evolutionary paradigm that, in general, behavioral traits *coevolve*, it appears that females, for their part, developed a psychobiology that at times welcomes "masculine hostility" because it is an honest advertisement of his ability to protect her and their offspring.

The work of Neil Malamuth, a principal architect of theories that integrate evolutionary biology and psychology with the facts of sexual coercion and violence and with modern feminism, also suggests that our ancestors' inability to "tell" with any certainty whose child a pregnant woman carries might explain not only why some men today behave in the ways they do with respect to domestic abuse, but also why "imagined" aggression against women among men who never actually hurt a woman is so widespread a fantasy. Malamuth has conducted studies of men's own accounts of their sexually aggressive fantasies and discovered that men are indeed aroused by such fantasies and by their admissions that they would force sex more often on women *if they could be certain there would be no punishment.*

For critics of any effort to link evolution, adaptation, and human behavior, this presumed tolerance for coercive sex is an especially bright red flag, symbolic of the many abuses and distortions of Darwinism perpetrated by racists and sexists. Paleontologist Stephen Jay Gould of Harvard, for example, scoffs at scientists who say any widespread behavioral traits are by definition adaptive, that they enhance survival. They might just be random leftovers of other evolutionary processes. But again, the nature-nurture disconnect is illogical. Darwin, remember, was perhaps the ultimate environmentalist, seeing the outside world as the principal shaper of traits and biological destiny. The devil, as they say, is in the details, in just how nature and nurture might interact. The issue is not whether something we do now directly leads to more successful mating or survival, any more than wearing thong bikinis is adaptive among teenage girls because it attracts potential mates. The issue is whether something we do might have indeed had an adaptive purpose at some time and such adaptations left their footprints on our psyches. Other evolutionary scientists scoff at such a conclusion, noting that we have no idea what the realities of the Stone Age might have been, and that there is just as much possibility that evolution made behaviors like rape maladaptive.

What can be said is that sexual coercion, however "natural" it might or might not be, is costly to females. They are often severely wounded. Some die; and the high rate of infanticide, up to 40 percent in some monkeys and 37 percent among mountain gorillas, can in extreme cases threaten the survival of the entire troop. Females also pay a high price in "anxiety," in energy expended to ward off male aggression and to form bonds that can help them with their offspring. "When we look closely," says Smuts, "we find that in many primates, hardly an aspect of female existence is not constrained in some way by the presence of aggressive males." The strategies female primates use to resist such aggression sound achingly familiar to their human cousins: physiological responses that change the timing of her estrus, forging strong bonds with other females, especially female relatives and "chaperones," to protect themselves. These sororities also mean that females control, to some extent, each other's fertility and influence male-male competitions for their favors. In rhesus macaques and vervet monkeys, for instance, no male can get to high-status females without the help and cooperation of a group of high-ranking females, so such males are reluctant to attack the highest-status females. Female savanna baboons, moreover, make male "buddies" who are not their "lovers" or sex partners but are recruited to protect them from aggressive males and to protect their infants; eventually, in many cases, these females will give their friends sex.

All male sexual aggression against women can be viewed as reflections of the varying costs and benefits to males in search of sex and to females in their use of counterstrategies. So, far from being an immutable feature of animal or human nature, male aggression toward women varies dramatically depending on circumstances. Even among chimps, rape remains relatively rare as a sexual strategy, and such strategies as courtship, orgasm, and long-term pair-bonding emerged as having an equal or greater overall *impact* than violence to secure and keep a mate.

Balance is key. In any animal society, individuals over time have much more to lose than gain if the cost to females demanded by males is too high. Balance must be kept, and given the escape of most people from the dangers of prehistoric society, it's hard to make a case for *any* need for violent sex as a tolerable sexual strategy. Sex, concludes Parker, is many things, but notably a war of "attrition," with men insisting and women resisting until one gives in or gives up. Men "win" often enough to fuel the feminist movement, but the only lasting male "win" is when women "win" something, too—successful offspring.

University of New Mexico researcher Jane Lancaster and others consider this fundamental need for both men and women to benefit from sex, mating, and marriages to be a key element in human social evolution, the kind of leap that argues against rape as a "natural" occurrence. At some point, male aggression toward females of the orangutan variety became obviously less effective in advancing reproductive fitness than gender cooperation and "love."

In support of this need is the fact that in nonhuman primates as well as humans, males tend to form alliances, too, their version of buddies and allies that "stick together." In Lancaster's view, evolving males began to "get it," in the sense that those males who developed and enhanced their physical and mental capacities to cooperate with other animals and not just fight all the time became our ancestors, while those who fought all the time, against each other and against females, tended to die out at a faster clip.

Our ancestors may have originally become cooperative during hunting, in order to bring home bigger kills. Cooperation required alpha males, the most dominant ones, to tolerate some mating activity by lower-status males, so that lower-status males would help stock the larder—in other words, when there was a little more "democracy" in the group with respect to access to sex. This turns out to be the pattern discovered among Goodall's chimps and de Waal's apes. Says Smuts: "Among hominids, the kind of tolerance we see among male allies in nonhuman primates became formalized as each male began to develop a long-term mating association with a particular female or females, a trend foreshadowed [in humans] by savanna baboons. This tolerance does not imply an absence of male mating competition within the group since some males would undoubtedly continue to have larger numbers of mates than others as is true among humans today." But things were loosening up until finally exclusive mating relationships were conserved because they also benefited females. "It is in this context that male sexual coercion becomes relevant," says Smuts. "As among many nonhuman primates that live in multi-male, multi-female groups, hominid females were vulnerable to sexual coercion, including infanticide by males. . . . This pattern was likely to change once males began to claim particular females as mates and to respect the mating relationships of their allies . . . including inhibition against attacking his mate and his mate's offspring, particularly when the ally was present. Similarly, as males attempted to develop long-term relationships with particular females, they would be likely to protect those females and the females' offspring against aggression by other males."

Several attributes of modern human social relationships—even harems, the veiling of women, and marriage itself—make sense in the context of reduced "rape" on the savannas. For one thing, once the male apes and hominids began to respect the other guy's "wives" and to offer protection to their own mates, any female who pursued a promiscuous strategy or who appeared to be "tempting" other males would become increasingly vulnerable to sexual coercion and infanticide.

Why? First, the promiscuous female would not have a protector against other males, and the "respect" that served to inhibit rape against an ally's mate and his children would not apply to her. For a female to risk "adultery" in such a context, the price would be potentially very high and it had better be worth it. "In other words," says Smuts, "as male efforts to establish pair bonds increased, females were forced to reduce promiscuous mating in exchange for male protection from harassment." Ironically, the females became *more* vulnerable at the same time to aggression from their "husbands," or mates, because other males would be less likely to interfere with the guy—owing to the costs of disrupting the male-male bonds or, worse, actually killing or maiming their buddies. While many animal societies and human cultures seem to bend the wrong way in permitting their female offsprings' submission to males who marry or pair-bond with them, few families not in dire straits want their daughters harmed lethally. After all, the purpose of courtship and mating is to *survive* and to perpetuate the family's DNA. Almost universally, as a result, families have found ways to allow male dominance in the marriage home, but to protect females as well.

One astonishing human parallel to what primatologists have discovered has emerged from an unusual study of family deterrence of domestic violence among daughters of Spanish aristocracy. This study, led by Aurelio Jose Figueredo and his coworkers at the University of Arizona and

University of Madrid, was dubbed the Daughters of El Cid. (El Cid, fabled in a poem as the hero of the Spanish wars against the Moors, was a twelfth-century nobleman, warrior, husband, father, and epitome of high culture properly named Don Rodrigo Diaz de Bivar.)

Figueredo set out to look at the role families might or might not play in protecting daughters whose courtships ended in marriage, and they chose to do so in Madrid, where, as epitomized by El Cid's credo, there is a strong cultural tradition against spousal abuse and where a husband who abuses his wife is not only considered a worthless coward but an affront to family honor and the crown.

Figueredo's study, conducted in 1995, was intended to test a contro-versial model used to predict domestic violence among another Hispanic population, Mexican-Americans in Tucson, a model that concluded that sex, money, and paternity were the key factors in wife beating and other abuse, particularly by lower classes.

To look for evidence that such a model was more universal, thus nature-based, Figueredo identified a population of Spaniards who, while Hispanic, come from a much different cultural tradition. In a telephone pilot survey of battered and non-battered women with children under twelve, he and his colleagues looked for patterns in such factors as paternity, income, number of children, educational level, social class, professional status, social power, extended family size, and emotional support.

If Figueredo's evolution-based ideas were right, and paternity, sex, and money were not the major or only factors that predicted domestic abuse, then he ought to have found in the Madrid study that the biggest predic-tor of such abuse was the absence or presence of some *male deterrence* factor. In other words, if the ape biological model has been conserved in humans in some way, then the families of the women he surveyed would have some effective means of *indirectly* deterring violence against their daughters and sisters, of allowing their husbands control, but not enough control to produce lethal effects, and of deterring violence rather than *intervening* in the daughters' and sisters' households directly, thus dis-rupting extended family bonds among the males.

Analyzing both the direct and indirect effects of a woman's kin on domestic violence, Figueredo found that the more male kin a woman had *living nearby* but not inside the woman's household, the less abuse hus-bands perpetrated. Moreover, it was only the nearby presence of male kin that deterred violence, not female kin. Clearly, El Cid, or at least the "noble" biological strategies his legend illuminates, still rules in Castille—and presumably elsewhere.

One other conclusion from this study seems obvious to Figueredo: It's not how much kin you have but where you put them that counts. Underpinning it all is the serious possibility that mutual deterrence is simply another way of describing mutual interest. They are two sides of the same sexual coin. In this context, human pair-bonds and marriage are a way in which cooperating men agree about mating rights and respect to some degree other men's wives and territory, giving each other in return husbandly "control" and the right to limited coercion of their own wives without interference from outside males.

Studies of primitive societies such as the !Kung and Yanomamo also support such conclusions, along with the presumption of some biopsychological foundations for both sexual violence and sexual cooperation. As with Old World monkeys, where female kinships or sororities are weak in these societies, males are more aggressive, and women are more vulnerable when females lack support from their own relatives. Where male alliances are very well developed—either because of hunting or military alliances—male coercion of females also is more likely. Gang rapes, for instance, are common among the Mehinakus, a South American tribe that uses this form of violence only to punish women who have violated men's customary rules of power.

Since cooperation and "love" cannot be coerced but are earned through mutual provision of benefits at a "price" that is fair to all sides, men and male monkeys and hominids are probably less likely to beat and rape their mates in societies in which exclusive matings are emphasized, and they are more likely to injure mates when marital bonds are subsidiary to male bonds.

Variations on this theme by evolutionary biologists and psychologists are plentiful if speculative. For example, as Laura Betzig notes in her panoramic survey of Western history, in societies where men's wealth and power are more equally shared, there is less incentive for them to cooperate in punishing female adultery. And as Smuts points out, such men will rely more on individual tactics to prevent real or imagined cuckoldry and will tend to support their female relatives in disputes with their husbands. As a result, women will have more sexual freedom and be less vulnerable to wife beating.

In the !Kung society, where the relationships between men is more egalitarian and wealth is somewhat evenly distributed, there is less male aggression toward women. But in other societies marked by enormous stratification of men, say few wealthy nobles and many peasants, powerful men will promote sanctions against female adultery among their women and not support female relatives in conflicts with their husbands over female adultery. The sexual behavior of high-status women in this case will be rigidly controlled, and beatings and other coercion will include sexual coercion and rape; even low-status men will support such punishment because of the benefits they get as "mimics" of high-status males.

A study led by Patrick Gray at the University of Wisconsin in the 1980s looked at marital power in 122 societies and found that high levels of female power in a marriage do nothing to hurt male sexuality, but did give more foreplay to women, less punishment for premarital sex, and far less fear of impotence, homosexuality, and transvestism. There were no increases in divorce, either.

From this perspective, it's not terribly hard to understand how in societies where men emerged with all the power (because of, for example, historic military might or good fortune in the hunt, or accumulation of wealth), they are loath to give it up or share with other *men*. It is in these very societies that women are the most subjugated physically and sexually. And the more resources that men control, the worse it will be for women. The women's movement is threatening to traditional male-dominated societies not so much because it might give women power, but because it opens the floodgates to more evenly distributed wealth among *men*. Some would argue that if this were the case, it's hard to see why men with lower economic and social status don't help women in their battle for equality and why they seem to be the most threatened by the women's movement. One possible answer is that such men perceive there is more to lose if they cross the alpha males.

It also is not far-fetched to see how prehistoric women and their hominid ancestors, faced with this situation, developed and evolved means of manipulating men and sexual situations to either get a bit of equity or protect themselves from harm. One of the more intriguing bits of evolutionary evidence was gleaned by Karl Grammer and Frank Salter of the Research Center for Human Ethology at Max Planck Institute in Germany and the Ludwig Boltzmann Institute in Austria. They had been interested in the fact that studies show people are less aggressive when they are sexually aroused, and they set out to see if men and women used "sexual signaling" or manipulations such as flirting, teasing, or other sexy displays to avoid or appease threatening situations. They decided the perfect place to do this was in nightclub entrances, where beefy bouncers screen incoming patrons for suitability of "dress and decorum." These are "excellent research posts," they discovered, "for studying dominance and submissions and social strategies." In their studies in Brisbane, Australia, and Munich, Germany, they found that indeed females used flirtation and sexy dress and males were much more likely to show submission to the doorman (with money or bowed heads and appeasements).

Feminists, both male and female, are often the most vigorous critics of any suggestion that bad sex, in the form of rape or abuse, can be traced to traits that coevolved between males and females in our ancestral evolution. Thus, when evolutionary psychologists like David Buss, Donald Symons, Martin Daly, and Margo Wilson—or Leda Cosmides, John Tooby, and John Barkow, who together edited *The Adapted Mind: Evolutionary Psychology and the Generation of Culture*—argue that men's control over resources can be in large part traced to women's sexual choices of men who are the "winners" in terms of power and status, and that men's control over resources have triggered their angry defense of such power and contributed to coercion and abuse of women, critics call this "blaming the victim twice." Once for being raped or beaten and a second time for having created the conditions under which she is raped or beaten. Even in chimp society this seems to happen. Dr. Craig Stanford, a University of Southern California anthropologist, has documented chimpanzees' thirst for meat and blood, almost to the point of what would be human barbarism. The willingness of males to kill and murder far beyond their food needs he attributes in part or all to female chimp sexual behavior. Female chimps will copulate with more than a dozen males a day during estrus, and the attraction of flesh, freshly killed, which is linked to the survival of her offspring, is the seduction dinner on the savanna. If lower-ranking males kill more they can give more meat and get more matings in to compete with the higher-ranking males. It is the difference between "getting lots of sex and getting lots and lots of sex." So it's the females' proclivity and preference for meat in exchange for sex that drive the barbarous killing. One can almost hear the male chimps telling the judge, "the she-devil made me do it."

But Stanford offers a measured and logical response to both sides of the debate. "Not only do women suffer from men's control over resources," he writes, "now it seems like they are getting blamed for it as well. The inference of blame, however, does not follow from the identification of women's participation in one aspect of the causal chain. From the vantage point of evolutionary psychology, neither sex deserves blame. Instead, we are all the end products of a long causal process that involved the coevolution of women's preferences and men's intra-sexual competition tactics. As products of this process, men and women are equally blameworthy or blame free. Issues of blame are irrelevant."

Like it or not, however, the evidence suggests that women's sexuality may well be at the "core" of the battle of the sexes and particularly so when it gets violent. "Over the course of human evolutionary history," writes Buss, "men who failed to control women's sexuality, for example by failing to attract a mate, failing to prevent cuckoldry or failing to keep a mate, experienced lower reproductive success than men who succeeded in controlling."

Commenting on the book *Demonic Males, Apes, and the Origins of Human Violence* by Richard Wrangham and Dale Peterson, British zoologist Matt Ridley agrees that "red claws and teeth" are part of the evolutionary contribution to human personality and biology. But he expresses "fear" that one should then conclude that violence is inevitable because it is biological and biological "causes" are more fixed and rigid than social ones. To the degree that there is convincing evidence that rape and domestic abuse, infanticide, and prostitution are part of the continuity and change that come with evolution, the idea that biological causes are more fixed than social ones does not wash. In fact, says Ridley, "biological systems are often flexible and social systems reactionary." The entire foundation of evolution is that attributes are *adaptations* to changing circumstances. Had our biologies not changed, and our psychologies not changed in our ancestors, we wouldn't be hear to tell the tale.

In his book *Good Natured*, which argues that humans are just that, Frans de Waal has made perhaps the best case for the state of the evidence about rape when he asks, "Isn't it likely that some rapes are mainly sexually motivated and others mainly acts of hostility and misogyny?" Rape, he explains, is "mechanically impossible in the absence of male genital arousal. Hence the view of rape as a hate crime pure and simple is silly. A penis is no fist. This doesn't imply, however, that rape rests on natural urges as Randy Thornhill and Craig Palmer want us to believe. As sexually reproducing animals, people have sexual urges. But to say that all men will rape under particular circumstances is like saying that all people will eat human flesh when stranded in the Andes." Even if that were true, de Waal argues, that is still not evidence that all people are "born" to be cannibals.

Even if men evolved to have some traits that support physical dominance over their sex partners, it's just as possible to read those traits as having other, socially useful and gender-neutral outcomes. Demonstrating an evolutionary predicate for a rare behavior (and rape is indeed relatively rare) is not very useful unless there is an equally compelling predicate for the more common behavior, in this case cooperative sex.

Stephen Emlen, in his studies of small birds called scrubwrens, shows that even among these creatures, when "divorce" or death occurs and new families are formed, the stage is set for incest, abuse, and other sex-based conflict, the sons competing with fathers for sexual access to stepmothers, for example. The stepchildren in these bird families are at increased risk of abandonment and sexual abuse; and "remarried" pairs with offspring are more likely to "divorce."

It's not known, of course, how the birds feel about all this, but among human societies—even those that have relativistic attitudes about rape and other forms of spousal abuse—almost all have rigid taboos and zero tolerance for incest and parent-child abuse. Overall, the incest taboo is so intense that in some societies, including our own, fathers are fearful of hugging their young daughters, teachers are afraid of touching other-sex children, and pop psychologists raise fearful specters about children who innocently climb into bed with Mom and Dad when they have nightmares.

Dr. Paul McHugh, chief of psychiatry at Johns Hopkins, calls the epidemic of "recovered memory" of father-daughter sex a "witch hunt" for incest and one, like most sexual obsessions, with little or no foundation in fact or reality.

Once again, there are feasible genetic and psychophysiological roots to both the abuses and our disgust over them. A genetic tendency to not have sex with mother, father, or sisters or brothers was identified more than a century ago by Edward Westermarck, who noted that children have an aversion to sex with those they grew up with; as we saw in chapter 10, studies on Israeli kibbutzim have borne this out. Children who play and even sleep together rarely if ever "fall in love" as teens or adults. People have substantial biological and neurological means of identifying kin and close relatives, mechanisms that enhance our ability to bond with our children, to guard our mates, to invest in the long-term survival of our offspring, and to protect the family's DNA as well as the individual's. Among birds, insects, and primates, too, "kin recognition detectors" are built in to their biologies and clearly have as one purpose to avoid inbreeding.

Paradoxically, perhaps, some of the same reproductive imperatives—and the biology and psychologies that attend them—undergird the grim statistics on sexual abuse and physical abuse of stepchildren by stepparents (usually stepfathers).

It is to no male's advantage to literally support or enhance the life of another man's children. In her book *Anatomy of Love*, Helen Fisher recalls Margaret Mead's statement about her three failed marriages. "I have been married three times and not one of them was a failure," said Mead, and in that dictum she bore testimony to the endless desire of males (mostly) for a variety of mates, and particularly for what would now be called "trophy" mates—(relatively) younger and younger women to bear them more children as they age.

But to that end, fathers—and mothers—must leave one mate (and children) to get another mate (and other children), leaving some offspring to the mercies of potential stepfathers; and men or women who leave their primary mate may be faced with handling their new mate's children by a previous relationship. Ancestrally, perhaps, when troops of hominids and Stone Agers did their mate swapping, villages, grandparents, and extended families were around to rear the children, and perhaps some stepparents in close-knit villages or troops considered their stepchildren as their own and protected them, in part because the biological dad may have been around to not only protect but also contribute to the common weal.

More recently, with nuclear families frequently cut off from extended families or community bonds, divorce and stepparenting are bringing record levels of abuse and disaster. Clearly most stepfamilies and blended families get along and don't sexually abuse stepchildren, testimony to the fact that people are not always victims of their ancient biology. But where a man's abuse of stepchildren occurs, it often is coupled with abuse of his spouse, as wives often forgo blowing a whistle on a stepfather who abuses her biological daughters or sons out of fear for her own safety or to protect the husband-wife relationship.

As Daly and Wilson wrote in *Sex, Power, and Conflict*, "evil stepparents are part of the folklore and legend of every society, and there is abundant evidence from a variety of animal and human studies that parents are more likely to neglect, assault, exploit, and otherwise mistreat their stepchildren than their genetic children." Study after study has found that neither poverty nor social status has any bearing on the level of sexual and other abuse by stepparents to stepchildren, and the only characteristic that consistently pops out is the degree of genetic relatedness as a predictor of violence.

Studies among many human populations show that statistically speaking, step- and blended families are at increased risk for instability, conflict, and abuse. And patterns apparent across many animal kingdoms suggest that there are at least biological predispositions for all of these problems in humans as well, as Emlen's studies of stepfamilies demonstrate (see chapter 10).

Emlen notes that in more than 90 percent of all animals known to live in "multi-generational family structures," there is cooperative care of the young. And in almost all animals that live in genetically unrelated groups, such cooperative care of the young "is virtually unknown." Similarly, in the "cooperative care" society, sexual competition is reduced because most of the potential sexual partners are close genetic relatives. Although there are high genetic costs to such inbreeding, social structures in which sexual competition is discouraged in other ways are rare. Without competition—and all of the problems competition brings—populations tend to die off faster than populations with more competition.

What the biological and evolutionary evidence tell Emlen and others who study incest and child abuse is that "some of these tendencies, particularly the conflict and strife seen in step families, really do represent behavioral predispositions that were selected in our ancestral path and that as a consequence, they're going to be very difficult to eradicate."

"We're basically seeing large increases in numbers of step families and in single parent families as the divorce rate goes very high in the United States," Emlen notes. Norman Rockwell doesn't live here anymore. And the single-parent child rearing situation has no antecedents in our ancestral family environments. "Whatever our heritable predispositions and decisions might be they were never adapted for single parent child rearing," he adds. "We've created a culturally new social situation for which our inherited predispositions have poorly prepared us. There are two take home messages in this. One is that we are creating an impossible situation for single parents, the other is that we are trying to socially construct alternatives, surrogates for that [relational] work force that is now missing."

Moving beyond the kind of stepfamily conflicts that lead to incest and other abuse, Daniel T. Fessler, an anthropologist at the University of California, San Diego, notes that parent-child incest may best be viewed as a special kind of adultery. Because parental investment differs across the sexes, "there are significant differences [between men and women] in reactions to adultery and likewise, because the reproductive concerns of men and women differ, mother-son incest is a significant threat to fathers, but father-daughter incest often goes unreported by mothers. Father-daughter incest in fact appears to be more common than mother/son incest."

From an evolutionary standpoint, says Fessler, this makes sense, because when you make a cost-benefit analysis, looking at what's in it for men and women, or what the biological rewards and penalties are, incest fits a scenario that would maximize reproductive strategies for each gender according to what it needs to survive.

Children, of course, generally are subordinate to powerful parents. Parents could always "get their way." So a likely explanation for the relative absence of incest outside of stepfamilies is that parents have both a natural and learned aversion to sex with their own children—mothers more than fathers possibly because being around a dependent, demanding being for five or ten years leaves Mom with the idea that caregiving is incompatible with sexual desire. That also could explain why stepfathers who "step" into the picture only after a daughter is through with childhood years are more likely to have sex with such a girl than is a natural father, or than to have sex with a female he has helped rear from infancy. Similarly, mothers who nurse infant sons and have most of the child-rearing responsibilities are far more likely to commit incest with a stepson, especially if she "steps" in when the stepson is already adolescent.

Incest also appears to be wound up in our often violent reaction to sexual jealousy. A son, for example, already knows that his mother, attractive and caring though she is, has already had sex (at least once, as the joke goes) with the son's father. And Dad may also be very jealous of anyone who could potentially cuckold him, so he is likely to watch his son very carefully, making mother-son incest the rare thing it is.

Though not nearly as vehemently condemned as incest, most modern societies condemn prostitution—particularly as it is practiced on urban mean streets. But the taboo is far from universal, and even some feminists acknowledge that selling sex is a rational alternative to starvation or abject poverty for women with few economic choices. In societies where prostitution is legal and more open, "some women choose prostitution because it provides a quick and lucrative source of income and . . . may be seen as a desirable option relative to a . . . job or a demanding husband," says David Buss. These women are far from "victims" or home wreckers, high-end call girls or, more famously, the "whores with the hearts of gold." (*Gunsmoke*'s Amanda and Margaret Mitchell's Belle Watling were women clearly in charge of their own destinies. If they were vilified by the "good people" of the town, they also "came through" for the heroes and the community when the chips were down.)

Belle even used her earnings to rear a child in relative obscurity and luxury, never wanting him to know how she financed his future, a meaningful evolutionary strategy for uneducated women in times of scarcity or war, when sending a relatively resourceful son out into the world would have been a woman's only means of fulfilling her DNA's desires—if not her own.

In this sense, prostitution would have emerged as a widespread phenomenon because, like ancestral promiscuity, it had more to do with the stark realities of reproductive biology than with wantonness. When it came to solving the problem of fetal attraction and the competing genetic interests of the two sexes, nature gave females as many options as it gave males. As a consequence, females would likely need to have other options for securing male resources. And among males, if one was a low-ranking male instead of a powerful or wealthy one, getting sex in the Stone Age— when women risked their genetic futures if they got pregnant without the certitude of resources—was not likely without a price.

Evidence for a biology of prostitution also appears in Craig Stanford's interpretation of "bad sex" described earlier in chimpanzee societies: The rate at which males kill animals to satisfy their thirst for blood is matched by the enthusiasm with which the females will trade sex for such food. Male chimps will hunt excessively to support their sex-for-food swap meets, with both sexes interested in the hunt and the grope long past their need for nutrition. In observation after observation, Stanford demonstrated that male chimps with a dead and bloody colobus monkey will share it only after the female gives him sex, and the females, sexually promiscuous even without the promise of meat, will put out even more if they get "paid." Stanford believes the biological foundations for such sex-food bartering probably exist in humans as well, because it might give an extra nutritional edge to any offspring that results, it could give lower-ranking males a bigger chance of getting sex at all, and it could offer alpha males and females "the difference between getting lots of sex and getting lots and lots of sex." As noted earlier, promiscuity in many animal societies rises and falls with male dominance, and, anecdotally at least, prostitution seems to be more common in patriarchal societies in which women's behavior and opportunities are seriously proscribed.

Monkeys, apes, and many other animals from insects to birds have mechanisms for sexual arousal that often appear in concert with giving and getting pain. They nip and bite and screech and coerce. They commit infanticide and they promiscuously mate with younger members of their troops, colonies, and swarms.

So far, however, the search for genes and psychobiological roots of pedophilia, paraphilia, sadomasochism, and other "perversions" has brought little to light. As a consequence, the more likely explanation at this point in time is that these behaviors are traceable to a combination of early experiences, crossed biological wires, and inadvertent genetic warping of biological and psychological mechanisms.

For one thing, these behaviors, unlike rape, violate a key theme in the story of sex by failing to confer *consistent* advantage to one or both sexes in the game of reproduction. As we've seen, even sexual aggression can produce babies. By that measure alone, many scientists believe, it's highly *unlikely* that sexual abuse of preadolescent children or the proclivities of the Marquis de Sade will turn out to have the biochemical and evolutionary correlates found in such behaviors as rape and domestic violence.

Instead, experiments suggest that with respect to paraphilias such as fetishism, as well as autoerotic asphyxia and pedophilia, the phenotype is a result of distorted lovemaps. These distortions are created by the confluence of brain circuits linking emotions, learning, and our limbic brain's sexual apparatus in response to uniquely human—and individualized—stresses. For example, studies by sexologist John Money show that pedophiles who are aroused only by male adolescents won't be aroused by prepubertal boys or girls at all. Most of the sexual activity is fondling and masturbation and only rarely intercourse, and some pedophiles not only lust after their victims but fall in love with them. Among the pedophiles treated by Hopkins psychiatrist Fred Berlin are those with hormone dysfunctions, genetic defects, brain damage, and mental illness, as well as those whose early childhoods had a history of abuse, repression, and psychological disturbance.

In his psychiatric evaluations of paraphiliacs, Berlin has found that their fantasies are even unique to the nature of their sexual deviance. "People," says Berlin, "do not decide voluntarily what will arouse them sexually. Rather, they discover within themselves what sorts of persons and activities are sexually appealing to them. Thus a man with conventional heterosexual interests tends to seek out adult women [in his fantasies as well as real life], just as the homosexual pedophile, who may be impotent with women, seeks out boys. The heterosexual voyeur repeatedly seeks out situations where he can peep upon unsuspecting naked or partially clad females in response to his sexual cravings, whereas the male transvestite repeatedly cross-dresses."

Regardless of the evidence for or against a biology of perversion, the fact remains that uniformly, human culture views most paraphilias, but especially sex with children, as predation, as unfathomable or at least bizarre and bad. This has led in part to societal ignorance of all the paraphilias, which sex scholars mean to include pedophilia, exhibitionism, sadism, and masochism. The trend is to punish, not understand or treat sex offenders, and the physicians, psychiatrists, and scientists who do study paraphilias are in some circles as unloved as their victims.

Money, who has written extensively on transsexuals and the origin and practice of paraphilias worldwide, and treated thousands of paraphiliacs, is renowned for his efforts and holds the record for grants from the National Institutes of Health for his research. But long before he became the focus of controversy over sex reassignment, Money was shunned by many at his home institution for his scholarly interest in "perversions." Berlin, who is head of the National Institute for the Study, Prevention, and Treatment of Sexual Trauma in Baltimore and has consulted widely on the biology and medical aspects of pedophiles and rapists and paraphiliacs of all kinds, has been forced to do most of his work in a private clinic outside of his principal academic workplace, also Johns Hopkins.

Both of these men and their coworkers worldwide insist that society has every right to shun and punish people who use sex to hurt others or themselves. Their research is in no way intended to stop or impede society's taboos. But they argue with equal insistence that the desire to punish behavior society deems harmful should not exclude a desire to understand the biological and psychological roots of such behavior. Only from knowledge can better interventions emerge, they say.

The level of ignorance is enormous, however. For example, it's widely believed by scientists that much of the biological and psychological foundation of bad sex occurs early in childhood or even prenatally, yet we know very little about childhood sexuality, and just about every effort to study it produces more heat than light. "Don't even think of going there," one physiologist said during an interview. "Even writing a grant proposal to study childhood sexuality can destroy a career or make you the target of death threats. And I'm not kidding."

Some headway has been made. In one study of fifty pedophiles, for instance, investigators found that many grew up in sexually repressed households, where sex was never allowed to be discussed openly. And contrary to popular belief, investigators have found no consistent evidence that children who are abused sexually are more likely to abuse other children sexually when they grow up. Just as many who are sexually abused are averse to committing such acts, and there is evidence that many who sexually abuse children have no history whatsoever of abuse in their own lives.

In addition, even the criminal justice system is coming to recognize that while pedophilia and other forms of exploitive sex must be punished in order to protect victims, the perpetrators may also be victims—not necessarily of any abuse but of their biological predispositions. Says Berlin, "Nothing in the research suggests that perversions are 'volitional' or that their expression is a failure of self-control." Evidence for this position, he points out, rests on the fact that it would be as extraordinarily difficult to persuade a conventional heterosexual male to want to have sex with children as it would be to teach a heterosexual male to enjoy sex with another male; and it would be practically impossible to "teach" a pedophile to want to have sex only with nice girls who liked him. The failure of most "treatment" programs of pedophiles and rapists, and their repeated arrests and recidivism, are grisly testimony to the biologically driven component that must be at work and to the fact that deviance, whatever its origins, are entrenched. "Aversion" therapies that use electric shock or other "punishments" to negate the positive rewards of sexual arousal when people are exposed to deviant sex images don't work very well either.

"In society and in the criminal justice system, there are the prevalent assumptions that pedophilia is a *voluntary* [italics mine] orientation and product of jaded depravity and that the next step will be sadistic assault and molestation ending up in lust murder," says Money. "These assumptions are faulty and are based on rare and sporadic cases in which there is an overlapping of pedophilia, which is a paraphilia of one type, with a paraphilia of another." In other words, bizarre, kinky behaviors develop independently in each individual as a response to that individual's experiences, brain circuitry, and genetic profiles, and not as a result of some willful "desire" or biological imperative.

Social scientists at East Tennessee State University demonstrated Money's conclusion in a study of 106 women involved with men who cross-dress (heterosexual transvestites). The transvestites' typical partner was a forty-year-old Protestant white woman who was a firstborn child, in her first marriage, more likely to be childless compared to her age mates, and with at least a two-year college degree. She was *no* more likely than other women to have had lesbian experiences or substance-abuse problems, and she had been married to her mate cross-dresser for thirteen years and known of his paraphilia for nine. A fourth of the women said they were occasionally aroused by their mate's cross-dressing. Most were certainly not happy but were coping, and everyone was functional.

By contrast, psychiatrists consider pedophilia to be dysfunctional, a mental disorder. Yet, according to Berlin, men who sexually abuse children do so for some of the same reasons that paraphiliacs cross-dress or have foot fetishes: they have seriously distorted mind-brain views of what is erotic, influenced by their psychological and biological development, and by their individual experiences in the world.

Where it gets mean is when fantasies that go with fetish get out of hand, and because so many of the fantasies have something to do with punishment, humiliation, uncontrolled lust, restrictions, and dominance, they often do.

Some social engineers have suggested banning material that nourishes fetishism. But what to do about the ads in the *New York Times* for five-hundred-dollar spike heels that make a woman feel sexy, but some man aberrantly and inadvertently turned on?

Berlin struggles not only with his patients' problems but also with the general queasiness most people have about any display of arousal. "We may wonder why we evolved in such a way that distorted lovemaps can happen to so many," he says. They may indeed be by-products of mechanisms nature developed to make sure we paid attention to every sexual cue possible. But if the process is a natural one, why the queasiness? That one's probably easier to answer, says Berlin. "We have evolved a biological imperative to have sexual drives and craving for satiety and these drives tell us 'if I don't have this, or eat this, or have sex now, I will die.' Nature wanted all of us to be sexual. Similarly, we evolved to be upset about any set of human behaviors that are terribly different from the mainstream. We see this in people and in all social animals. It's almost instinctual the way people come together in groups that support each other and quickly exclude and alienate those who don't fit. That is adaptive, too. My guess is that if a large or significant percent of people were pedophilic, they might have a bigger voice in what's acceptable." Indeed there are "man-boy" societies whose leaders argue for just that point. "On the other hand," Berlin warns, "mothers and their maternal instinct to protect their children is just as strong if not stronger." In a duel of such interests, it's likely mothers would win.

Explaining further, Berlin notes that "all people *must* choose to eat and have sex to perpetuate the species, thus we are born with powerful biological drives for both. A decision to neither eat nor have sex will therefore be opposed by such powerful biological forces. Whether paraphilias or hypersexual behavior is like obesity in that we cannot control it, is to draw the point much too fine. But it is a remarkable testimony to the power of a biological drive that so many are struggling so hard to lose weight when 'all' that is necessary is to eat a little less."

In these scientists' views, normal biological and biochemical "templates"—the testosterone, the estrogen, the serotonin and the nitric oxide, the AVP and the oxytocin—may be present in paraphiliacs, including pedophiles. But in bad sex, as in violence, neural circuits are likely to be distorted by postnatal stresses and experiences. The literature on the biology of aggression, for example, links such behavior not only to high testosterone and serotonin imbalances, but also to abusive early punishment and repressions that may sculpt young children's biological hardware and software, leaving them physiologically unable to tell the difference between situations that are truly threatening and situations that are not. A similar scenario may be at play in children who develop adolescent or adult paraphilias.

Whether these children come already vulnerable to such distortions because of prenatal hormone baths or deficits, or genetic susceptibilities that leave their brain circuits more fragile or vulnerable to deleterious sculpting, is not known. But there is evidence that some aspects of *temperament*, widely understood to be an overall inborn setup for experiencing the world, do not change very much. Some children are born shy or furtive, others more adventurous and daredevil. And while their behaviors can be tempered by training and education (shy kids can become actors and daredevils can learn to be more careful), there will probably be some individuals whose brains and bodies are more likely to lead them to bad sex.

One theory is that mammals and humans in particular evolved to be able to have sex under extreme conditions of environmental stress—with predators threatening and food scarce and all sorts of things that go bump in the night. That's because all mammals except humans have only brief, seasonal periods of fertility and because even human females are fertile only a few days each month, so males and females would have had to "seize the moment" regardless of the surroundings. It is unlikely that our ancestors were able to "plan" their families in the way we do today, waiting until there's enough money in the bank and a college trust fund. This may be one reason why sexual arousal and orgasm can result from mental or physical violence or stress. And pain and violence may have become a sexual stimulus to our hominid ancestors, who may have frequently faced painful stress and deprivation, but survived by continuing to seek sex and survival of the species anyway.

"In general," according to Berlin, "behavior, whether sexual or nonsexual, is a reflection of one's state of mind, as persons tend to act in response to their thoughts and feelings. Some states of mind can be considered pathological, for example, when an individual loses the capacity to determine whether heard voices are coming from the environment or are imaginary. This type of impairment can occur in a variety of psychiatric syndromes such as schizophrenia, dementia, delirium, or manic-depressive illness. Mentally ill people sometimes commit sex offenses as part of their mental distortion. But in other cases, some persons commit sex offenses in response to intense, unconventional sexual hungers."

The causes of conventional and unconventional erotic interests are multiple and complex, he notes, but there is evidence that biological factors such as hormone levels or chromosomal makeup sometimes play a major contributory role with respect to the nature of an individual's sexual desires.

Among the evidence Berlin has accumulated in more than two decades of research on sex offenders is the case of a forty-year-old white male referred by his lawyer for assessment after he was arrested for having sex with a thirteen-year-old boy and who was on probation for a similar offense five years earlier. He had been sexually active most of his adult life almost exclusively with young males, some as young as eight, since he himself was seven years old, although he had rarely been caught. He undressed, fondled, and participated in mutual masturbation and organ genital sex several times a month. While he usually persuaded the children rather than forced them, when drunk he had threatened his victims with a knife.

Typical of such offenders, he hated his behavior and felt ashamed and guilty, using alcohol to fire up the courage to seek other sex partners among children. He would try to resist but have irrepressible urges. Just watching a young boy in a TV commercial would bring on strange urges to focus his attention on the child's genital area. Here is what this patient told Berlin:

"If I have seen an exceptionally nice-looking boy I get aroused. I want to go over there but then again I don't. I see him and I want to get out of there because I know I am going to start fantasizing. I have noticed that the first thing is I drop my eyes to his genitals. It gets more intense, the fantasies, that I dream about a South Sea island, nothing but boys on the island. I know it is wrong. I know what the legal issues are. All I can think about is getting the boy. I want to keep doing it and doing and doing and doing it. No matter how."

Berlin found a number of "biological" pathologies in this patient. He may, for example, have had the genetic disorder known as Klinefelter syndrome (where sufferers have an extra X chromosome), which has links to sexual deviation. Such ambiguous sex translates into people with enlarged breasts developing at puberty, small testicles, abnormally long arms and fingers, low sperm counts, and elevated LH levels and FSH levels. Clearly, these are not hormonally normal individuals.

From the Greek for altered love, paraphilia, says Money, must always emerge from a distorted or mismatched lovemap, the template that exists in all of us at birth and that develops in the mind and brain to depict our idealized lover, love affairs, and program of sexuoerotic activity in our fantasies or actual acts of sex. It is comprised of our genetics, our hormones, and our experiences. About the only certainty Money writes about in his lengthy expositions of paraphilias over his sixty-year career is that during early childhood development, normal sex play and sex rehearsal are necessary if children are to have normal lovemaps. "During the developmental years of juvenile sexual rehearsal play," he says, whether in monkeys or in humans, "appropriate matching of the lovemaps of playmates for age synchrony and image reciprocity has a healthy developmental outcome more often than does mismatching and this kind of sex takes place predominantly between infants or juveniles of the same age." The very young may learn from the older people having sexual play (as children do with parents or teens), but their best shot at normalcy is if this observation is from a distance. And in any case, such rehearsal play is "essential as an antecedent to normal lovemap development."

Thus, even assuming that some bad sex is influenced by Klinefelter syndrome's chromosomal and hormonal rearrangements of the body and brain, bad sex is also a consequence of what a Klinefelter child is taught. Some Klinefelter children, while usually "assigned" the sex of "male," may feel a gender identity with a male or female and, depending on how that is reinforced or repressed, may develop fantasies, urges, and desires for different sexes or different age groups. When enlarged breasts occur at puberty, many Klinefelter patients have the breast tissue removed to look more like men, but others do not. Some will want to dress in girl's clothes because their hormones are saying "you're a girl." Whether their families allow it or not, or what kind of response they get from others when they do or do not dress one way or another, can have a profound influence on how their sexual behaviors—kinky or otherwise—will play out.

In humans, says Berlin, structural and functional differences in the brain between males and females depend on exposure to various sex hormones during embryonic development, and there is every reason to believe that such exposures, normal or abnormal, play a role in the sexual deviancies of those who are already predisposed by other factors—chromosomal, environmental, social, and psychological—to have sexual ambiguities.

Berlin has a long list of such biological and neurological findings. In a patient with erotic sadism, he has found oculomotor abnormalities suggestive of basal ganglion dysfunction. In dozens of cases of homosexual pedophilia, he has found not only Klinefelter but also cortical atrophy, grand mal seizures, schizophrenia, elevated testosterone, dyslexia, and other chromosomal disorders. In a patient with voyeurism, he has found elevated LH levels, and in cases of paraphilia rape, elevated FSH and LH as well as testosterone.

Another kind of link between the brain or mind and bad sex emerges from the painful lessons of psychosurgery, which damages the brain in order to try to save the minds of sex offenders. These operations range from castration to making lesions in the preoptic area of the hypothalamus, a rich source of sex hormone receptors. Animal studies have shown that ablating such structures can reduce sexual impulses; the grim, dark history of prefrontal lobotomies in the 1940s, 1950s, and 1960s includes hundreds of documented cases in which men and women either lost their sexual drive altogether or exhibited bizarre sexual behavior as a result of surgery in the amygdala and hypothalamus.

Ordinary house cats, first described by Kluver and Bucy in 1939 to have hypersexual activity when their temporal lobes were removed, experienced a reversal of this hypersexuality when specific sites in the ventromedial nucleus of the hypothalamus were destroyed surgically. And the first reported attempt to do this in a human in 1966 was successful, on a homosexual pedophile who reported after the operation that he had lost his pedophilic urges.

Stereotactic hypothalamotomy, lesions made in the middle of the hypothalamus in attempts to control violent behavior and sexual disorders, have been reported to be 95 percent successful in studies in Japan, although there are serious side effects, including hyperactivity and lack of bladder control. One case in 1977 involved a twenty-nine-year-old man sentenced to six years in prison for multiple rapes. Six months after his operation, he said, "I can converse with women, without thinking of sexual intercourse. I feel considerably safer and freer. I only regret that this approach could not be used much earlier." In 1979, neurosurgeons reported substantial sex-drive reduction in thirty-four sex offenders undergoing a similar effect after voluntary surgery. But psychosurgery created so many other side effects and was so patently abused in some treatments of the mentally ill, women, and minorities, as well as some sex offenders, that only very recently has any serious interest in neurosurgery for such problems even been discussed in scientific circles. It's unlikely that the therapeutic promise of psychosurgery will ever be genuinely resuscitated.

Taking all the evidence into consideration, Berlin continues to view pedophilia and other paraphilias as "variants" in human sexual behavior. That is not to say that behavior that hurts others, or conflicts with societal expectations, should be tolerated or allowed to continue unchecked. It is to say that it is not moral turpitude. Almost without exception, psychiatrists and psychologists and therapists who work with pedophiles, rapists, and paraphiliacs emerge over the years with a great deal of sympathy for their patients even while they are corresponding with them or treating them in prison. To these physicians and investigators, the perpetrators of sexual crimes are also prisoners of their minds and bodies, whose punishments are as often self-inflicted as they are inflicted by a society that, justifiably in most cases, wants them gone.

For most of us, there is little sympathy for practices that use sex and sexuality to harm others, to coerce or to cause pain. Nevertheless, it also is true, paradoxically, that what causes pain may also bring pleasure to many people, suggesting a link between the two.

One possible explanation is that the pain and pleasure centers in the brain operate in an overlapping and integrated way so that when pain is perceived, natural mechanisms can instantly react and attempt to neutralize or overwhelm the painful stimuli.

Pain signals are carried by nerve fibers along the spinal cord to two key switching stations in the brain. The first, located at the top of the spinal cord, or lower end of the brain stem, is called the reticular formation. The second is the thalamus, the walnut-sized organ deep inside the base of the brain that functions like a relay station. From here, pain information is further processed and signals are relayed to the highest level of the brain, the cerebral cortex, where pain is "felt," or perceived, and its emotional consequences experienced. Pain signals continue to travel to areas around the thalamus. Here, the brain has a large supply of opiate receptors and a rich supply of cells that produce endorphins, the brain's opiates. This system sends "stop pain" signals back down the track.

While these signals are traveling up and down the central nervous system, neurotransmitters—which cause nerve impulses to jump from the ending, or axon, of one nerve cell across a tiny gap, or synapse, to the dendrite of another nerve cell—are being released to regulate the speed with which signals fire across synapses, or to turn signals on and off. These neurotransmitters operate alone or in combination to produce a variety of chemical "messages" at the same time. Tomas Hokfelt, a Swedish scientist, has found that each single neuron can release more than one transmitter and two or three in combination may be able to convey more subtle sensations and information than any single one can.

Some neurotransmitters, including norepinephrine and epinephrine (aka noradrenaline and adrenaline) are released not only in the brain but also from the adrenal glands near the kidneys in the form of hormones. In their hormonal form, they increase the heart rate, blood pressure, and flow of sugar into the blood—all part of the "fight or flight" response to danger and injury, but also part of sexual arousal.

And then there is the limbic system, that strip of brain tissue around the thalamus responsible for emotional responses, including suffering as well as love. As the thalamus interprets the chemically regulated electrical impulses we perceive as pain, these messages pass to the limbic system and to various parts of the cortex, the outermost layers of the brain responsible for memory and physical movement, as well as pain assessment. Finally, the impulses reach the pituitary and the hypothalamus, parts of the brain that also release endorphins and enkephalins to ease pain from any source, physical or psychic. In this complex, integrated system, pain can even seem to occur in an amputated limb, because nerves that once served the missing part continue to send pain signals to the brain. And it's likely that, influenced by early experiences, the brain's neural connections can create the perception in some people that pleasure, rather than reflecting the absence or diminution of pain, *requires* pain.

While it may not be S&M, nevertheless couples frequently, perhaps universally, do things in the course of lovemaking that they wouldn't want their closest friends to know, practices they consider well within the bounds of good sex but that hug the edge of injury.

While everyone has a favorite body part that is sexually arousing— neck, breasts, genitals, ankles, and feet (there are an estimated 1.5 million foot fetishists in the United States alone and there are even Web sites devoted to it), the point at which an obsession or "fetish" over a body part or some other item involved in arousal crosses the line to sexual injury is a psychologically and physically blurry area.

What all such "kinky" sexual practices, from foot fetishism to sado-masochism, may have in common is their apparent origins in childhood, in the plasticity of the developing brain and perhaps in the biological vulnerabilities of some brains more than others. All of them, moreover, are compulsive and exclusive. As Fred Berlin puts it, if you don't have foot fetishism, which probably emerges in response to something sexually comforting involving feet, no amount of exposure to feet or high heels will turn you on.

The complexities of paraphilias, as a behavior triggered by both body and mind stimuli, are illuminated by studies comparing rapists and non-rapists. Scientists in Montreal assessed the sexual arousals of a group of ten rapists with a history of low-level physical violence and compared them with ten non-rapists. Stimuli consisted of audiotapes with such themes as mutually consenting sex, rape with violence, rape with humil-iation, physical aggression without sex, and neutral scenarios. While the subjects listened to the tapes, penile responses were recorded using a special gauge that measured the strain against it by erection.

The results showed no differences in the level of erection between the groups for rape themes involving violence. However, the tapes depicting rape involving humiliation found the rapists, far more than the non-rapists, responding, suggesting that for the rapists, it's the humiliation they inflict, not the pain, that is arousing. It doesn't take much imagination to conceive of how young children subjected to physical or psychological humiliation might work hard to turn that pain into some kind of "perverted" pleasure and might grow up making pain and injury, plus or minus humiliation, prerequisites to sexual arousal and orgasm.

Perhaps the truest thing that can be said of human sexual arousal and performance is that they manifest in endless variety. Some of the displays may have persisted because they continue to confer some ancient advantage in the reproductive wars. Others may persist because the wiring to facilitate them evolved incidentally or coincidentally to other aspects of our biology and psychology, making large numbers of people vulnerable to their occurrence. Still others emerged as powerful symbols of sexuality whose impact died with their civilization. Virgin sacrifices among the Aztec and Maya peoples come to mind.

And perhaps the only *sure* thing to say about "bad sex" is that none of it was likely to have emerged full blown in the dawn of recorded history to plague us with "immoral" images. They likely had their origins in some behavior that conferred some advantage to someone or some society at some time. Nor is it likely that sexual perversion as we modern humans define it is a matter of "bestiality," as if we had pushed some evolutionary rewind button and wound up more like animals than we used to be. If indeed sexuality and sex are so strongly ingrained to assure that humans can't or won't avoid sex, even under such daunting circumstances as war, famine, or disease, then the fact that we have developed cultural, psychological, and biological vulnerability and fragility where bad sex is concerned may be nature's way of reinforcing and rewarding some survival behaviors and punishing others at various times.

In any case, our cultural biases may blind us to whatever "good" may be echoed in "bad sex," and to what we might be able to alter about sexual behaviors that harm ourselves or others.

Dr. Marcy Lawton of the University of Alabama in Huntsville once gleefully described an example of this in a story about how male scientists were duped by their studies of a group of jays whose males were decidedly unwilling to be aggressive, compete for females, or dominate any member of the flock. It took a long time for the scientists to get over their shock and their convictions that something was "abnormal" about the birds' sexuality, and finally turn their attention to what the females were doing. When they did, they discovered that the females, not the males, engaged in sometimes mortal and often ferocious combat with one another just before breeding. But it got worse. Instead of theorizing that there was perhaps something about the biological and "psychological" evolution of these female birds that determined their ferocity, the male scientists assumed these females were "abnormal" and had some feathered equivalent of PMS. In fact, further study demonstrated that the female jays are simply the ones who regulate dominance and hierarchy within the group and that the group is a matriarchal one.

Other scientists eventually revisited the data and conducted more field-trip observations to look for explanations beyond their old assumptions. They've begun to see that the paradigm for sexual behavior, between sexes and among sexes, in competition for mates and in competition with one another for mates, involves cooperation, pair-bonding and tenderness, and romance, as much as competitiveness, aggression, coercion, and manipulation. It involves honesty as well as deceit; what's more, these traits coevolved and did not occur in some vacuum of evil male or female design. If males fight one another to strut their genetic fitness for mating, then females had to have had a hand in preferring that males do so. If females fake orgasm in order to manipulate their reproductive fitness and success, then males evolved mechanisms for detecting the fakery at least some of the time, or of overcoming the advantage it gives to females.

And consensus is forming around the idea that whatever the relative contributions of nature and nurture, humankind's big brains and ability to learn from experience can modify and even overturn traits that were thought to be so hardwired they were intractable. Fortune 500 company CEOs in power suits don't split the skulls of their subordinates who fail to show submission by lowering their eyes the way subordinates to alpha males did in the ancient past, or the way wolves and bull elephant seals still do.

The boys-will-be-boys and snakes-and-snails-and-puppy-dog-tails version of male psychosexual development is no more verifiable or informative than the women-are-all-gold-diggers or sugar-and-spice-and-everything-nice version of female psychosexual development. If boys have a biological susceptibility to attention deficit hyperactivity disorder (ADHD) that girls do not, and if it makes them more adventuresome, more willing to take risks, and more naturally aggressive and violent toward girls, that is an interesting phenomenon—but it does not make boyhood, or maleness, what one writer has called a "protodisease."

In an essay written for the Carnegie Corporation of New York, Carnegie's president, David Hamburg, mused on the subject of war and ethnic conflict and aggression, but his words speak eloquently if unintentionally to the issue of sexual violence, to bride burning and battering, to Lorena Bobbitt's attack on her husband's penis, to judicial tolerance of wife beaters, to Nicole Simpson's cut throat, and to the profound and predictable possibilities for negotiated peace between the sexes on the whole.

"The capacity for attachment and the capacity for violence," he said, "are fundamentally connected in human beings. We fight with other people in the belief that we are protecting ourselves, our loved ones and the group with which we identify most strongly. Altruism and aggression are intimately linked in war and other conflicts. . . . My lifetime has witnessed terrible atrocities committed in the name of some putatively high [moral] cause. Yet there have also been vivid examples of the reconstruction of societies, major reconciliations and real enlargement of opportunities for substantial segments of a population. . . . If we could understand [the conditions under which the outcome can go one way or the other] maybe we could learn to tilt the balance in favor of a stable, enduring peace. . . . [T]here is certainly the possibility that we humans can learn to minimize these tendencies. . . . The ancient propensity toward narrow identity, harsh intolerance and deadly intergroup conflict will confront us with new dangers. . . . In the realm of scientific research, the interactions of biological, psychological and social processes in the development of human aggressiveness leading to violent conflict must constitute an important frontier in the decades ahead."

There is absolutely nothing in the scientific data banks to suggest that an individual's level of a brain chemical—or any other factor, including a mother who beat him, a father who was absent, and poverty—justifies rape or any other act of bad sex or violence. But as the more thoughtful philosophers of sexual science suggest, information about the genetic and biochemical roots of sex behavior—good and bad—provides a heuristic framework, a provocative set of hypotheses, a way of thinking about the interactions of biology and learning, of brain and mind and body, that will lead us as a species to ask the right questions of one another. And certainly to begin to approach the important issues of our romantic and most intimate relationships with more intelligence in the future than in the past.

12

ALTERNATIVE SEX

"The most remarkable thing about sex is its diversity."
—SIMON LEVAY

Plato, undoubtedly influenced by Greek mythology's story of Hermaphroditus, described the case for a third sex, called androgynes. Hermaphroditus was the son of Hermes and Aphrodite, who, with a nymph called Salmacis, formed a single body that merged male and female characteristics. Plato argued that an angry Zeus split them in half and doomed them forever to seek their "other half," which he believed neatly explained heterosexual and homosexual love and sex.

In some ways, scientific studies have not substantially altered Plato's prescientific view of things. There is growing and substantial evidence that however much we may use the shorthand of "masculine" and "feminine" to describe the entirety of normal animal and human sexual behavior and biology, sexual behavior fails entirely to be neatly binary. There appear to be many ways to "be" a male or female. No one can know how animals feel about their love objects, but at the level of behavior, "homosexuality" is hardly uncommon in nature. If female cats are deprived of male company, that's OK, some of them will take on the "male" role and arousal will take place. Bonobos have routine same-sex rendezvous, including genital-to-genital rubbing, probing, and orgasm. Male gorillas will seek other male gorillas for sex play. In fact, "homosexual" behavior is so commonplace among animals that its relative infrequency in people—best guesstimates are about 4 to 5 percent of men and 2 to 4 percent of women—is the genuine surprise.

Clark Gable and Gore Vidal can both be accurately described as men and Marilyn Monroe and Ellen DeGeneres accurately described as women, but not the same kind of men and women. To be a man or a woman is first to have been a boy or a girl, and before that to have been XX or XY in the egg-sperm unity. And if we marvel at the variety that meiosis and mitosis and hormones and genetic switches can bring to something as socially inconsequential as our fingerprints or our complexions, how much more so must we marvel at the variety these processes may bring to our sexuality. We cannot claim infinite diversity in one trait and fail to allow for it in another.

To be sure, there are predominating patterns that are recognized as reproductively fit male and female, heterosexual and homosexual, under-sexed and oversexed. Of an estimated 235 to 245 existing species of primates on our planet, only mandrills and orangutans are entirely disinterested in sex with females—or anything having to do with females—except during the five-month breeding season. Western culture especially has tended to put rigid lines between categories of sex and gender, but every human being's biological sex, brain sex, and sexual orientation lie along a continuum that is physiologically fluid and dependent on the context of their lives. A person may *feel* one way and act another; act one way this time and another way that time. As sex neurobiologist Roger Gorski put it, "A man who has anal intercourse with his wife isn't likely to consider himself homosexual." Just as people adapt their appetite for food and drink to what is available, so do they adapt to their sexual circumstances. Just as nature offered options for our temperaments and preferences, it offered the potential for options about how we perceive and play out sexuality.

Behavioral geneticists, endocrinologists and neuroscientists, and psychologists are in an almost frenzied search for physiological confirmation, affirmation, or refutation of homosexuality and bisexuality as genetic or biological constructions. Despite the effort, the jury remains out. "I do not know—nor does anyone else—what makes a [particular] person gay, bisexual or straight," says Simon LeVay, whose controversial but respected studies on the biology of sexual behavior and orientation guide much of the research. "I do believe, however, that the answer to this question will eventually be found by doing biological research in laboratories, and not by simply talking about the topic, which is the way most people have studied it up to now."

One reason for LeVay's optimism about finding at least some biological foundation for alternative gender orientation and preference must rest in the evolutionary rule that anything widespread and persistent among human societies makes a compelling case for having some biological predispositions to account for it. Various studies have measured the rate of homosexuality in human populations at a fairly consistent 2 to 5 percent of all males in every generation.

Given that most of these men don't have children and thus don't pass on some "gay gene" to the next generation, there remains the likelihood that somewhere along the line, nature endowed us with the capacity for sexual flexibility and organized our genes to allow for the development of variations on themes at the least. In nature, some species can literally alternate sexes, depending on conditions and the needs of the community. Some creatures can produce both male and female gametes and produce offspring without benefit of a male and female "parent." There is increasing evidence of specific genes linked to same-sex preferences. And evidence is growing that the brains of heterosexual and homosexual human beings are different in measurable ways. "This startles a lot of people," says Roger Gorski. "But anyone who thinks about it for 30 seconds will find the concept logical."

The hard evidence for it is still relatively scant, but there is the possibility that homosexuals, persistent and consistent as they are in human and apparently some other animal societies, have a valuable role in a particular population's sexual success, perhaps as a control mechanism, perhaps as "altruists" who sacrifice their reproductive fitness for some other purpose. Homosexuality, from the standpoint of evolution, some argue, could also be a "progressive" or a transitional phase en route to more than two genders.

In fact there *are* societies with three genders, known to anthropologists as berdaches, hijras, and xaniths, all of whom are biological males who live, act, and are treated as "social women," or "male women," according to sociologist Judith Lorber. Similarly, she says, "there are African and American Indian societies that have a gender status called manly hearted women—biological females who live as 'female men.'" In Western culture, transsexuals and transvestites approach such status, although they are not accepted as normal by the wider society. But clearly, Lorber notes, "genders . . . are not attached to a biological substratum. Gender boundaries are breachable."

The breach has been particularly visible in traditional and non-Western societies that seem to have more ease in shuffling the traits—both biological and psychological—that make up genders, and less devotion to tying genitals to either sexual orientation (how the individuals feel themselves to be and to whom they are attracted) or gender status (how the larger society views them and how they want the larger society to view them.)

Among the strongest examples of such societies are some Native American cultures that have a formal, socially acceptable gender known as the berdache, validating the lives of men who do women's work. Unlike male-to-female transsexuals or intersexuals, the role of the berdache can be a sacred one, and they enjoy a safe and even revered status in some tribes, a haven for boys who from early childhood display what we in the West would call effeminate behavior. Berdaches are given the tribal status of educators of children and singers and dancers at tribal events. Historically, they took care of the sick and helped the women in provisioning war parties. In her book *Paradoxes of Gender*, Judith Lorber reports that among the Navahos, berdaches do women's "craft work" but "also farm and raise sheep, which are ordinarily men's work." They are believed to be lucky in business, act as heads of family, and are entrusted with family business.

The surprising thing about berdache and other social genders of this kind is that homosexual sex may occur between non-berdache males, but the cultures emphasize the berdache as the designated person that a man would seek out for male sex. The men in the tribe are not required to "come out of the closet" or make a choice about their sexual orientation. They can choose a berdache one night and a wife another. But in many of these tribal cultures, a man who does want to have sex exclusively with a berdache can "marry" him.

What this accomplishes in the society is that the berdache satisfies the sexual needs of the tribe's homosexuals without being any threat at all to the institution of marriage, which is sacred to the tribe. And since homosexual sex *acts* are not what define berdache status for a man, the berdache is very unlike the designation of "male homosexual" in Western culture. Rarely do berdaches marry or have sex together, and the berdache's "husband" is not labeled a homosexual.

Another example of such gender-expansion are hijras, a group in northern India who define themselves as intersexed males who have become women; they may in fact undergo voluntary castration. In some ways they are like transvestites, female impersonators, or "drag queens," but they more often have the status of European castrati, who first emerged as a special class in the seventeenth century because the Roman Catholic Church would not allow women to sing in public. Hijras live in their own households and create extended families of "adopted" mothers, daughters, sisters, aunts, and grandmothers. They sing and dance at weddings and births, work as cooks and servants. They are required to dress like women but do not try to "pass" as women; instead they are openly deviant, almost a caricature, as is the drag queen of *Torch Song Trilogy*. They wear skirts *and* a beard. This legitimatized function is coupled at times with prostitution with men, and many of these people become "mistresses" to homosexual men. They worship a mother goddess and sometimes Shiva.

A third intermediate gender, xaniths, are found in Oman, a highly sex-segregated Islamic society in which women's purity is strictly controlled by the heavy veil and by relatives. Xaniths are male homosexual prostitutes who openly dress in men's clothes but in pastel colors. They are steered to women's circles during celebrations but still have men's legal status. Their social role is to be the sexual toys for unmarried or separated men, a means for this society to protect the "purity" of their women; the men who have sex with them do not consider themselves homosexual, nor does the society in which they live. In some families, xaniths are a "family tradition" and can revert to full manhood when they marry and "deflower their bride," according to Lorber.

A few years ago, some newspapers carried a story about a women in Lepurosh, Albania, Sema Brahimi, who at age fourteen apparently decided to become a "man." She cut her hair short, put on trousers, went to work in the fields, changed her name to Selman (the masculine form of Sema), and insisted her mother and sisters refer to her as "he" and "him." Now in her late fifties, she is completely accepted as a man in her traditional Muslim village in the Dinaric Alps, a population of three hundred so isolated it has no telephone, car, or house with indoor plumbing.

Remarkable as her individual story is, what is more remarkable is that Selman is one of a legendary group of "Albanian virgins," first described by British writer Edith Durham in a 1909 book. These were women who swore never to marry in order to keep order in a society where males were scarce. They often become fierce chieftains of villages in the area bordering Montenegro. Those who chose the life also escaped the strict

Kanun law that essentially defines women as childish chattel, with no property or inheritance rights and at the mercies of husbands, fathers, sons, and village men in general. Women who vowed virginity and took on the whole life of a man could get the status and respect of men, making such status a haven for women who did not want to enter into an arranged marriage, who had no brothers to protect them, or who were needed to replace male family members who were killed in blood feuds or wars.

Established social scientists (most notably gender expert John Money) consider these "extra" genders to be "odd" or "deviant," whose learned, nontraditional sexual behaviors are the exceptions that essentially prove the rule that humans must ultimately "learn" to be males or females regardless of their genital legacy. But there also is the possibility that they continue to be nature's experiments with sexual reproduction and mating strategies. If the only purpose of sex were to reproduce, then the persistence of sexual desire long after reproductive fitness fades would be a "leftover" phenomenon without a purpose. Perhaps it is; or perhaps desire beyond our ability to reproduce is a way to encourage psychological denial of death and extinction and promote sexy instruction to our grandchildren. Or perhaps homosexuals and the transgendered are echoes of some other natural phenomenon evolved to meet challenges that no longer exist or that we don't yet recognize.

Whatever one's views about the mutability of gender, these individuals bear witness to the variety of ways that human beings orient themselves in a sexed world and to the fact that genital sex and procreation are not the only defining characteristics of sexual activity. When it comes to the biology of sexual *preference and orientation*, the picture is even fuzzier. It might be that biology predisposes people to their orientation, or it may be that their behavior and socially defined roles influenced their biology. Prenatal hormones and genetic influences may determine or predispose people to one sexual orientation or another, but it's possible that in some cases the reverse is true. Not all that is biological is genetic and not all that is genetic is biological.

Money and some of his colleagues, after an extensive review of the literature on "gender identity," or a person's primary identification with one or the other sex, concluded that prenatal hormonal environments had little to do with gender identity. They based their conclusions on case histories that show that in children born with ambiguous chromosomal and genital genders, gender identity can be largely whatever the parents rear them as, even when their genitals "disagree" with what sex they are "assigned."

Other unusual cases—most notoriously in the last few years, one of John Money's—just as easily argue another way. In his book *As Nature Made Him*, John Colapinto dramatically describes the accidental surgical mutilation of an infant boy's penis more than thirty years ago; his parents' desperate search for help; their consultations with Money, who recommended female gender assignment; and the boy's subsequently successful struggle to reclaim his male identity after decades of female hormone treatments and enforced living as a girl and woman. Money's critics, most notably a Canadian psychiatrist, point to this case as solid evidence against both the theory and practice of gender assignment, and have gone so far as to denounce Money's handling of the case as unethical experimentation in the absence of any theoretical foundation for such gender manipulation. For his part, Money has pointed out that the family removed the boy from his care during early adolescence, and that it's impossible to know whether the family was truly able or willing to comply with the rigors of complete gender reassignment during their son's childhood. Insisting that he cannot publish any reevaluation of the case because he neither has continued treating the individual nor has permission from him or his family to write about him, Money at this writing has remained silent on the subject.

Whatever the true "nature" of the person around whom the firestorm erupted, it *is* true that he was a completely normal chromosomal XY male, one born not with a genetic or genital ambiguity of any kind. Thus, a more compelling case for the biological determinism camp can be made with a group of genetic (XY) males, first identified in the Dominican Republic, called *guevedoces*, or "penis at twelve years of age."

These children were raised as girls for most of their childhoods because they were born with female-appearing external genitals. Because they had an inherited enzyme defect called 5-alpha-reductase deficiency syndrome, they had severely deficient dihydrotestosterone production but normal capacity for making and being sensitive to testosterone in their tissues. Suddenly at puberty, when the hormones turned on, they became virilized as testosterone levels rose. Their penises grew, their testes descended, and their voices deepened. And despite the fact that they were raised as girls for twelve or thirteen years, seventeen of the eighteen examined in one study made an easy segue into being completely male.

Several cases of males with ambiguous genitals assigned a sex in childhood, with or without cosmetic or hormonal treatment to support the assignment, have also undermined the "social learning" theory of gender identity to some extent, by making decisions at some point in adulthood to abandon their "assigned" sex and adopt the lifestyles of their chromosomal sex. Nevertheless, some have maintained their assigned roles, and some studies of "ambiguous" gender identity that have been completed again suggest only that the human sexual experience is nothing if not flexible. One of these involved children exposed prenatally to diethylstilbestrol (DES), a synthetic estrogen given to women for medical reasons, became pregnant. DES, already known to exert masculinizing or defeminizing effects in the central nervous system of animal species, reportedly increased the incidence of bisexuality or homosexuality in female offspring of the mothers, even the offspring of the 75 percent of the mothers who described themselves as entirely heterosexual. In another of these studies, genetic males exposed to an androgen-deficient prenatal environment because of a genetic insensitivity to androgens were found to be female in appearance and attracted to men as they grew to adulthood.

Also investigated were genetic (XX) females with congenital adrenal hyperplasia, an overgrowth of the adrenal gland, who were genitally masculinized to varying degrees because they were exposed prenatally to excess androgen caused by a deficiency in an adrenal enzyme that breaks down androgen. Despite the fact that these girls were reared as girls and developed a female gender identity after genital surgery, they all had "masculine" toy preferences and were often "tomboys." They tended to not be very maternal or interested in feminine behaviors or clothes.

THE BIOLOGY OF HOMOSEXUALITY

Whatever the lessons from such unusual genetic disorders, they stand in contrast to what scientists are beginning to learn from larger numbers of people who are free of disease and genetically, at least at the sex chromosome level, completely normal. This is the group that comprises the vast majority whose genital and chromosomal sex mismatches their sexual *orientation*, homosexuals whose romantic and erotic attachments are for those of their same genital and chromosomal sex.

It has taken seventy-five years to substantially undo the damage that psychodynamic theories of sexuality perpetrated on homosexuals, transsexuals, and their families—theories that blamed weak dads, domineering moms, and homosexuals themselves for their orientations. Only now have we begun to resuscitate the remarkable insights of Havelock Ellis and Magnus Hirschfeld, who suggested that homosexuality and heterosexuality were equally normal aspects of human nature determined in some part during early development. Some of the old thinking persists, marked by such initiatives as the Christian conservative movement's gay conversions, but, as with most expessions of human behavior, it is difficult to lift proof of biological principle out of faith-based convictions or rare examples. Behaviorally, at least, just as with heterosexuals, homosexuals are stable in their preferences, and where scientists have tested the biology and psychology behind the behavior, a consensus is emerging that their lovemaps, fantasies, and attractions are as well, whether or not they *act* on their orientation, whether or not they have sex. Sexual orientation, in fact, is stable even when brain function is temporarily interrupted by such things as injury, memory loss, and sleep. People don't wake up in the morning or from anesthesia and become a different gender, even after years of coma.

It needs saying that rigorous scientists have yet to identify any gene, or group of genes, or discrete cascade of biological events that conclusively and entirely *determines* or *predicts* homosexuality, or for that matter heterosexuality. But to ignore the contributions that distinct physiological and evolutionary events are being shown to make in *all* facets of human sexuality is to fail anyone's silly test. Worse, it is to inflict harm. In Simon LeVay's opinion, past and even some present "treatments" of homosexuality by the medical and psychiatric professions are the sexual equivalent of the Tuskegee experiments on syphilitics. "The physical and psychological damage done by psychoanalysis, castration, testicle grafts, hormone treatments, electric shock therapy and brain surgery in attempts to cure homosexuality must be counted among the most serious crimes ever committed by the medical profession," he says, because homosexuality is not a disease. At worst homosexuality is an unwelcome (by some) condition, but at its foundation, it simply *is*. How could it be otherwise? To ascribe to numberless generations of homosexual children and adults some otherworldly capacity for self-delusion, self-destruction, or conspiracy would require the heterosexual world to take on just those characteristics. To insist there is no continuum of human sexual expression requires inflexibility in all other facets of human existence. Logic and universal human experience of vast variation in behavior demand that if nothing else, we take homosexual existence and feelings at face value. To do that means homosexuality exists as an alternative to what most men and women may experience as the dominant expression of their sexual lives, but one that is no more natural or unnatural than any heterosexual preference or orientation.

One pot of evidence is epidemiological. Homosexuality often "runs in families." James Weinrich and Richard Pillard of Boston University have discovered that having a gay brother, for instance, significantly increases the chances of being gay. About a fourth of all brothers of gay men are themselves gay, compared to the rate of male homosexuality in the general population of 3 to 5 percent. Dean Hamer of the National Institutes of Health and an expert on the sex chromosomes and the genetics of behavior has found that among gay men, there is an increased incidence of homosexuality in their maternal cousins or uncles as well.

Among women, the figures for lesbians are equally significant. But there is no evidence at all that having a gay brother will increase the likelihood that a woman will be lesbian or vice versa, strongly suggesting that whatever biological factors are at work, they are different for men and women.

Not surprisingly, studies of twins, both identical and fraternal, offer insights into a genetics of homosexuality. Studies show that having a gay identical twin makes your own chances of being gay a little better than even, while having a gay fraternal twin makes your chances only about 25 to 30 percent, still higher significantly than the "random" chances in the general population, but not as high as sharing the entire genetic load with an identical twin. These studies reveal the influence of prenatal and afterbirth factors—including stress and even the position of the fetus in the uterus—at work against a background of inborn potential. In twin pregnancies, for instance, although they may be genetically identical, one twin may get more than half of the mother's blood supply or hormone bath, leading to differences as obvious as birth weight, but perhaps also subtle differences in sexual development and orientation. "All in all," says LeVay, "these twin studies point to a strong but not total genetic influence on sexual orientation in men and a substantial but perhaps somewhat weaker genetic influence in women." For Gorski, the evidence is strong enough to say that "genetics contribute 25 to 75 percent of the variance we see in human sex orientation, although not sex behavior."

Yet another store of evidence comes from individuals with known chromosomal abnormalities, such as Klinefelter syndrome. Born with an extra X chromosome, Klinefelter babies have a chromosomal address of XXY instead of XY. They are, from one genetic standpoint, unambiguously men because of the SRY genes on the Y chromosome that confer maleness via the triggering of male sex hormone secretion at critical stages of development in the womb. But the extra X, along with the visible stigmata of Klinefelter—large breasts, tallness, some retardation—also appear to influence sexual *orientation*, if not genital identity.

In his book *The Sexual Brain*, LeVay reports results of a U.S.-Danish study in which scientists counted the chromosomes of every resident of Copenhagen who matched certain age and height criteria and identified sixteen XXY men. They then compared their sex histories with controls matched for height, age, weight, IQ, and so on and they found a "highly significant excess of homosexuality" among the XXY men. How the extra X might work is unclear, but LeVay hypothesizes that testosterone levels were lower than normal in the prenatal hormone environment.

The prenatal hormone theory of homosexuality gets some backing from a variety of other pieces of evidence, pieced together by a persistent cadre of sex scientists. Among them was psychiatrist Richard Green, who several decades ago began a famous study of "sissy boys," boys who as children were effeminate and who as adults declared themselves gay or bisexual.

The sissy-boy study, for all of its deprecatory labeling, was one of the first solid studies demonstrating that within the first year or of life, homosexuals already were exhibiting what Green and others still call "sex atypical traits."

As early as 1979, John Money and one of his graduate students at the time, Anthony J. Russo, reported on an unusually long-term study of eleven boys whose early childhoods already were marked by sex atypical behavior. The investigators looked at makeup, jewelry, hobbies, clothes, perfume, dreams, self-image, parental relationships, careers, romantic relationships, friendships, aggression, and so on in long (twenty-seven-page) surveys conducted over twelve years. They concluded that nine of eleven boys with sissy-boy tendencies before adolescence grew up to live a homosexual or mostly homosexual life. None was a transvestite or transsexual, and all nine had at least some college education and were "well achieved" occupationally. They all seemed well adjusted, a fact that Money and Russo attribute principally to their parents' refusal to second-guess their children, to allow them to be the people they felt they were.

Weinrich and coworkers at the University of California at San Diego have further reported that gay men who have a strong preference for receptive anal intercourse also have the histories, as reported by themselves and their families, of sex atypical childhoods, while gay men without this preference had a much more typical pattern of boyish behavior in young childhood.

The take-home message from such data is that the appearance of sex-atypical behaviors in such young children, children who haven't had time to develop a sexual "will" or the experience to "choose" their attractions, is powerful indication that adult sexual orientation is at least in part determined or influenced by the early biological circuitry laid down in the brain. Behavioral patterns shown by sissy boys or tomboys could not all have been "taught" or "learned" after birth. They are too consistent, too early to appear, and too entrenched.

Delving into the mind-brain connections to homosexuality, a large number of scientists, including LeVay, Gorski, and Laura Allen, have gone looking within the brain's structures and chemistry for differences between homosexuals and heterosexuals. The research has been successful—there are some real differences—but the results have brought controversy, in part because it remains impossible to say with certainty that the differences are the *cause* of anything. They could well be the *result*. Or incidental.

After all, homosexuals, just like heterosexuals, are not wholly defined by whom they want to have sex with, or by what gender they identify and feel comfortable with. They have many other differences, too, so finding differences between their brains and the brains of heterosexuals could be a reflection of social learning, family upbringing, and myriad other things that influence neural development.

But again, the persistence of certain traits across all cultures, ethnic groups, time periods, and socioeconomic levels makes it very unlikely that all of the parents—or environments of these people when they were children—acted the same way toward their children or created the very set of circumstances, prenatally or postnatally, that created same-sex preferences. A much more likely scenario is that because male and female brains *do* develop differently (see chapter 4), they respond differently to intra-uterine and postnatal challenges involving hormones, neurotransmitters, experiences, and genetic expression. In other words, because the sexual brains of men and women are known to be different in a variety of ways, and because most of what makes us sexual beings is between the ears and not the legs, it is likely that what creates or influences sexual orientation happens in the brain, too. Moreover, given that there are vast brain differences between the behaviors of men and women, it is not surprising that neuroscientists are discovering brain structures and functions that are literal templates and road maps for human—and animal—behaviors that vary markedly by sex.

Gorski, Arthur Arnold, Michael Gazzaniga, and others all have demonstrated that men's brains are lateralized differently, that they do best on such tests as spatial ability and eye-hand coordination and do worse relative to women on verbal ability tests. Their brains are also more specialized for different tasks. And it turns out that in studies that compare gay and straight men on such tasks, gay men tend to score more like heterosexual women.

Consider the trait of left-handedness. Handedness, as studies of right and left brain have long shown, is determined by the opposite side of the cerebral cortex, so that the right brain controls the left hand and side of the body and vice versa. Most men and women have a "dominant" hand. If it turned out that gay men and lesbian women were more consistently left-hand dominant or at least more ambidextrous, or less lateralized for handedness, that would form a powerful clue to how the brain might organize itself with respect to sexual orientation. And indeed, several studies have found that to be the case, although, as LeVay points out, this difference does not show up in the one aspect of handedness most influenced by learning and family: which hand a person eventually writes with.

Other traits that distinguish between gay and straight people are more elusive but no less intriguing. There's a widespread perception that gay men are more artistic on average than heterosexual men and that they are overrepresented in the performing arts, such as dance and theater. One explanation for the seeming preponderance of gay and lesbian actors, artists, and performers is that in the performing arts, open tolerance and acceptance of homosexuality is greater than in the general population. But it's also possible that people who grow up with sex-atypical behaviors are drawn to certain occupations. As LeVay notes, jobs that seem to require lots of organization skills, as well as physically demanding occupations, seem to be especially attractive to some lesbians while those jobs requiring "creative or caring" traits such as design, writing, dance, theater, and nursing, he says, are attractive to some gay men.

In the last few years, several avenues of investigation have pointed to distinctive differences in the anatomy and chemistry of homosexual and heterosexual brains. The news of these discoveries has been so widely reported that in one gay enclave, West Hollywood, California, legend has it that T-shirts bearing the legend "My Brain Is Smaller" have sold well. The allusion is to research by LeVay and the Netherlands' Dick Swaab (who tediously counted individual neurons and found more and bigger ones in males than females in a cluster of cells similar to the sexually dimorphic nuclei, or SDN, found in rats), as well as Roger Gorski and Laura Allen, who, Gorski says, has identified more differences in the brain anatomy of men versus women than any other investigator to date.

LeVay's investigations, to which he turned to help him through his grief after his lover of twenty years died of AIDS in 1990, has been fruitful in these areas. A highly trained neuroanatomist who worked at Harvard and the Salk Institute on the brain's visual cortex, he has identified the medial preoptic region of the hypothalamus as one site of differences between homosexual and heterosexual men.

As described earlier, this part of the brain, in humans and other animals, is involved in the regulation of male-typical sex behavior and contains those four clusters of nerve cells known as the interstitial nuclei of the anterior hypothalamus, or INAH, the whole thing only the size of a teaspoon full of liquid. LeVay learned that INAH 3 is bigger on average in men than women, but INAHs 1, 2, and 4 show no major sex differences overall.

In a move that brought some good out of the tragic deaths of young men with AIDS, LeVay obtained the brains of gay men, the brains of heterosexual men who also had died of AIDS (intravenous drug abusers), and the brains of heterosexual men who died of other causes, as well as the brains of several women who were presumably heterosexual on the basis of the law of averages, since most women are heterosexual.

He studied the hypothalamic tissue in a "blind" analysis, not knowing which brain tissue came from which group, and found consistently that INAH 3 was on average up to three times bigger in the heterosexual men (whether they died of AIDS or not) than in the women, a finding first made by Allen in Gorski's lab at UCLA.

LeVay also found that in gay men, INAH 3 was on average the same size as in the *women* and up to three times *smaller* than in the "straight" men. Admittedly, some of the women and gay men had a large INAH 3 and some of the heterosexual men a small one, but on average, the findings showed that the homosexual population had the smaller. "There's no question," says Gorski, that INAH 3 is consistently smaller in gay men."

What this suggests to LeVay is that gay and straight men "may differ in the central neuronal mechanisms that regulate sexual behavior.... To put an absurdly facile spin on it, gay men simply don't have the brain cells to be attracted to women," because it is likely that the smaller INAH 3 reflects the fact that there are fewer neurons there in gay men.

The research has its critics and skeptics, who note that because most of the men in the study died of AIDS, their brains might have been unusual as a result of the HIV's impact. But since he included brains from men with AIDS who were not homosexual but IV drug abusers and found the same results, LeVay feels his data are valid. He also, however, shares some of the skepticism about claims he has found a proof for a genetic basis for homosexuality. "It is not possible, purely on the basis of my observations, to say whether the structural differences were present at birth and later influenced the men to become gay or straight, or whether they arose in adult life, perhaps as a result of the men's sexual behavior."

But animal research *does* lend weight to the former rather than the latter. Recall that in rat studies, for instance, scientists have located a brain structure called the sexually dimorphic nucleus of the medial preoptic area, which might be analogous to the INAH structures but in any case is exquisitely involved in rat sex behavior. If it is altered hormonally or surgically during prenatal growth or shortly after birth, it can have profound effects on the rats' sexual behavior as males or females, but if experimenters wait more than a few days, then nothing changes the size of this structure at all. Even castrating the adult males, removing the source of androgens, does little to affect this structure. "If the same is true for INAH 3 in humans," LeVay has written, "it would seem likely that the structural differences between gay and straight men come about during the initial period of sexual differentiation of the hypothalamus (and) that these differences play some role in determining a person's sexual orientation."

In the rats, it turns out, the major sculptor of the MPOA is circulating androgen, suggesting that the INAH 3 size differential between gay and straight men might be due to the same thing in fetal life in humans. Some critics argue that hormone studies often have shown that the level of androgen is the same in gay and straight men, which they claim undermines the whole INAH-androgen hypothesis. But neuroscientists on the other side of the issue note that it's quite likely that the difference has nothing to do with the total amount of androgen, only with how *sensitive* neurons in INAH 3 in the fetal brains are to the hormone, with the variation between gay and straight individuals accounted for in the way the neurons respond to the hormone.

LeVay is convinced that there are inborn genetically determined differences in the brain's hormone receptors or neurotransmitter receptors that influence how much androgen or other hormone gets access to neurons, gets to sculpt brain circuitry during fetal development, and therefore determines how the hormones are put to use. The number of neurotransmitter receptor "keys" may be the same in gay and straight men and women, but in gay people some of the locks may either be missing or their tumblers changed just enough to let the keys only partway in.

Laura Allen and Roger Gorski, who conducted studies on autopsied brains from AIDS and other patients, found gay-straight differences in another brain structure called the anterior commissure. This structure, like the corpus callosum (the coaxial-cable-like band of fibers that forms an arch over the thalamus and originates as the axons of millions of individual nerve cells all over the cerebral cortex), connects the halves of the brain, but is smaller than the corpus callosum and a more primitive connection between the hemispheres. Before there were mammals with placentas, there was no corpus callosum at all, just the anterior commissure.

Both of these structures are larger in women than men. But the commissure was found by LeVay and Allen to be on average *larger* in gay men than straight. Along with the corpus callosum, the anterior commissure is one of the brain structures that also varies greatly between men and women in general, straight or gay.

According to Gorski, the differences in the commissure may explain some of the other differences perceived in gay men compared to straight men: their less masculine, more feminine-like diminished brain lateralization, for one. If such things as visual-spatial relations, verbal abilities, and other higher brain functions are less lateralized in gay men as they are in straight women, then the anterior commissure would have had to be bigger to connect the hemispheres and integrate these functions.

But perhaps the most important insight the anterior commissure work reveals is that because this structure is not directly involved in any way in *regulating* sex behavior the way the hypothalamus is, it's unlikely the size differences Gorski and Allen found are a *result* of different sex behaviors, but instead originated during the initial development of the structure by sex hormones or other circuits.

Dutch scientist Swaab and his coworkers have added still more to the growing evidence in favor of a gay biology with reports that the supra-chiasmatic nucleus, or SCN, important for circadian rhythms is larger in volume and has more neurons in homosexual men than heterosexuals. Swaab and others found this nucleus by measuring its vasopressin receptors, proteins that react immunologically to certain probes.

In a study of male-to-female transsexuals, Swaab also identified some possible brain structures associated with the fact that these individuals feel female although they are genetically male. Again using brains taken at autopsy, they found that the size of the BSTc, which plays a role in rodent sex behavior, was in the male transsexuals the same size as in women and smaller than in gay or heterosexual men. Sex-change treatments involving heavy doses of female hormones are common among transsexuals and could account for the change, so the jury is, once again, out.

"I've been working in the field of sexual dimorphism in the brain for thirty years," Gorski says, "and while there is no definitive evidence of a biological foundation in the brain for homosexuality, there are increasingly clear differences. In 1973, English neuroanatomists published information on the differences between the male and female brains of animals in the preoptic area. In 1976, Arthur Arnold showed extraordinary differences in the brains of male and female song birds in the same preoptic areas and to date, more than 30 such sexually dimorphic areas or connections have been identified in rats alone. The evidence is so good that in neurobiology, we can assume sexual differences in the brain unless proven otherwise. Such anatomical differences, coupled with the fact that the brain is an integral part of the reproductive system, that there is neurological control of sex, and that the brain of mammals is in the earliest stages of development an essentially female brain, it's hard to argue against the hypothesis, at least from a solid theoretical base. If you are a scientist in this field, and if you accept the huge amount of data that show sexual differences in the brain, it is not hard to leap to the idea that there is a biological root to all sex behavior, gay or straight."

With high praise for LeVay's work, Gorski, a pioneer in the field, says that before LeVay published his evidence in 1990, little attention was paid to efforts by neuroanatomists and biologists to identify biological roots of homosexuality. Since then, "it's been wild."

Gorski is ambivalent about the attention, in large part because of what he considers to be premature political fallout on all sides of the debate over sexual behavior in Western societies. "People become polarized on the basis of partial information and people get hurt. What so many fail to understand is that if there *is* a gene for homosexuality, it still may only make you behave as a homosexual, or have homosexual sex, if you have opportunities to do so." Mozart may have had rare musical talent in his "genes," but without a father who bought him a piano, we may never have had those forty symphonies. In a similar way, it almost certainly takes more than a gene to make human beings express any behavior. In a sarcastic swipe at the anti-gay factions, Gorski, a father and grandfather, said that "homosexuality *is* a choice of sorts. Gays have to decide to swim upstream and say 'f—k' society if they want to live in accordance with their orientation."

He also is distressed to some extent by the relative lack of interest in pursuing the biological foundations of lesbianism, although there have been some advances. In the spring of 1998, for example, Dennis Mc-Fadden and a team at the University of Texas reported in the *Proceedings of the National Academy of Sciences* the first physiological difference between heterosexual and homosexual women: echo-like sounds made by the inner ears of homosexual and bisexual women are weaker than the same sounds made by their heterosexual counterparts. "The findings," the team said, "suggests that the inner ears and some unknown brain structures responsible for sexual preference are masculinized in homosexual and bisexual women because men also exhibit weaker echo-like sounds in their inner ears. The inner ear may be a valuable window into events that occur during brain development and sexual differentiation.

"A big part of our problem with conducting research into the biology of homosexuality is that our society can't deal with sex at all. One outcome is that we define masculinity and femininity via cultural/social means, not biological ones. To say that masculinity equals aggressive behavior is silly, just as silly as to say femininity equals passivity."

There is, he concludes, an entire cascade of biological processes that determine masculinity and femininity, and quite probably homosexuality and heterosexuality, including those mediated by brain structure, genes, and hormones. "There is not only the induction of male specific characteristics onto an initially female background, but also defeminization, the suppression of female-specific characteristics that must take place to produce a sexually intact male as we commonly define such a person. And vice versa. To produce a feminine female, there must not only be feminization, but also demasculinization. And within each of these major components, there may be sub-levels that are operating independently."

In this context, even a temporary or brief alteration of usual events in the process can alter the sexual brain's development and the outcome of any individual's sexual orientation.

That's because in humans, just as in the rats he has studied, Gorski believes, the process of sexual differentiation of the brain occurs mostly and most powerfully *after* the genitals are formed in the fetus, not before. Scientists can deduce this in great measure by the fact that in homosexuals, the external genitals—penis, vagina, breasts, and testicles—are normal.

A final storehouse of evidence for a biology of homosexuality holds that a collection of genes acts singly or in combination to apparently produce gay behavior.

Principal among those adding to this collection is Dean Hamer, the molecular geneticist at the National Cancer Institute, who has become one of the premier trackers of behavioral genes. Hamer, who trained at Harvard in biological chemistry and genetics, is the coauthor with Peter Copeland of *The Science of Desire: The Search for the Gay Gene and the Biology of Behavior.* He too turned to homosexual biology for partly personal reasons but also because he has long been intrigued by the twin and family studies that suggested that homosexuality runs in families. But Hamer took a different tack: He launched his search for a gay gene by looking critically at the DNA of his subjects.

Preliminary studies led him to search first on the X chromosome of forty pairs of gay brothers to see if they shared any DNA that would parallel their shared sexual orientation. If so, that would strongly suggest that a gene plays some role in expressing the trait, although he was cautious enough to insist that this commonality would have to occur in many more than half of the brother pairs, since in the general populations, brothers, gay or straight, would get the same X from their mother half the time.

He found in fact that thirty-three of the forty gay brother pairs had the same piece of DNA at one of the tips of the X chromosome called Xq28. Geneticists map chromosomes with geographic regions by first identifying whether a site is on the q or long arm of the chromosome. So, in this case, it's on the long arm of the X at level 28 in a horizontally banded map of the chromosome. The chance of that occurring by chance was less than one in ten thousand, Hamer reported in the journal *Science*; but ever cautious, he denied that he had found a gay gene, only that he had found that "one subtype of male sexual orientation is genetically influenced."

Critics of Hamer's work—and they are numerous and vocal—vigorously point out that he had never looked at Xq28 on the straight male brothers of his subjects to see if they had inherited or not the same peculiarity, which would have knocked the hypothesis into a cocked hat. It's also possible, of course, that whatever influence Xq28 exerts may be very indirect, on, say, personality or impulsivity or risk taking or optimism or pessimism or some other aspect of temperament.

Hamer might well agree with that. He has repeated his findings and affirmed them, and although he defends not checking out the brothers of the gay men whose X chromosomes he studied—such brothers would possibly have been highly motivated to *not* tell him their true sexual orientation under the circumstances, particularly if they were living straight lifestyles—he later told a small group of journalists at a genetics symposium that he has indeed studied these heterosexual brothers. "We have checked the heterosexual brothers of the gay men and they generally got the mother's other X chromosomes, but only 75 percent had this opposite marker, meaning that a fourth of them had the same marker their gay brothers had, but stated they were heterosexual.

"We don't know what the gene is or what it makes. The region is in the 10 to 20 centimorgan range, very large and it could make a transfer RNA, a hormone or who knows what. All we can say with any certainty is that in the population at large, no fewer than five and no more than 67 percent have it and it's probably in the range of 10 to 15 percent, the very high end of the 'guesstimated' percentage of gay males in the population." (Most estimates are in the 4 to 5 percent range, as noted earlier.) He adds, "This gene or genes is a factor but not THE factor in sexual orientation. We would need thousands of sib pairs to narrow down the region to even one centimorgan and it's probably not worth going after it."

Hamer says what's more important is that the more we learn of "our diversities and differences, the more tolerant we will become of these and of all behaviors, including sexual behaviors." The point is to tolerate, not to establish some cause and effect so that we can "prevent or change" orientation.

In a separate family-based study of female sexual orientation genetics, Hamer has found that the trait in women is more flexible and diverse, less strongly distinctive. "Maybe women are just more honest in talking about their sexual orientations, fantasies and so on. Men, even if they are homosexual, will deny it, even to themselves." In short, in women, the genetic load for sexual orientation seems to be less heritable.

Perhaps the most intriguing new evidence for a homosexual "gene" comes not from studies in humans, or even mammals, but in investigations of that old standby in genetic studies, fruit flies.

One late winter day in the early 1990s, neurogeneticist Ward Odenwald, who like Hamer works at the National Institutes of Health, looked at some genetically engineered drosophila in his lab and saw that the males were actively "courting" one another.

In an astonishing piece of videotape that he makes available to other researchers and uses to dramatic effect during scientific presentations, the male flies, in Odenwald's lab at the National Institute of Neurological Disorders and Stroke, are seen in a courtship "dance" called "chaining," in which the males literally line up and chase each other around a cage in distinctive patterns that form lariats and rings, as well as straight lines.

And as the story unfolded for Odenwald and his team, it turns out they did so because of a single mutant gene—a red-eye color gene ironically called "Drosophila white" (*w*) that was transplanted into their bodies. Male-male courtship had also been shown to occur in other mutants of drosophila, notably by Jeffrey Hall at Brandeis in 1994, but nothing up to that point had suggested a single gene to be totally responsible. The lab gods of serendipity had indeed smiled upon Odenwald and his team, who had not set out to look for a homosexual gene, but instead had begun their work examining some of the incredible and elaborate repertoire of gender-specific actions that go on when drosophila, or fruit flies, court and mate.

Their instruments of study were the tools of molecular biology, specifically the ability to transport in or knock out single genes, manipulations that were already known to have dramatic effects on the flies' courtship and other behavior.

"When we first saw this phenotype, however, we knew something was very different, very unusual, but we didn't suspect just how much until later," Odenwald recalls. "It was a completely serendipitous observation. We had set up a family of flies in a bottle to collect embryos, as we do for many studies, and there was a huge blizzard, with 17 inches of snow that kept us from getting back to our lab for four days. When I finally got in to the lab, there it was. We thought it was due to the fact that the flies were stressed by the lack of food, attention and constriction in the bottle and that the stress had activated or misactivated a heat shock gene we had transplanted into their bodies. This is a standard type of gene transplant for doing studies in these flies, because it allows us to heat the flies and activate particular genes to see their effects. In our case, we wanted to use the heat shock gene to allow us to heat the flies and in this way activate the typical red eye color gene so that it would express not just in cells that make the color red (in the eye) but also all over the flies' bodies, what we call ectopic expression.

"So for the first six months or so, we thought this unusual activity was simply due to some 'misexpression' of our transplanted gene."

To understand what changed their minds requires a bit of background about fruit fly genetics and bedroom habits.

First, all fruit flies have the red eye color gene, but it generally is only expressed in cells that make red pigment. But by injecting fly embryos with the gene together with a bit of DNA called a heat shock promoter gene (that turns on the red eye color gene in response to heat), scientists can get the red color gene to turn on all over the flies' bodies, so-called ectopic expression. Odenwald and his colleague Shang-Ding Zhang also knew that the red eye color gene's protein products are used not only in making red pigment but also in forming serotonin.

Second, in normal courtship conditions, fruit flies transmit, accept, or repel an assortment of pheromonal chemicals, along with their display of wing flicking, body tasting, antenna waving, and other signals of astonishing variety. Each of the signals is depending on feedback from a partner, which accepts or rejects him in response to his signals and then transmits other signals.

Thus, it is well known to fruit fly geneticists that when flies initiate courtship and mating, any attempt by one male to court another male is instantly *suppressed* by emissions of an antiaphrodisiac pheromone, and other rejection signals that males elicit in response to such inappropriate advances. Clearly in the case of their flies, this suppression was not working.

"But we were relative novices at all this, so we did a bunch of things to try and explain what we saw." They even manipulated the genetic sequences of the carrier or vector they used to deliver the gene before transplantation, but *still* got the same result. "The only thing that we didn't change in all our experiments was that each vector had this so-called white gene.

"Given the nature of the behavior—the 'homosexual' nature of it—

we didn't want to speculate about a single gene being responsible because if we were wrong, we'd be wrong with bells on. So we set up another experiment in which we altered the flies with chemicals and got rid of the homosexual behavior by knocking out the white or *w* gene. Then we knew we had the answer."

What induced the homosexual behavior was indeed a misexpressed form of the *w* gene, which is more specifically a gene that transports tryptophan and guanine across membranes. (More properly, it's a gene for both homosexual and bisexual behavior courtship in fruit flies, as the flies will go for heterosexual males as well as other *w* mutants.)

In repeated experiments, when Zhang and Odenwald transplanted a mini *w* gene into drosophila embryos, the mature but virgin males began to vigorously court other mature males. Even when there were plenty of females around, most males formed male-male courtship chains, doing elaborate homosexual courtship dances. Every time they knocked out the *w* gene, the behavior went away.

What is also intriguing, Odenwald says, is that while female sex behavior does not appear altered by the expression of *w*, even males who have not been genetically transformed by *w* will, when exposed to an active homosexual courtship environment, actively participate in the male-male chaining. Odenwald concludes that in fruit flies at least, both genetic and environmental factors play a role in male sex behavior, since the non-gene-transformed males in their studies were clearly "recruited" to the homosexual lifestyle.

In a paper published in 1996 in the *Journal of Neurobiology*, Audrey Liang Yin Hing and John R. Carlson of Yale University reported follow-up studies to Zhang and Odenwald's experiments. They explained that the *w* gene encodes a small protein, or polypeptide, that plays a role in transmembrane transport of essential biochemicals. "Specifically," they wrote, "the products of the *w* and another gene called scarlet form a combination that transport tryptophan while the products of *w* and brown genes form a combination that transports guanine. Both tryptophan and guanine are precursors to the eye pigments essential to normal eye color. The role of the *w* product in transport and in courtship suggests that possibility that ectopic expression of the mini-*w* gene might lead to abnormal transport of a molecule that modulates behavior, conceivably a pheromone or a visual pigment."

In their experiments they found that because they were able to reproduce the homosexual behavior in isolated pairs of flies, it's unlikely that *w*-induced male-male courtship depends on sensory cues long suspected in male-female fruit fly matings, or on pheromones or sounds made by wings or antennae. Indeed, the only sensory deprivation they found to stop the male-male courtship was when they tested the flies under dim red light instead of normal light; that reduced the bisexual or homosexual behavior by three-fourths compared to controls.

Put another way, the Yale team found no evidence that taste, sound, or smell had anything to do with the flies' altered sexual preferences. The researchers did microsurgery, clipping their antennae or body parts or damaging their ability to smell or give off sex pheromones, and still got the same male-male behavior. In a test of the role of olfactory input, for instance, flies were found to carry out male-male courtship even after microsurgical removal of olfactory organs. Nor did cutting their wings off do anything to alter it, a sure sign that the sense of touch is not involved.

Hing and Carlson concluded that the homosexual courtship they see is not due to any "sense" these flies have that males are suddenly more attractive or that other males suddenly stop rejecting them, but probably is due to some inability to process information in general. Because the males with the *w* gene will also mate with the wild type normal males, it's not that normal males stop giving off the right signals, it's that these signals aren't interpreted in a normal way by the mutant.

What they were left to conclude, as Odenwald's team had, is that the *w* gene alone is responsible for this male-male courtship behavior. And remember that serotonin connection? As it turns out, other studies in rabbits, cats, and other mammals show that manipulations of serotonin are linked to homosexual behavior. In rats, for instance, scientists have been able to alter serotonin levels and induce one male to mount another in a parody of intercourse. Because the chemical products of the *w* and scarlet genes form a chemical combination that transports tryptophan, and because tryptophan is a precursor chemical to serotonin, it's likely that some misexpression of the mini-*w* form of the gene may affect serotonin levels. It could be that the gene alters brain areas that make them richer in nerve cells that produce serotonin. Studies are currently under way to investigate areas of the brain rich in serotonin receptors and test the serotonin-bisexuality hypothesis further in the flies and other animals.

Odenwald says the gene sequence for *w* bears striking resemblance to a known human gene, but its function, if any, in humans, is unknown. "This does not mean that what happens in a fruit fly happens in a human, but it's certain that many basic mechanisms in life" have been "highly conserved from day one from fruit flies right on up to man.

"The general theme is for nature to build something and then modify it by natural selection," he adds. "My gut feeling is that while a fly is surely not a man, and there are gross structural differences between them to say the least, when dealing with a behavior so essential to survival of species, it's easy to argue that many of the molecules and the genes that make them are highly conserved by nature."

Odenwald says similar things have been reported in other model systems. In one study on rats and rabbits with tryptophan removed from their diets, investigators found a marked increase in hypersexed males and males mounting other males. "It's pretty striking to see a photograph of four or five rats all humping each other in a chain. Astonishing and too much of a coincidence." He also notes that the experimenters had done the study because they had noticed similar behavior in animals being tested with drugs that inhibit serotonin synthesis. What may be important is not so much absolute levels of serotonin, but the ratio of serotonin to dopamine. Guanine, the other biochemical involved in *w* transporter function, is a precursor of dopamine.

"I can't tell you how many sleepless nights I've had," Odenwald says, "because it was so remarkable to look into our lab bottles and see what we thought we were seeing. It's a jaw dropping experience. If we're right, this is one of the very first genes identified for a specific sexual behavior and the first time that misexpression of a single gene can be shown to have such an impact. Geneticists tend to think in terms of altered genes or non-expressed genes when they look at dramatic behavioral outcomes. But here is a gene that is working, it's pretty normal, but is misexpressed. We don't yet know why the misexpression occurs," he says, although some have suggested that the misexpression is more a matter of where in the brain the genes are expressed than anything else.

If, in fact, there are genes that code for homosexuality, why would they persist over the eons of human evolution? Obviously, homosexual men and women don't reproduce themselves or their genomes very vigorously, and the evolutionary process, it would follow, would have long ago ended their nonreproductive sexual behaviors along with them. Why, for instance, would the Xq28 region that Hamer has found persist and be conserved?

One possibility, says Hamer, is that the gene was conserved to help the female somehow, the original holder of the X chromosome, and that males are simply "innocent bystanders" in this game and randomly get the gene from one of Mom's chromosomes. This kind of phenomenon has been shown in fruit flies and possibly in sheep, but not in mice.

Another possible explanation is that "gay genes," if they exist, might be kept around by nature because of some other beneficial trait linked coincidentally with homosexuality, perhaps one of which we are not yet aware or one that might promote cooperation among men or among women. But even if this were the case, it wouldn't explain why it persists, because homosexual men don't reproduce as often as heterosexual men and natural selection would not favor the retention of such a trait or its genes.

What if gay genes are "unselfish" genes, the contrarians to the selfish genes that drive heterosexual behavior. Several sociobiologists propose that while gay genes might indeed reduce the reproductive fitness of its bearers, it could mean that such men or women are kept around to help provide resources to their brothers and sisters who are reproductively fit. At times of scarcity in our evolution, such help would have meant the survival of the line, of at least some of the homosexual's DNA, at the expense of his direct line, much the way the drones (all males) in a hive never have sex with the queen.

But as LeVay points out, the problem with this explanation is that "it does not account for homosexuality, it only accounts for the lack of heterosexuality." The gay men and women in this model don't sexually reproduce, but why would this model cause them to love and want sex with members of their own sex? "To put it crudely," says LeVay, "why do gay men waste so much time cruising each other, time that according to this theory would be better spent baby-sitting their nephews and nieces?"

There also are theories that suggest that, like the spandrels of San Marcos that Stephen Jay Gould writes about, homosexuality is a "leftover" trait from a time in evolutionary history when it was necessary, but nature never completely managed to unload it. LeVay likens this theory to the fact of sickle cell anemia. SCA is an inherited disorder in which a person who has it had to have inherited recessive genes for the sickle cell. Such people have grossly distorted red blood cells (they look like sickles) and they not only fail to carry their allotment of oxygen, as red blood cells are supposed to do, but they also get hooked into capillaries and larger blood vessels, causing excruciating pain to victims.

Now, it happens that carriers of the sickling gene who have only one copy of the gene do not have sickle cell anemia, although they do have something called sickle cell "trait," with some minor alterations in their red blood cells. And it further turns out that SC trait occurs in high frequency in areas of the world where malaria is endemic. It seems that the slight alterations in red blood cells in sickle cell trait populations offer resistance to malarial illness, probably by making it more difficult for the mosquito-borne malarial parasite to invade the red blood cells and produce more parasites. The benefit of this trait is obvious, and for it the population at large was "willing" to pay the price of killing off some of their numbers in an extremely painful bout of sickle cell anemia in those who got a double dose of the gene.

Similarly, suggests LeVay, it may be that gays and lesbians are "the losers in a genetic roulette game," their gay genes kept around by the human race because overall they conserve some element that benefits the heterosexual community. LeVay considers the idea unappealing "but nevertheless plausible."

And finally, he says, there's the possibility that gay genes are bad for us in that they fail to get our DNA to survive in the next generation, and the reason they stay with us so consistently in 5 to 10 percent of humans is that they are easily re-created out of "normal" genes. That is, they occur in parts of the genome that are easily mutable, possibly because of nearby events or breaks in the genome. To determine that, however, the genes must be found first; only then can their genomic environments be discovered.

Meanwhile, as Judith Lorber notes in her book *Paradoxes of Gender*, our polarized notions of male and female, although likely to be a "human invention," continue to result in inequalities of one sort or another, usually, but not always, favoring males.

The widespread insistence that "men are men and women are women" is more about social status and power than procreation, Lorber concludes, because biology and physiology are telling a much more flexible story. The concept of distinct genders, says Lorber, is one way society has developed of compensating men with some power over women's fertility. Appropriate or not, such compensation may or may not have anything to do with physiology. Gender is "so pervasive," she says, "that in our society we assume it is bred into our genes." But a growing body of facts suggests that like any cultural phenomenon, gender also is learned and its shape and meaning can be altered. "[G]ender and sex are not equivalent and gender . . . does not flow automatically from genitalia and reproductive organs, the main physiological differences of females and males. . . . Whatever genes, hormones and biological evolution contribute to human social institutions is materially as well as qualitatively transformed by social practices."

If Lorber is right, it may also be true that mating and procreation are not the only *natural* or *normal* goals of sex. No sex scientist denies that gender is principally about reproduction. But even those who do not procreate, including homosexuals, the celibate, or, historically, younger sons who don't inherit enough wealth to marry, can enhance family strategies such as resource conservation across generations to retain power and maximize the success of the offspring that are produced.

That legendary observer of the sexual scene, Gore Vidal, has added at least literary support to this case. In a somewhat notorious article in *The Nation* in 1991, he thought it was time that he "explained sex" and concluded that given our planet's overpopulation and over-breeding, gays and lesbians "should be considered benefactors" of planet Earth and breeders should be "discouraged."

THE COMPROMISE BETWEEN THE SEXES

"Roses have thorns, and silver fountains mud;
Clouds and eclipses stain both moon and sun,
And loathsome canker lives in sweetest bud."
—WILLIAM SHAKESPEARE, POET

"We don't know exactly why sex evolved."
—R. J. REDFIELD, ZOOLOGIST

It makes sense to me, as someone with a reporter's temperament and (it is hoped) skill, that a five-year foray into the science of sex would yield some novel, or at least more coherent, picture of sex. And so it did. What surfaced, for me, as most important and perhaps most true about the material was the endlessly *compromising* nature of this intensely intimate behavior. Moreover, it seemed fitting that my judgment was informed not principally by the compilation of theories, experiments, and facts, but by human imagination, culture, and emotion, which scientists, like the rest of us, use mainly to help make sense of our biology. The eminent and controversial sexologist John Money once was asked to speculate what life would be without sex. In his response (to editors of *Discover* magazine) he concocted "The Fable of the Whiptail Lizard Kingdom."

In the story, Whiptail the Tenth, the ruler of the kingdom, was in despair because the greenhouse effect was warming up his kingdom to desert proportions and he alone knew that new males required cold sand in which to hatch. The queen was driving him crazy, in a constant uproar when no males hatched season after season. How, the king asked himself, could his kingdom survive without sex? The answer came in the form of a "miracle," the birth of Unitail, whose belly was filled with eggs that carried all the genetic material necessary for fertilization. "They were the eggs of a virgin sovereign who would rule over an entirely new race of whiptails, a race in which there would be neither male nor female, only unisex members."*

At the end of Money's fable, Unitail issues a proclamation in the memory of Whiptail the Tenth and his queen: All egg laying will begin with a ceremonial marriage with another lizard. "We shall never forget how it used to be in the days of our forebears, when every whiptail lived with sex," Money wrote, making two points: first, that in almost all cases where we now see asexual reproduction, sex was certainly the ancestral way of begetting and the move to asexual reproduction was clearly a backslide wrought of catastrophe and unpredictably disruptive events; and second, that the physical wonders of sex, vital as they are, and even if only in memory, are tightly linked to the social, cultural, and emotional ceremonies, the rituals, the *mutual benefit*, the cooperation between and among individuals that accompanied the evolution of sex.

Money's tale addresses perhaps the single most important question begged by all suggested histories of sex and its earliest origins to this point, namely: What was likely to have characterized the series of events that transformed some one-celled creature's trial and error episode with conjugation and proto-sex into the protracted, complicated, subtle, and enduring system of sex we find today in people, penguins, and peacocks? The trip to sexual procreation may have started simply enough, but it now winds through a perverse maze of back roads, detours, traffic jams, and dead ends.

One answer to the question may be found in a new reading of the old story of the battle of the sexes. Pop psychology and public TV fang-and-claw mating specials belie the accumulating evidence: The principal feature and goal—or at least *a* principal feature and goal—of our sexual

---

*In computer simulations of outcrossing and recombination of DNA, R. J. Redfield, a University of British Columbia zoologist working in Vancouver, claims to have found evidence that makes Unitail's way look like the evolutionary winner. He says that females' high cost of splitting and exchanging genetic material in the egg before and during fertilization, and the fact that the most deleterious mutations arise in the male germ line (ten to one male-to-female rates in humans) make parthenogenesis a pretty good deal for females. Unitail seems to have had it right because "male gametes may give progeny more mutations than . . . sexual recombination eliminates." On the other hand, when scientists genetically engineered mice to have chromosomes that came only from one parent, Dad or Mom, these mice died very early in their development. One lesson is that genes, at least after fertilization occurs, act differently in the offspring depending on whether they came from Mom or Dad. They leave a sex-specific "imprint" on the genetic blueprint, making dads absolutely and generally indispensable, whiptail lizards notwithstanding.

design is cooperation and compromise, not competition. The wages of sex, in this view, are the capacity for negotiation and problem solving, not dominance, submission, or aggression. Seen this way, successful sex, defined as reproduction, pleasure, and nurture of offspring to sexual maturity, is a win-win situation, a balance of gender interests and power that turns out to reflect the very essence of accommodation. Wham-bam-thank-you-ma'am might give you offspring, but it will not get them optimally fed, clothed, and enrolled at Harvard, thus likely to successfully rear a grandchild or two. To get that requires commitments, resource sharing, and mutual support.

The battles and conflicts are what have attracted so much of the attention in the evolution of sex. They're universal, easy to spot, and fun to watch. But humans are not merely animals; they are animals with big brains, with emotions as keenly honed by time and trial and error as our genes. And as scientists have accumulated more sophisticated means of going beneath the surfaces of bodies, minds, and behaviors, they have seen a very different paradigm that enriches if not substitutes for the old nature-versus-nurture and male-versus-female view of things. Instead of, or at least along with, competition and "war" between the sexes, there is enormous evidence that our natures run to another kind of sexual sport: nurturing, romance, love, extended family bonding, parenting, altruism, trust, sacrifice, commitment. The cooperative nature of our sexual selves is more subtle, perhaps. The signs are harder to read and measure, but they are there in the everyday life and dominant social contracts of almost every mature human society: protection of the young and the weak; a desire for peaceful coexistence; and reinforcement of fairness, romantic love, and commitment to the common good. None of this obliterates such equally prominent features as ambition, violence, and greed. But we have the capacity for both and the apparent desire for less of the latter.

What we have come to think of as the softer sides of sex never quite fit the old every-gender-every-individual-for-itself story; they never made complete sense in just that context. Without a drive to get past conflict and to cooperate in forming intense relationships for some periods of time, what point would there be to our psychologies, to our unique human capacity for learning, for understanding, for *wanting* to understand others, and to change and adapt our behaviors to accommodate others? What would be the point of our having evolved to have conversations with our minds and feelings, to be what Robert Wright has called the "moral animal"?

If sex is survival and thus the fundamental driving force of life, moreover, it makes sense that all of the things we do to get, keep, and use it would make use of the full repertoire of our mental, physical, and emotional capabilities. In a natural world where *not* getting sex is a one-way ticket to eternal genetic oblivion, it only makes sense that nature lets us bring *everything* we have to bear to get sex done right. If this is so, then how could the sexual process not be about negotiation and a sexual contract?

Business or any other kind of contracts, of course, at least good ones, require that both parties be satisfied, each getting something and giving up something to guide a relationship to mutual goals. Can there be a better definition of the goal and purpose of sex?

Such a view explains a lot. Under the make-love-not-war model, the most successful sex would rarely, or only under stressful circumstances, favor rape or a zero-sum game. Such matings would create significant handicaps for the children and the parents. Rape plus abandonment in favor of more sexual pleasure would have doomed the species long ago by keeping us from becoming interested in parenting or accumulating and protecting resources needed for survival.

The newer view of sex also helps explain not only why we are so conflicted about sex and sexuality, but also why we work so hard to resolve our conflicts. The sight of an erect penis or an engorged vagina can at some times in our lives be scary or even repulsive, yet incredibly attractive and arousing at other times. One reason we may have such ambivalence is that we have both inherent and learned capacities to *time* our sexual activities and appetites. At some time in our evolution, it was vital to align our interest in sex with the realities of our environment and the likelihood that we could take care of a pregnancy, a child, and ourselves. And over time, our hormones, our bodies, and our brains conserved the best collection of tools for the job. Here is an essential paradox of sex made rational: It's a messy, intrusive, time-consuming, sometimes painful, and always complicated process, translated by our minds and moods into pleasure and purpose. This is precisely *because* we need a lot of experimentation and wiggle room to accommodate the unpredictable world around us. We need to make sure we do sex right and with the right mates, at any point in time and in tune with whatever may come.

Even the painful aspects of sex, such as jealousy, heartbreak, betrayal, and childbearing itself, might be seen in part as ways to make us think hard and choose carefully before we engage in sex.

Only, perhaps, in the context of sex as a means of enhancing conflict *resolution* does it make sense that animals, especially people, court slowly and make long-term commitments with some trepidation. Only, perhaps, in this context do some of the weirder cultural constructs surrounding sex, including laws against anal sex, incest taboos, chaperones, harems, dowries, chivalry, prenuptial agreements, and the Kama Sutra, make more sense.

If indeed the best theories are those that most rationally explain the most, then this newer and often-ignored proposition about the goal of sex stands up well. It seems to explain a great deal about why sex, and not just in humans, is so all-consuming an enterprise, so overwhelmingly complicated and such a bag of contradictions. Why it informs every aspect of our lives, both *nature* and *nurture*. Why it requires so much of our behavioral, cultural, emotional, intellectual, and physiological repertoire to achieve it in its total grandeur and to experience the contentment and rapture of which poets write. Why we don't just copulate, but *love*. Why indeed there is such an incredible diversity of sexual expression, why it's *natural* that a cat's penis should be barbed, why love lost literally, physically, hurts, and why along with the capacity for orgasm, we universally evolved the emotional capacity for sexual shame, anger, jealousy, dominance, submissiveness, passion, passivity, flirtatiousness, adultery, modesty, inhibition, and exhibition. All of this so that we have better chances of making good choices in the race to send our DNA into the next generation.

Evidence is plentiful that when it comes to sex and our reproductive apparatus, nature and nurture are inseparable and equal partners and that how we *behave* can alter our *biology*—as individuals and as a species— just as what we are *biologically* can alter how we *behave*.

"Just because something happens after you are born doesn't mean it was not prenatally ordained by genes or the environment of the womb. And just because something is prenatally ordained doesn't mean it can't change after you are born," says John Money. Yes, hormones (nature) are what make a pregnant woman weepy or hungry. But equally dramatic, stress and hard physical labor (even exercise) can render a woman infertile by sabotaging her sex hormones. We almost certainly evolved this sublime interaction between biological destiny and practical reality so that in times of hardship, disease, injury, or other stress, women and their mates might not be compromised by additional burdens. The stressors have changed—

getting into an MBA program and Olympic gymnastic tryouts instead of coping with famine—but the connections remain.

One answer, then, to why sex looks and acts the way it does—why there are different sexes, but almost always *two*, not three or nine or forty-seven; why there are sperm and egg cells so different in size and requirements; why sexes ever battle at all—rests in observations and experiments designed to reveal the interactions of biology *and* psychology, of inborn traits and acquired ones that together render men and women able to mount practical, mutually beneficial responses to the world around them.

The evidence is indeed abundant for the notion that our sexual legacy includes war, but mostly peace; conflict, but predominantly conflict resolution.

Sir Ronald Fisher, a British statistician and biologist whose monumental 1930 work, *The Genetical Theory of Natural Selection*, published just four years before he died, introduced the notions that the rate at which an organism increases its "fitness"—that is, its ability to mate, reproduce, and send its genes into the future—is equal to its genetic difference from other members of its species. He would have argued that he wants a girl just like the girl that married dear old Dad—but not exactly just like her—and she wants to marry someone just like her dad but not exactly like him. Too different and the offspring could be rejected by others in the community; too alike and the collection of deleterious mutant genes could bump the offspring off the fitness wagon forever.

Fisher believed that the reason there are two sexes is that nature was a pretty fine and practical mathematician. "No practical biologist interested in sexual reproduction would be led to work out the detailed consequences experienced by organisms having three or more sexes," he wrote. That is, beyond two sexes, reproduction just gets too complicated.

It took more than sixty years—until 1992, in fact—for another group of British biologists, led by Laurence Hurst and William D. Hamilton, to offer another cohesive answer. Hurst and Hamilton suggested that the whole evolution of separate sexes was a way for nature to manage conflicts between "selfish" genes. Any organism, including humans, that reproduce by fusing two cells not only fuse the DNA of their nuclei, but also the genes that roam around manufacturing essential proteins and chemicals in the rest of their cells. In their view, while the powerful nuclear genes fight for supremacy, another battle takes place among mitochondria and other organelles in the cytoplasm of cells outside the nucleus, duking it

out, looking to conserve and preserve their sources of nutrition and survival. So when two cells fuse, the non-nuclear troops from each of the
parent cells know that one army has to go. They're not needed. But if all-
out war occurs, the whole cell, including those precious gametes, could
die as well.

Enter nature's more or less diplomatic solution to "genomic conflict."
Sexes. In a two-sex system, one, call it the male if you want, sacrifices
itself, surrenders its non-nuclear genes, so that its nuclear genes may live
on in the fusion with the other sex, call it female, which will forever be
able to pass on her non-nuclear organelles. (Incidentally, the fact that only
females can pass on mitochondrial DNA has been used by Emory University's Doug Wallace, L. Luca Cavalli-Sforza, and other evolutionary
gene hunters to track the origins of different races and their migrations
across the planet. Every woman alive today has mitochondrial DNA
passed on to her *only* by her female ancestors, making it possible to link
her genetic footprints and blueprints to similar blueprints anywhere in
the world they might be. Studies by Cavalli-Sforza and Mark Seielstad at
the Harvard School of Public Health using a set of Y chromosome studies
of Africans and Europeans have measured the degree of similarities and
differences in shared Y markers. They found that over millennia, women,
not men, dispersed their DNA most widely, possibly because of our
female ancestors' habit of joining their husbands' tribes and traveling with
the male leaders of their marital families. These "patrilocal" societies persist in some areas of the world even today, such as among the Bedouins
of the Sinai Peninsula.)

In humans and other animals, males thus make tiny sperm, which
contribute no mitochondria to the new offspring, and females make huge
eggs, which hold on to all her non-nuclear contributions. But she pays,
too, after all, because it is her eggs that need to be sound enough to
nurture the whole developing new offspring, and the resources she alone
must bring to that enterprise cost her in terms of time and energy.

Hurst and Hamilton found more than enough evidence for their theory in the annals of observations about creatures great and small that
reproduce. They knew that living things that have sex come in two styles:
those that fuse their cells during actual mating and those that don't fuse
but pass on nuclei between two of them. If their model of genomic conflict
is right, only the first category evolved into two true sexes, because only
they were at risk of war between their cytoplasmic genes. The other group
never has to worry.

From algae to gorillas, this turns out to be the case, while some organisms, like paramecium, never fuse. The latter come close, bore a hole between them, exchange nuclear material, and blissfully roam around exchanging with an endless array of other paramecium. As long as the exchange involves nuclei that aren't too closely related, no problem.

A few life-forms, it seems, including some tail-flicking ciliates, can switch from a paramecium-style fusion of nuclear material to sex in which everything is exchanged. When the latter occurs, there are only two sexes that scientists have identified. That would seem to nail down the idea that the availability of two sexes is a way of managing genetic conflict in the cytoplasm.

One seeming fly in the ointment is the thirteen-sex slime mold, *Physarum polycephalum*. On closer inspection, however, it also helps reinforce the notion that two sexes is the optimal number for managing gene wars. Whenever slime molds meet and mate, they appear to recognize a particular hierarchy of the thirteen sexes; they're not all equal. If slime mold gene number 8 wants to get it on with number 9, 8 gives up its organelles to 9, marries "up," so to speak. Only if you mate with a partner beneath you do you get to keep your mitochondria and other cellular genes.

Why, then, don't humans have thirteen sexes, or eighteen or forty? Hurst believes that with more than two sexes, there is a lot of incentive to "cheat" and not unilaterally give up your mitochondria. The "Balkanization" of sex, in which it's every selfish mitochondrial gene for itself, can too easily lead to war and ultimate devastation of the whole species. Two is an easier number to manage.

The "war" between the sexes appears to go on, then, at the biochemical level in an almost steady-state fashion, but a fashion designed to keep things negotiable and balanced. Biologist William Rice of the University of California at Santa Cruz proved the case in his studies of fruit flies, which, at first blush, would appear to be a society that gives males a huge advantage over females. As described in detail in chapter 6, however, Rice made a kingdom of females that he kept from evolving, bred them, then killed all the offspring except males that had been given only paternal genes. After forty generations of such breeding, normal males were more often able to re-mate with females that had already mated with competitors, and they made semen that gave far more male offspring but killed a lot of females. In this way, Rice has demonstrated a leveling of the reproductive playing field and convincingly argued that sexual conflict, at least in insects, provides a strong biological incentive for making new species, a hallmark of vigorous reproduction.

As we've seen in chapter 10, Australian geneticist David Haig not only believes Hurst and Hamilton are right, but says management of gene wars during reproduction was so successful a strategy for maintaining lively species that this kind of "battle of the sexes" keeps on going right into the human fetus developing in its womb, in what he calls "fetal attraction." For evidence, he points to a variety of complications seen in pregnancy, and to molecular behaviors and traits that first became apparent when scientists focused on the phenomenon of "genomic imprinting." This, remember, is the genetic series of events that, contrary to Mendelian "rules," results in a gene behaving differently depending on whether it has come from Mom or Dad. (Mendel demonstrated that a gene is a gene. Whether from Mom or Dad, it does the same thing. As it turns out, that is usually the case, but not always.)

One example of genetic imprinting is the absence of a gene on chromosome 15 in humans that causes a condition marked by mental retardation and small hands and feet if inherited from Mom, and a very different disease, marked by worse mental retardation and a large mouth and red cheeks if it came from Dad. Linked to these differences are inherited patterns of DNA chemistry and structure, along with gene complexes that influence or regulate the genes involved in the disease. But if you ask *why* this non-Mendelian situation evolved in the first place, Haig says it's all about intra-genomic conflict and the competing interests of Mom's and Dad's genes. Backed up by a growing army of molecular geneticists, including Shirley Tilghman of the Howard Hughes Medical Institute, Thomas DeChiara of Columbia University, Davor Solter and Andrew Feinberg of Johns Hopkins University, Haig says imprinting's rules, which state that the same gene in some cases acts differently, depending on whether it came to junior from Mom or Dad, are a reflection of that old intra-genomic warfare.

Working on his theory that Dad's genes want *big* fetuses and Mom's want *more* (smaller) fetuses, these scientists went hunting for nontraditional behavior of genes in those genes involved in whatever might influence big and small; in other words, growth factors. And they found that genes for a growth factor called IGF-II, which manages fetal growth in mice, told a nice story, indeed.

The gene that makes Insulin-like Growth Factor II, or IGF II, binds to receptors made by another gene for Type I IGF and holds the blueprint for the signal "grow." More IGF II means more growth, and it turns out that paternal DNA indeed manages the biggest production of IGF II. DeChiara manipulated the genes from Mom and Dad and found that only Dad's gene for IGF II operates, while Mom's copy is silent. Mouse embryos that inherit a mutant IGF II gene from Dad grow up smaller than usual, but if they get the same mutant gene from Mom, they are fine.

Haig also found that a second receptor, Type 2 IGF, also binds with IGF II but does not stimulate growth the way the Type 1 receptor does. And when scientists looked at whether it matters if the gene for Type 1 comes from Mom or Dad, it turned out that the picture is the reverse of what happens with the IGF II gene. The gene for this receptor is expressed only in DNA from Mom, not Dad, because in this case, Mom's DNA wants to increase production of the receptor as a way to absorb all that IGF II made by Dad's DNA, and in this way keep growth at bay.

Although Feinberg, among others, says none of this work yet *proves* that genomic conflict or the growth genes so far identified cause imprinting—the process of silencing Mom's copy of them—there is, he says, "good evidence that there was evolutionary pressure" to deal with complications that arose from competing genes and genes that might crowd one another on the biochemical highway.

The genius of nature avoids a costly fight. By favoring two sexes and no one sex above the other, moreover, nature makes sure that there's always a gene that selects "mates" of the opposite type to prevent the possibility that both sides will surrender and the new fused cell will have *no* support. More than two sexes would so complicate the alignment of allies and enemies that we'd wind up with a reproductive process the UN couldn't handle. That thirteen-sex slime mold and some other creatures are exceptions to the basic notion that this so-called uniparental inheritance of cytoplasmic genes is the cause of sex and two sexes. But they rarely get far in life.

There is more going on in the evolutionary name of sex and it would surely boggle Darwin's mind. For example, consider a species where the fathers also contribute to the well-being of the young. If it's true that the best theories *predict* events and behavior, then any theory developed for this trait would have to predict that females hotly compete with one another for access to males (to make sure their offspring get the best nutrients) and that males would be the choosers in picking a mate.

Studies in prairie voles confirm those predictions and more, for this trait might also explain why whole communities of these rodents stay in another kind of sexual balance, gender equity, in which there are rarely more males than females. If the male contributes nutrients to offspring, he as well as she need time between matings to restore nutritional status and energy. In a series of experiments, scientists have shown that by manipulating the relative time each sex needed between matings, they could change the degree of female-to-female competition and male mating choice—literally turning on or turning off the battle of the sexes. Because among voles the larger males are on the bottom during mating (to avoid crushing the females), experimentally creating male shortages in the population means the largest males (who normally may not be the nimblest cross-country racers) are more in demand. They get pooped and even *more* choosy, and their size makes them better able to boot off the smallest, least competitive, albeit sexy females and check the population that way.

In the end, sex—the visible consequence and outcome of sexual selection and all of the strategies that go into mating—potentially affects the behavior and survival of both sexes when males invest something of themselves and their resources into their offspring, as human families can attest. It's pretty clear that interactions between the sexes are so far beyond the simpleminded popular models of a "battle of the sexes," "chemistry," or "love at first sight" that only a perpetual adolescent might buy them. Compromises and cooperation are more characteristic of sex than battles.

Humans are products of biological evolution but also of culture, so that while we share much with our closest biological relatives (chimpanzees and *Homo sapiens* are separated by a mere 5 million years or so of natural selection on the evolutionary tree), the precepts of competition for scarce resources on the savanna of Africa are far from the only kind of evolutionary pressures driving us. There were also 5 million years of using what else nature gave us—our problem-solving brains, our psychology—that define our approach to sex and everything else. We have also inherited mental mechanisms from our ancestors, and mental limits, in which we compensate for things no chimpanzee or other animals can. Among the psychological traits that have biological roots are altruism and cooperation, perhaps even charitableness and love, even though out there on that African plain, nice guys may have finished last in the scramble for food and mates.

Thus, when it comes to sexual behavior, there is at least as much possibility that we've left behind hard-scrabble competition—winner-take-all mentalities, male aggression and promiscuity, female dependence and submissiveness—as there is that we haven't completely divested ourselves of it. Just as kinder circumstances may have led the bonobo pygmy chimps—though only 1.5 million years away from nasty, male-dominated, aggressive, standard chimpanzees—to practice egalitarian sex and use it to relieve all kinds of social tensions within and between generations, so too is it possible that some early human traits may have vanished under the influence of our ability to reason and manufacture. "Unlike most other animals," says David Papineau, professor of philosophy at King's College in London, "we aren't just bags of reflexes; we are able to reflect on our circumstances and to figure out the best thing to do." Noting that human children easily learn language but not math, Papineau says we make arrangements to compensate: "Maybe the right course of action doesn't always come easily, but often we can manage it. When we need to do multiplication . . . we can make the effort, or better, build a calculator."

If there's a problem scientists have with all of this, it's that most of us forget that in nature and nurture, nothing is ever one-way or simple. It's certainly true that on *average* males, human and otherwise, are more indiscriminate in their search of sex, and women are more interested in how much a guy is willing to invest in her *offspring*. At least that's been the case until easy, cheap, and safe contraception made pregnancy a less likely outcome of sexual liaisons for women.

But studies also strongly show that men, once they *decide* to invest in a relationship, in a romantic relationship, want someone who has two brain cells to rub together as well as two gorgeous thighs or breasts. Although each sex may use similar mating tactics, the tactics take on the coloration of each sex's particular priorities. According to University of New Mexico's Gangestad and Thornhill, for example, both males and females display similar wares, in part because both sexes have faces, bodies, hearts, and genitals, but "men display resources, status and athleticism more than women do. Women display attractiveness and sexual restraint more than men do." These different traits also play out in how each sex operates to keep their mates, fend off sexual competitors, and gain other sexual advantage.

These days, evolutionary psychologists are discovering how ancient are the *social and cultural* lessons and signals passed on from one generation to the next, signals that shape how the young learn to be sexual adults.

What seems clear is that the learned lessons that become deep-rooted traits are those that lend stability to the individual and the society. For instance, in today's society, a girl who grows up in a fatherless household is more likely to develop an earlier interest in sex and adulthood. From an evolutionary standpoint, such girls would have understandably linked the father's absence (through death, for example) with the likelihood that she had better learn to fend for herself at an earlier age if she wants to survive, and sex would gain her the lost "fatherly" resources and protection.

What evolutionary psychology predicts is that boys will grow up interested in holding on to information and behavior that is likely to get them wealth, status, and girls, while girls are likely to grow up learning those lessons that give them an edge in parenting and a rewarding use of their sexuality—playing hard to get, for instance, to get resources from men and giving her favors only to those men who are likely to invest in her and her children. And each of these evolutionary trends will be played out against a backdrop of what's happening in their immediate lives. Is there war or peace? Famine or plenty? Does each of them have family resources to fall back on? Or none at all? Is there a bumper crop of fierce competitors around? Or did disease give them a buyer's market? As University of Maryland geneticist Christine Hohmann puts it, "Genes are the brick and mortar, but the environment is their architect."

Nature and nurture do not conspire to move us in some "advanced" or "more complex" direction. Just as sometimes dreaming of a cigar just means you want a cigar, some sexual behaviors and differences may not be adaptive at all, just . . . there. Natural selection may be continuous, but changes may come only when individuals in a given population *cannot* adjust to the stresses of their environment. As Darwin and his interpreters and disciples all agree—but lots of people don't realize—evolution does enable living things to change progressively and nonrandomly through natural selection, and new mutations do bring a new supply of hereditary information to the task of survival. But there is nothing about this that suggests there is some "predestined design" or creeping "perfectionism" that drives the process. Random mutations happen all the time but are routinely put to rest by individuals who have learned to overcome adversity. Such mutations may spread and become part of our biological makeup. When genetic change and new species do occur, it is rare and, in a way, the result of environmental and biological phenomena that push some envelope. To be a "living fossil," as one scholar put it, is the very essence of biological success because it means that the animal involved has learned to adapt to whatever its biological load is by either altering its environment or escaping it.

Humans, for instance, clearly evolved to fear strangers, possibly because they were a real threat when food was scarce. But social customs and cultures have equipped us to take some risks to get to know "strangers," an essential strategy if we're to avoid inbreeding.

Consider the gray tree frog. When they are ready to mate down at the pond, the females with their eggs ready to be fertilized come out of the trees around midnight for one annual clutch in the frog sack. The female selects a male and pulls him onto her back, but then hops away quickly, with her love hanging on for dear life, to find the perfect place to release her eggs. If the purpose of sex is to get reproduction done, why after thousands of years haven't the male tree frogs found a way to call to their loves from the ponds the females want to release their eggs in, instead of being dragged around all over creation by her need to find the perfect honeymoon cottage? It turns out that environmental factors are even more important than sex drives in tree frogs. The females are mostly worried about finding a place to release their eggs where the tadpoles are least likely to be eaten by something higher up on the food chain, or where they are more likely to find more food. The males, for their part, want to avoid contact with other perhaps more dominant males.

Lending more support for (or fuel to the fire over) the idea that among humans nurture can overcome nature are some intriguing experiments suggesting it may be possible to so ingrain a behavior in our offspring that it persists for generations to come in much the way the genes are inherited and persist. In what a pair of Israeli scientists call phenotypic cloning, parents can train their children to act as they wish them to act so that their children do the same despite new environments, new conditions, and even different biological temperaments. DNA has nothing to do with it. And if different families have different enough behavioral styles, and one is more adaptive than the other in any given environment, the "best" behaviors or strategies will be the ones to survive and get passed on, just as DNA is passed along. These so-called memes, remember, are culturally derived traits that pass vertically as well as horizontally among individuals. They aren't necessarily reflected in any particular genetic makeup, but ultimately they *might* in some fashion come out in our genomes because those that survive travel along with certain genetic makeups, while those that don't survive take all that failed gene pool with them.

For evidence, there are some roof rats in Israel, which have learned how to remove seeds from pinecones by opening them. That skill can be learned only from another mother rat caretaker (Dad doesn't participate) who can open pinecones. If that gives some rats an evolutionary advantage—more food—then that trait persists and nature may select for whatever it was that made the first mom a pinecone opener, eventually making her DNA match her socially learned skills. Social and cultural evolution is also faster than the genetic kind and more likely to preserve your genes. Neither Darwin nor modern biologists ever believed that people or animals just have to wait around for an "advantageous" mutation in their DNA to crop up and confer some progress on an individual or the species. Nature *and* nurture count. They work together.

Israeli scientists Aden Vital and Eva Jablonka of Tel Aviv University also propose as evidence of memes the altruistic behavior of some animals to take care of orphaned newborns who are not their kin and offer no genetic incentive to the mother. Why would such a situation exist? Because it gives Mom an opportunity to put her own behavioral stamp on more offspring. It may also be why humans want to adopt infants—they know from others' experience that infancy makes it more likely that parents can impose their behavioral shape and choices on the child so that they're more like biological parents. Why might females in many animal species not care very much if the male ejaculates and runs, leaving her with all the work of upbringing? Maybe because her payoff is that she can master the mind and behavior of the child in an animal version of "she who rocks the cradle rules the world." She gets to pass on not only her genes but her worldview, her gestalt, her beliefs and opinions. And if they're adaptive gestalts and opinions and views, her daughters and sons will have a better chance of ruling the world than will some other female's offspring.

Could this mean that the postmodern family in which dads as well as moms are responsible for child rearing might be the source of some sex-based cultural wars? Do the children of such couples pick mates differently, have different views of sex? Choose different occupations? Have more or fewer Oedipal complexes? Evolutionary psychologists and anthropologists believe it's quite possible.

Still other research suggests that our minds and brains, our psychological makeup and temperaments, also come with both built-in genetic equipment and the flexibility to learn new tricks. Animals and birds, for instance, seem to be able to "discount" in ways that would make an economist proud. Behavioral ecologist Marc Hauser of Harvard University and biologist Alex Kacelnik of Oxford have looked at rhesus monkeys that can do simple arithmetic and animals that can trade off the value of future opportunities for immediate rewards, depending on the amount of food available. "Play today and pay tomorrow" strategies are used by animals and birds, says Kacelnik, when it means maximizing your chances of mating and keeping your genes alive. Human personalities and psyches, notes biologist John Tooby, come "factory equipped," but options are available. Moreover, Tooby says, to know how sexual reproduction evolved, you need to understand the psychological mechanisms that underlie cultural behaviors and how they are generated and transmitted. As we've seen, men and women differ in some parts of their brains that are intensely involved in how the mind deals with and parses opportunities for courtship, mating, and sex. "Evolutionary psychology [no less than evolutionary biology] jettisons the false dichotomy between biology and environment," says Michigan's David Buss, "and provides a powerful . . . theory of why sex differences exist, where they exist and in what contexts they are expressed." Social, cultural, and economic contexts always can override propensities to do anything and the differences, the great variation *between* people is far greater than the similarities in response to dealing with a given situation. "Genes," says Cornell anthropologist Meredith Small, "are passed along in the real world where attraction, love, mating and marriage are not always linked. In reality, people choose mates (or have them chosen for them) in myriad ways. Our brains are not designed specifically to desire (this ideal or that ideal), but to weigh options and then get down to the business of having babies."

All sorts of reproductive and mating strategies evolved to *initially* at least confer advantage on either men or women, but were countered by strategies to equalize things. "Each sex," according to Buss, "possesses mechanisms designed to deal with its own adaptive challenges, some similar and some different." But always, those adaptations appear to coevolve with the other sex. Even "men's control over resources can be traced in large part to the co-evolution of women's mate preferences and men's competitive strategies," he says, and to the capacity for both sexes to mutually adapt in the face of change.

Under the old interpretation of sex biology, evolutionary psychologists and biologists suggested that it was only the battle of the sexes that played out on many levels of body and mind, including the fact that because men and women have different reproductive needs and agendas, as a rule men want lots of sex partners, while women prefer one stable relation-ship—a well-recognized field and formula for gender wars. This account predicted, therefore, that women will always want men with wealth, status, and stay-at-home qualities, while men will always want lots of women of beauty and youth. And indeed, Buss, whose studies of ideal mates are the most prolific in the literature, has discovered that among college-age men and women, this seems to hold true. When subjects are given a list of traits to rank in assessing potential mates, supermodel status appeals more to men than Tom Cruise looks do to women, and Fortune 500 CEO status appeals more to women than wealthy widows do to men. But as Small and other feminist biologists have noted, lots of men choose older, less fertile, less beautiful women as life mates and mothers of their children; and lots of women choose starving actors and farmers without a penny in their pockets. When Small looks at Buss's list of traits, she finds it more telling that both sexes rated such traits as "kind and understanding" and "intelligent" even higher than traits on which men and women differed. "Men and women surprisingly want the same thing," Small concludes.

Overall, it's important to keep in mind that DNA and nature have no "winners" or "losers" except in the species-wide race to reproduce. Thus, coevolution and adaptation have no inferior or superior, only successful strategies. "Over the course of human evolutionary history," Buss notes, "men who failed to control women's sexuality . . . by failing to attract a mate, failing to prevent cuckoldry or failing to keep a mate, experience lower reproductive success than men who succeed. We come from a long and unbroken line of ancestral fathers who (got) mates, preventing their infidelity and providing enough benefits to keep them from leaving. We also come from a long line of ancestral mothers who successfully secured investing mates, acted to prevent the siphoning of a mate's resources to (other women or prostitutes) and granted sexual access to men who pro-vided beneficial resources."

Lest we undermine the role of evolutionary biology by concluding that only humans get into this game, studies of bees and ants have shown that the number of reproducing queens are linked to differences in a variety of social traits and behaviors, including the size and shape of the queen and the colony's setting. By manipulating the social environment, inves-tigators have been able to shift the number of queens at will, which in turn has enormous effects on the genetics of the offspring in relation to which ones do best in the face of dwindling food supplies or more crowd-ing. Even something as simple as birth order can alter biological destiny, with a bit of evidence that firstborns, at least among humans, have some traits more similar with one another than with their own siblings. If behavior has a major genetic component, it's clear that for genetic reasons some are better than others at getting resources if they are firstborns, and that there may be some behavioral component that reinforces this in the parents' behavior toward their firstborns that doesn't do so in subsequent offspring. The whole cultural phenomenon of primogeniture, in which only the firstborn inherits wealth, can be seen as both genetic and social in its origins, at least by some theorists. In fact, sophisticated, learned culture in general is emerging as a characteristic of many animals, including dogs and nonhuman primates.

Physiologists and other biological scientists have put other new spins on old observations about the difference between the sexes and the alleged battle between them. Just one example: Among many animals, males are highly ornamented and females are not, suggesting that nature invested most of its creative energy in producing attractive males that females would find irresistible.

But wait. Such a mating strategy is very costly to males. They have to find and eat more food, grow bigger and stronger so that they can channel a lot of biological resources into strutting their stuff. All that color and frippery makes them easy targets for predators. And then there's all that *defending* they have to do to ward off even finer-looking fellows.

What we're left with is robust reaffirmation that cooperation is the point, if not the origin, of sex. We see again and again in these pages that nature invested at least as much, perhaps more creative energy in co-evolving females with extravagant demands for the males' ability to garner resources; and nature has done so with impressively clever means, using those fine feathers to indicate a male's commitment and parenting potential. In other words, it's *her* tune *he's* dancing to. While his agenda is to mate, hers is to keep a roof over the heads of her brood and in a good neighborhood, too. As the reproducing partner in the duo, the resources she invests are internally heavy and last a long time, so it's *her* choice that counts, at least as much as his—and arguably more—in ensuring the survival of *his* genes as well as hers. Females, not males, determine ultimately how a species evolves. "A one-night stand for a man can result in a 20-year investment for a woman," quips Mary Batten in *Sexual Strategies*. At the end of the day, the battle between the sexes is one with either two winners or two losers.

How do we know? Studies in ducks, for one thing, have found that if a male is forced on a female, she won't lay an egg—she didn't choose him, and physiologically, her reproductive system asks what part of "no" he didn't understand. Pigs and cows are much the same, according to David Crews. "A male is not just a bag of sperm. Animals, in order to be really fertile, need to sort themselves out. Breeders do best when they arrange things so that the female gets to make an active choice about whom she wants as a mate."

"In any living system I've looked at," says Crews, "including some plants, togetherness, cooperation and sexual collective bargaining always results in higher fertility than separation. I believe behavioral facilitation of reproduction came first, giving adaptive advantage to pairing up. Two sexes evolved because of those advantages."

From the catalog of sexual variety, scientists increasingly draw the conclusion that the diversity of sexual traits, mating rituals, physiology, and behavior is a tribute to nature's tendency to balance its act. The formal title of this hypothesis is called optimality and its influence over the past twenty years has grown, along with a lot of criticism. What optimality theory, like any good theory, tries to do is predict what animals might be expected to do. While there is considerable evidence to explain one or another idea about how sex originated, why it did, and perhaps even why it persists, there also is a bundle of data related to why and how sex exists in the almost endless variety of versions that it does.

Optimality theory holds that sexual selection is a trade-off in which constraints and freedoms are traded around in response to the external world to bring about some approximation of genetic and survival equilibrium. Game theorists who've played with this, including biologist Peter Mamerstein and Nobel laureate Reinhart Zeltson, show that the outcomes of natural selection, when manipulated in mathematical models, for instance, do indeed produce something like optimality solutions. By spotting patterns in a vast diversity of behaviors, notes University of Liverpool's Geoffrey Parker, a leading proponent of optimality theory, theorists can't show animals or their genes in perfect balance, but in a state that is more or less balanced to deal with exigencies.

For example, if a female produces very big eggs, having a large number of them is probably a bad idea because of the practical limits of laying, protecting, hatching, and feeding them, so the big egg producer will probably have fewer eggs than a small egg producer. Brought down to the level of that essential characteristic of sex—to send as many copies of one's genes into the next generation and assure the survival of that generation to carry them on again—it's hard to measure the precise number of copies of a gene replicated by a sexual strategy, so optimality theorists measure the number of offspring ideally, which is often close to that number. Even more directly, they have to measure use of available food or some other "indirect" measure of fitness.

Borrowing from John Maynard Smith's commonsense approach to evolutionary stability strategies, Parker and his colleagues look for sexual strategies that when used by most individuals in a given population of animals are pretty resistant genetically to rare mutants or other deviant strategies.

After studying the yellow dung fly for twenty-five years, Parker has found an example of simple optimization, specifically, how long a yellow dung fly would need to copulate to confer optimal fitness, or put as many of his genes into the future as possible.

Human beings, men and women, often put a lot of emphasis on "staying power," lengthy bouts of foreplay and intercourse that prolong arousal, pleasure, intimacy. Other species, however, seem to copulate almost casually and very rapidly. What was nature "thinking" of in either case? What, if anything, is the meaning of the vast variation in the length of time animals copulate?

Starting with what he could observe, Parker saw substantial differences in duration of copulation among these tiny, sexy creatures. He also saw that males struggle to get a female and that males frequently "flynap" females and remate them. Following the notion that natural selection for these traits produced something close to optimal solutions, he set out to find the optimal patterns in the diversity: Given the struggle, the mate stealing, how long *should* a dung fly copulate to gain any evolutionary advantage?

Presumably the yellow dung fly had no math whizzes or efficiency experts among them to figure it out, but figure it out they did at the level of their genes and biological fitness. What happens is that the last male to copulate with any given female gets 80 percent of her eggs covered by displacing the other males' sperm. (Every male must drop onto the female, copulate, and guard against displacement.) "What you can do is calculate the optimal copulation duration for a male for fertilizing the most eggs and getting his DNA to survive," Parker notes.

In the duration marathon, it turns out that the smaller males can copulate longer than the larger males, because only with this kind of setup can displacement rates level out somewhat. In dung fly society (not to mention our own, many would say), things sexual, it seems, are optimized from the male standpoint. What happens is that because the larger males win the get-a-girl struggles and get to the females *first*, their sperm is more likely to be displaced by the smaller flies' sperm, which get deposited last. This means the bigger guys must copulate more often and therefore for shorter times to get their DNA spread around for the long survival haul, while small males do just the opposite—copulate less often but for longer periods. The *very* small males lose out entirely more often. (What the female is doing all this time to optimize *her* sexual strategies is the subject of upcoming research. But if any of this resonates with the human experience, few biologists would be surprised.)

If what we can see of sexual behavior represents nature's way of optimizing, for both sexes, the strategic games of sex, then much of how we interpret what's going on biologically also makes new kinds of "sense."

For instance, females across all species—from insects to macaques to women in a medieval village in southwestern France—philander, eagerly pursuing multiple males serially or simultaneously, and the best philanderers win the survival game. A death blow to theories about male sexual aggression and female sexual passivity? Probably. A number of studies over decades suggest that fewer than 5 percent of mammals form sexual partnerships for life, notes anthropologist Helen Fisher in her book *Anatomy of Love*. From humans' earliest days on Earth, several million years ago, the pattern has been "monogamy with clandestine adultery," by both males and females. The flings may well represent the "simple optima" that Parker and others write about, nature's way of assuring something that works will help a species survive. In this case, the philandering increases the chances that new combinations of genes will be passed on to new generations and keep dangerous, recessive genes at a statistical disadvantage in the survival-conservation game. Suggests Fisher: "As long as prehistoric females were secretive about their extramarital affairs, they could garner extra resources, life insurance, better genes and more varied DNA for their biological futures. . . . Those who sneaked into the bushes with secret lovers live on, unconsciously passing on through the centuries whatever it is in the female spirit that motivates modern women to philander."

Once viewed as a straightforward power-struggle strategy, battles among males during mating season's competition for mates turns out not to be a simple matter of might and weaponry. Scientists have evidence that antlers, biting teeth, and plenty of testosterone-fueled strength may be little more than the visible outgrowths of the (literally) seminal struggle that takes place between competing males' sperm *after* copulation, inside the sex-hungry female who agrees to a more-the-merrier lifestyle. "Conception," writes Small, "once thought a dance of simple beauty between one female's fertile egg and a solo male's hopeful sperm, is now seen as a do or die cellular brawl that determines which males will pass on their genes to the next generation. All other forms of male rivalry, from the youngsters' first bickering over scraps of food to the adults' fierce struggles for dominance, lead up to these sperm-size moments of truth." And it's the "sperm wars" that really count.

What this means, says Meredith Small, a disciple of feminist biologist Sarah Blaffer Hrdy, is that "many characteristics that we think of as male across a wide range of species—the distinctive body with a projecting penis; the willingness to spend staggering amounts of energy on a thousand and one courting, mating and cuckolding strategies; perhaps even a tendency to struggle . . . for territory and status, may have evolved as so many weapons to aid a male's . . . sperm. The concept of competitive sperm may thus explain the basic source of maleness itself."

British scientists Robin Baker and Mark Bellis suggest that some sperm sacrifice themselves to help their "fitter" brother sperm hit pay dirt—team competition as it were. Baker and Bellis, using the same kind of working-back-from-observation method as Parker, came to this "kamikaze" conclusion by noticing the vaginal plugs left behind by male rats after copulation. The plugs, made from seminal secretions, are found in fruit flies, snakes, mice, bats, monkeys, and even marsupials. Once, they were widely believed to prevent leakage, giving the male who mated *first* an advantage by denying later males entry. However, it seems that the rat plugs are full of hundreds of dead sperm woven together like a web and these sperm are largely abnormal. So, instead of being nature's way of doing quality control and letting the fittest sperm through to the egg, the scientists now suggest that these sperm were destined to prevent competing males' penetrating thrusts from removing plugs and letting *their* sperm get through, *by producing enzymes that destroy the newly invading sperm.* These altruists aren't good enough to make it to fertilization but are good enough to let their fitter relatives win.

Apparently, a male may also vary his sperm counts to suit the need for competitive advantage when his mating situation changes. Baker and Bellis showed this by comparing one group of female rats housed near the one and only male they were allowed to mate with and another group of females housed near a single male but mated with another one. Results:

The male rats who *didn't* know their new mate ejaculated almost twice as much sperm at each mating as males who *knew* their partners were "faithful" in the sense that they were not mating with any other rat.

Rats aren't humans, of course, but both nature and these scientists are more interested in common pathways than exclusive boundaries between species. In a test on ten human couples with an "average" sex life, the scientists asked the men to use condoms during intercourse or masturbation, and had both the men and their mates keep track of their time together. Again sperm counts for ejaculations during intercourse decreased the more time they spent together, by about the same percentages as those in rats. During masturbation, though, the only thing that increased the amount of sperm was the amount of time since the last masturbation—the longer it was, the higher the sperm count. Put another way, when males can't be sure that their female partners aren't getting it on with another male, they are more likely to worry, or at least their sperm are, about sperm competition and the need to release more egg-seeking gametes. But when they know their chosen female is close and presumably faithful under their watchful eyes, they can afford to conserve their sperm supplies. This may not exactly explain why husbands forget anniversaries, but it raises some interesting questions about priorities.

We humans make a lot of assumptions about how much "control" we have over our sexual natures, prerogatives, preferences, and performances. But even our most vaunted and valued ones turn out to be evolution's necessity rather than our invention, and, as the growing amount of observation and experimentation seem to demonstrate, with no one sex having the ultimate edge in the battle between selfish genes. Think the penis evolved at least in part to wow the ladies, increase pleasure for a couple, or make copulation possible? In fact, most birds don't have penises, but sex is prominent among them. Think size counts? It also seems certain that natural selection gave advantage to males who could deliver sperm high in the female reproductive tract with a direct deposit capable of fertilizing more eggs than less well endowed males. But for championship-class male genital advantage, look to insects, not humans. Remember the dung fly's displacement strategy? Many female insects store sperm deep inside themselves for months before their eggs are fertilized, so the DNA competition is won by the male who can displace his rival's sperm. This would suggest that nature gives advantage to males with penises that are truly nimble, large, and adorned. And in fact, there are examples of such. The damselfly penis has two hornlike appendages with scooped-out flaps on each section that literally shovel out other males' sperm. Another species, a form of dragonfly, has a spongy penis that soaks up other sperm. A tiny fruit fly called *Drosophila bifurca* produces coiled-up sperm more than twenty times its body size, a thousand times longer than that made by humans, and the flies have huge (by fly standards, anyway) testes to produce them. Their size makes them so sluggish that they must rely on females' reproductive organs to get them to the eggs, giving the females a decided choice in the matter.

The nature of sex, then, is to dance, but in a pas de deux, not parallel solos. No, we don't know (yet) exactly why sex evolved, or why there are thorns among the roses. But if we pay attention to our imaginations as well as our observations, we'll see it's likely that evolution has cut our egos down to fit our true sexual natures.

BIBLIOGRAPHY

Abramson, Paul and Pinkerton, Steve, eds. *Sexual Nature/Sexual Culture*. University of Chicago Press, 1995.

Ackerman, Diane. *A Natural History of Love*. Vintage, 1994.

Alexander, Richard D. *The Biology of Moral Systems*. Aldine, 1987.

Allman, William. *Stone Age Present*. Simon and Schuster, 1994.

Angier, Natalie. *The Beauty and the Beastly*. Houghton Mifflin, 1995.

Axelrod, R. *Evolution of Cooperation*. Basic Books, 1984.

Baker, Robin. *Sperm Wars*. Basic Books, 1996.

Barkow, Jerome. *Darwin, Sex, and Status*. University of Toronto Press, 1989.

Barkow, Jerome, Leda Cosmides, and John Tooby, eds. *The Adapted Mind*. Oxford University Press, 1992.

Bateson, P., ed. *Mate Choice*. Cambridge University Press, 1983.

Batten, Mary. *Sexual Strategies: How Females Choose Their Mates*. Putnam, 1992.

Becker, Jill, S. Marc Breedlove, and David Crews, eds. *Behavioral Endocrinology*. MIT Press, 1993.

Bell, G. *The Masterpiece of Nature: The Evolution and Genetics of Sexuality*. University of California Press, 1982.

Betzig, Laura. *Despotism and Differential Reproduction: A Darwinian View of History*. Aldine, 1986.

Betzig, Laura, M. Borgerhoff Mulder, and P. Turke, eds. *Human Reproductive Behavior: A Darwinian Perspective*. Cambridge University Press, 1988.

Bishop, Jerry E., and Michael Waldholz. *Genome*. Iuniverse.com, 1999.

Botting, Kate and Douglas. *The Art and Science of Sexual Attraction*. St. Martin's Press, 1996.

Bowlby, John. *The Making and Breaking of Affectional Bonds*. Routledge, 1989.

Bullough, Vern L. *Sexual Variance in Society and History*. University of Chicago Press, 1976.

Bullough, Vern and Bonnie. *Women and Prostitution: A Social History*. Prometheus Books, 1987.

Buss, David M. *The Evolution of Desire: Strategies of Human Mating*. Basic Books, 1994.

Buss, David M., and Neil M. Malamuth, eds. *Sex, Power, Conflict: Evolutionary and Feminist Perspectives*. Oxford University Press, 1996.

Cairns-Smith, A. G. *Evolving the Mind*. Cambridge University Press, 1996.

Calvin, William. *How Brains Think*. Basic Books, 1996.

Cavalli-Sforza, Luca. *History and Geography of Genes*. Princeton University Press, 1994.

Cavalli-Sforza, Luca, and M. W. Feldman. *Cultural Transmission and Evolution: A Quantitative Approach*. Princeton University Press, 1981.

Cherlin, Andrew J. *Marriage, Divorce, Remarriage*. Harvard University Press, 1981.

Chisholm, J. *Romantic Passion: A Universal Experience*. Columbia University Press, 1995.

Clark, William. *Sex and the Origins of Death*. Oxford University Press, 1996.

Comfort, Alex. *Sex in Society*. Penguin Books, 1964.

Crow, J. F., and M. Kimura. *An Introduction to Population Genetics Theory*. Burgess, 1970.

Daly, Martin, and Margo Wilson. *Sex, Evolution, and Behavior*. Wadsworth, 1983.

Damasio, Antonio. *Descartes' Error: Emotion, Reason, and the Human Brain*. Grosset Putnam, 1994.

Darwin, Charles. *The Descent of Man and Selection I Relations to Sex*. Collier, 1972.

———. *The Expression of the Emotions in Man and Animals*. 1872. Reprint: University of Chicago Press, 1965.

———. *On the Origin of the Species by Means of Natural Selection or Preservation of Favoured Races in the Struggle for Life*. Murray, 1859.

Dawkins, Richard. *River Out of Eden*. Basic Books, 1995.

———. *The Selfish Gene*. Oxford University Press, 1976.

Dennett, D. C. *Darwin's Dangerous Idea*. Simon and Schuster, 1995.

de Waal, Frans. *Peacemaking Among Primates*. Harvard University Press, 1989.

———. *Chimpanzee Politics: Power and Sex Among Apes*. Johns Hopkins University Press, 1982.

Durden-Smith, Jo, and Diane Desimone. *Sex and the Brain*. Arbor House, 1983.

Eberhard, W. G. *Sexual Selection and Animal Genitalia*. Harvard University Press, 1985.

Edelman, Gerald. *Bright Air Brilliant Fire*. Basic Books, 1992.

Eible-Eiblesvelt, I. *Human Ethology*. Aldine, 1989.

Ellis, Havelock. *Studies in the Psychology of Sex*. Vol. 1. Random House, 1936.

Fausto-Sterling, Anne. *Myths of Gender: Biological Theories About Women and Men*. Basic Books, 1985.

Fisher, Helen E. *Anatomy of Love: The Natural History of Monogamy, Adultery, and Divorce*. W. W. Norton, 1992.

Fisher, Ronald A. *Genetical Theories of Natural Selection*. Dover, 1958.

Forsyth, Adrian. *A Nature History of Sex*. Chapters, 1986.

Foucault, Michel. *The History of Sexuality*. Vol. 2, *The Use of Pleasure*. Pantheon, 1985.

Fromm, Erich. *The Art of Loving*. Harper and Row, 1956.

Gazzaniga, Michael. *Nature's Mind: The Biological Roots of Thinking, Emotions, Sexuality, Language, and Intelligence*. Basic Books, 1992.

————. *The Social Brain, Network of the Mind*. Basic Books, 1985.

Givens, D. B. *Love Signals: How to Attract a Mate*. Crown, 1983.

Goodall, Jane. *In the Shadow of Man*. Houghton Mifflin, 1988.

Gould, J. L. and C. G. *The Ecology of Attraction: Sexual Selection*. W. H. Freeman, 1989.

Green, Richard. *The Sissy-Boy Syndrome and the Development of Homosexuality*. Yale University Press, 1987.

Hamer, Dean, and P. Copeland. *The Science of Desire: The Search for the Gay Gene and the Biology of Behavior*. Simon and Schuster, 1994.

Harlow, Harry. *Learning to Love*. Ballantine, 1973.

Hass, Jonathan. *The Evolution of the Prehistoric State*. Columbia University Press, 1982.

Hatfield, E., and R. L. Rapson. *Love, Sex, and Intimacy: Their Psychology, Biology, and History*. HarperCollins, 1993.

Hayflick, Leonard. *How and Why We Age*. Ballantine Books, 1994.

Hite, Shere. *Women and Love: A Cultural Revolution in Progress*. Knopf, 1987.

Holcomb III, Harmon R. *Sociobiology, Sex, and Science*. State University of New York Press, 1993.

Hrdy, Sarah Blaffer. *The Woman That Never Evolved*. Harvard University Press, 1981.

Jacob, Francois. *The Logic of Life: A History of Heredity*. Pantheon, 1973.

Johanson, D., and M. Edey. *Lucy: The Beginnings of Humankind*. Simon and Schuster, 1981.

Kaplan, Helen Singer. *The New Sex Therapy: Active Treatment of Sexual Dysfunction*. Quadrangle, 1974.

Kaplan, Louise. *Female Perversions*. Doubleday, 1991.

Kinsey, Alfred C., W. B. Pomeroy, and C. E. Martin. *Sexual Behavior in the Human Female*. Saunders, 1948.

————. *Sexual Behavior in the Human Male*. Saunders, 1948.

Konner, Melvin. *The Tangled Web: Biological Constraints on the Human Spirit*. Holt Rinehart and Winston, 1982.

Laquerer, Thomas. *Making Sex*. Harvard University Press, 1991.

Lawson, A. *Adultery: An Analysis of Love and Betrayal*. Basic Books, 1988.

LeDoux, Joseph. *The Emotional Brain: The Mysterious Underpinnings of Emotional Life*. Simon and Schuster, 1996.

LeVay, Simon. *The Sexual Brain*. MIT Press, 1993.

Levine, Joseph, and David T. Suzuki. *The Secret of Life: Redesigning the Living World*. W. H. Freeman, 1998.

Liebowitz, Michael. *The Chemistry of Love*. Little, Brown, 1983.

Lorenz, Konrad. *On Aggression*. Oxford University Press, 1973.

Low, Bobbi. *Why Sex Matters: A Darwinian Look at Sexual Behavior*. Princeton University Press, 1999.

Margulis, Lynn, and Dorian Sagan. *Mystery Dance: On the Evolution of Human Sexuality*. Summit Books, 1991.

————. *Origins of Sex*. Yale University Press, 1986.

Masters, William, and Virginia Johnson. *Human Sexual Response*. Lippincott Williams & Wilkins, 1966.

Maynard-Smith, J. *The Evolution of Sex*. Cambridge University Press, 1978.

Michod, Richard. *Eros and Evolution: A Natural Philosophy of Sex*. Addison Wesley, 1995.

Michod, Richard, and B. R. Levin, eds. *The Evolution of Sex: An Examination of Current Ideas*. Sinauer, 1987.

Moir, Anne, and David Jessel. *Brain Sex*. Michael Joseph, 1989.

Money, John. *Love and Love Sickness: The Science of Sex, Gender Difference, and Pair Bonding*. Johns Hopkins University Press, 1980.

Money, John, and Anke Ehrhardt. *Man and Woman, Boy and Girl*. Johns Hopkins University Press, 1972.

Morris, Desmond. *Intimate Behavior*. Bantam, 1973.

Perper, Timothy. *Sex Signals: The Biology of Love*. ISI Press, 1985.

Raff, Rudolf A. *The Shape of Life: Genes, Development, and the Evolution of Animal Form*. University of Chicago Press, 1996.

Ridley, Matt. *The Red Queen, Sex, and the Evolution of Human Nature*. Penguin Books, 1993.

Rodgers, Joann E. *Drugs and Sexual Behavior*. Chelsea House, 1988.

Sanday, P. R. *Female Power and Male Dominance: On the Origins of Sexual Inequality*. Cambridge University Press, 1981.

Small, Meredith. *Female Choices*. Cornell University Press, 1993.

Smith, John Maynard. *Evolution of Sex*. Oxford University Press, 1978.

Smuts, Barbara B. *Sex and Friendship in Baboons*. Aldine de Gruyter, 1985.

Springer, S. P., and G. Deutsch. *Left Brain, Right Brain*. W. H. Freeman, 1985.

Steele, Valerie. *Fetish: Fashion, Sex, and Power*. Oxford University Press, 1996.

Symons, Donald. *The Evolution of Human Sexuality*. Oxford University Press, 1979.

Taylor, Timothy. *The Prehistory of Sex*. Bantam Books, 1996.

Thornhill, Randy, and J. Alcock. *The Evolution of Insect Mating Systems*. Harvard University Press, 1983.

Tiger, Lionel. *The Pursuit of Pleasure*. Little, Brown, 1992.

Trevathan, W. R. *Human Birth: An Evolutionary Perspective*. Aldine, 1987.

Trivers, Robert. *Social Evolution*. Benjamin/Cummings, 1985.

Weinrich, James D., and John Money. *Sexual Landscapes: Why We Love Who We Love*. Scribner, 1987.

Williams, George. *Sex and Evolution*. Princeton University Press, 1975.

Wilson, E. O. *On Human Behavior*. Bantam, 1979.

————. *Sociobiology: The New Synthesis*. Harvard University Press, 1975.

Wood, James W. *Dynamics of Human Reproduction, Biology, Biometry, Demography*. Aldine, 1994.

Wright, Robert. *The Moral Animal*. Pantheon, 1994.

ACKNOWLEDGMENTS

Writing is a solo and solitary enterprise; producing books, however, is seriously collaborative. I owe much to many.

To begin at the beginning, there is John Michel, my hugely talented editor, and my friend. The original idea for this work, in fact, came from him during one of our lovely long Manhattan lunches on a blistering day in late summer of 1995. Ever since, even during his protean wrestling matches with my manuscript, and to his everlasting credit, he only once wondered aloud whether it took a "madman" to attempt such a book.

So my first and remaining debt of gratitude is to him for his creative, patient, and firm shepherding of my meandering ideas; and for his inexhaustible store of hilarious stories, witty salvos, gorgeous poems. His hand and mind show on every page. I hope I have been a worthy student of this remarkable teacher. This work has *two* parents with joint custody, conveniently a lifestyle we both know intimately as single parents. Andrew is a lucky boy.

Another huge bolus of debt is to the person who arranged that Manhattan lunch (and over the last fifteen-plus years, countless other happy meals), my indomitable agent, mother confessor, and pal, Elaine Markson. In my altogether too frequent and too long hiatuses between books, she persists in sticking by me, and in applying relentlessly cheerful prods to my professional procrastinations. I look forward to many more years as a devoted client and swapper of grandchildren stories.

Scores of scientists gave generously of their time for interviews, sent materials, and provided guidance; notable among them are David Crews (diversity and the sexual brain), Mark Tramo (neurobiology and brain organization), Greg Ball (avian sexual behavior), Davor Solter (imprinting and meiosis), Andrew Feinberg (imprinting and growth), Solomon Snyder (neuroscience), John Money (sexual anthropology and psychology), Jan Eppig (oocytes and embryogenesis), Barry Komisaruk (orgasm), Sue Carter Porges (monogamy and oxytocin), Arthur Arnold (the brain and sex), Richard Axel (olfaction), John Bancroft (The Kinsey Institute), Fred Berlin (pedophilia), James Boone (gender roles), Adrian Dobs (endocrinology), Eva Eicher (genetics of sex determination), Peter Fagan (sex therapy), Aurelio Figueredo (domestic violence), Roger Gorski (the brain and homosexuality), David Haig (evolutionary biology), Dean Hamer (behavioral genetics), Christine Hohmann (sex hormones), Barbara Migeon (X chromosome), Ward Odenwald (drosophila genetics and homosexuality), Godfrey Pearlson (the brain), Nicholas Pound (fidelity), Devendra Singh (mate selection), Randy Thornhill (sexual attraction), and Laura Betzig (history and anthropology of sex).

Others provided a sounding board at critical moments in the research and writing process, read parts of the manuscript, and offered clarity and insight to help me through foggy patches. Thank you especially Jon Franklin, Sol Snyder, Elaine Markson, James Weinrich, and Sue Carter Porges.

Finally, there are those to acknowledge because they willingly indulge my writing habit and necessary seclusions at the expense of time forever lost. Among these are Elaine Freeman, my friend of four decades and colleague at the Johns Hopkins University Office of Communications and Public Affairs, who carried my workload during two leaves of absence and many more distractions; my parents, Dorothy and Max Ellison, who sanctioned my hermitage and always came when I called; my partner, lover, and now husband, Robert Nath, with whom romance endures; and my remarkable sons, Jared and Adam, whose journeys in manhood and rigorous pursuit of difficult dreams mystify but always inspire me. Their love redeems my shortcomings and my past.

INDEX

ABOUT THE AUTHOR

JOANN ELLISON RODGERS is deputy director of public affairs and director of media relations for the Johns Hopkins Medical Institutions. For eighteen years, she was a reporter, then national science correspondent for the Hearst Newspapers, winning a Lasker Award for Medical Journalism. She is past president of the Council for the Advancement of Science Writing and of the National Association of Science Writers, a fellow of the American Association for the Advancement of Science and one of the few non-scientist members elected to Sigma Xi, the Scientific Research Society. She is a lecturer in the Department of Epidemiology at the Hopkins School of Public Health, a freelance writer, and author of six books. Among these are *The Academics' Guide to Media Relations* and *Psychosurgery: Damaging the Brain to Save the Mind*.